Springer Climate

Springer Climate is an interdisciplinary book series dedicated to climate research. This includes climatology, climate change impacts, climate change management, climate change policy, regional climate studies, climate monitoring and modeling, palaeoclimatology etc. The series publishes high quality research for scientists, researchers, students and policy makers. An author/editor questionnaire, instructions for authors and a book proposal form can be obtained from the Publishing Editor.

Now indexed in Scopus® !

More information about this series at http://www.springer.com/series/11741

Judith Kreuter

Climate Engineering as an Instance of Politicization

Talking Tomorrow's Technology—Framing Political Choice?

Springer

Judith Kreuter
Institute of Political Science
Technical University of Darmstadt
Darmstadt, Hessen, Germany

ISSN 2352-0698 ISSN 2352-0701 (electronic)
Springer Climate
ISBN 978-3-030-60339-7 ISBN 978-3-030-60340-3 (eBook)
https://doi.org/10.1007/978-3-030-60340-3

This Springer imprint is published by the registered company Springer Nature Switzerland AG
The registered company address is: Gewerbestrasse 11, 6330 Cham, Switzerland

This book is dedicated to my grandfather.

Preface

This book would not be what it is without the support and interest of a large network of people that accompanied me, in diverse ways, along this strenuous and rewarding journey. I am very grateful to Prof. Dr. Markus Lederer for his extensive professional and personal guidance, and for providing me with an example of how to be, not only an excellent researcher and supervisor, but also team leader and mentor. Further, I am grateful to Prof. Dr. Johan Schot who joined this endeavor at a crucial moment.

I had the pleasure of working with great people during the progression of this research project. I thank my colleagues at Westfälische Universität Münster and Technische Universität Darmstadt, in particular Sinja Hantscher, Dorothea Schoppek, Sebastian Schneider, Janina Obermeyer, Michael Verspohl, Heike Böhler, Björn Egner and Timo Richter, as well as all members of 'Team Lederer' for invaluable discussions, feedback and exchanges. I met a number of exceptional researchers during this project, many of whom have, by now, become my friends. I would like to express my thanks to all these researchers for insightful discussions and engaging interactions. Further, I am particularly indebted to everyone at the DFG Priority Programme 1689 for offering many opportunities to interact and network with researchers from around the world. Last but not least, I wish to express my deepest gratitude to my family and my friends.

This book was written with the aims to increase knowledge and understanding of the social construction of Climate Engineering as an emerging set of technologies to address specific climate change effects, and to learn more about the role of social construction for the socio-technical reality of climate technologies. It is written for everyone interested in political and social aspects of the processes of emergence of new technologies and development of existing ones, in the role of academia in political decision-making and in technologies in the climate policy discourse. I hope that it offers insights to students of climate policy and governance, of science and technology studies, of global governance in general and climate engineering specifically.

Darmstadt, Germany — Judith Kreuter

This book is the publication of my PhD thesis at Technische Universität Darmstadt, Germany.

Contents

Abbreviations

1.5 °C Report	Report "Global Warming of 1.5 °C. An IPCC Special Report on the impacts of global warming of 1.5 °C above pre-industrial levels and related global greenhouse gas emission pathways, in the context of strengthening the global response to the threat of climate change, sustainable development, and efforts to eradicate poverty"
AR5	IPCC Assessment Report 5
BAU	Business as usual
BECCS	Bio-energy with Carbon Capture and Storage
BMBF	Bundesministerium für Bildung und Forschung (German Federal Ministry for Education and Research)
C	Celsius
C2G	Carnegie Climate Governance Initiative
C2G2	Carnegie Climate Geoengineering Governance Initiative
CBD	Convention on Biological Diversity
CC	Climate change
CCS	Carbon capture and storage
CDR	Carbon dioxide removal
CE	Climate engineering
CM	Cloud modification
CO_2	Carbon dioxide
CS	Carbon sequestration
DAC	Direct air capture
DACCS	Direct air capture with carbon storage
DFG	Deutsche Forschungsgemeinschaft (German Research Foundation)
EC	Epistemic communities
ENMOD	Convention on the Prohibition of Military and any other Hostile Use of Environmental Modification Techniques
EU	European Union

EW	Enhanced weathering
GHG	Greenhouse gas
IPBES	Intergovernmental Science-Policy Platform on Biodiversity and Ecosystem Services
IPCC	Intergovernmental Panel on Climate Change
IR	International Relations (sub-discipline of political science)
JK	Judith Kreuter
MAUD	Military Application of Uranium Detonation
NAS	United States National Academy of Sciences
NDRC	National Defense Research Council
NET	Negative emissions technology
OF	Ocean fertilization
OIF	Ocean iron fertilization
OW	Ocean welling (upwelling or downwelling)
PA	Paris Agreement
ppm	Parts per million
RM	Radiation management
SA	Surface albedo
SAI	Stratospheric aerosol injection
SM	Space-based modification
SPP-CE	German Research Foundation (DFG) Priority Program 'Climate Engineering'
SRM	Solar radiation management
STS	Science and technology studies
UK	United Kingdom of Great Britain and Northern Ireland
UNEA	United Nations Environment Assembly
UNFCCC	United Nations Framework Convention on Climate Change
USA	United States of America
WMO	World Meteorological Organization
WWII	World War II

List of Images

List of Tables

Chapter 1
Talking Tomorrow's Technology, Framing Political Choice?

Abstract Kreuter provides a concise discussion of the role of emerging ideas of Climate Engineering in the broader context of global climate governance. She discusses 'wickedness' in the sense of the non-existence of a clearly right solution, with regards to both challenges—anthropogenic climate change on the one hand and Climate Engineering as a potential measure to address this problem on the other. Kreuter explores the empirical case of the current academic discussion of Climate Engineering as an instance of politicization which offers the opportunity to open up the topic to deliberation and political decision-making.

United Nations Secretary-General António Guterres announced in September 2018 at the UN headquarters that "[c]limate change is the defining issue of our time—and we are at a defining moment" (UN Secretary-General 2018).[1] Multilateral climate policy-making presents a remarkable global governance endeavor involving numerous state and non-state actors, for example in the forum of the United Nations Framework Convention on Climate Change (UNFCCC), to "prevent dangerous anthropogenic interference with the climate system" (UN 1992: Art. 2). In a ground-breaking multilateral achievement in 2015, 195 states collectively decided, through the Paris Agreement, to reduce their greenhouse gas (GHG) emissions significantly in order to limit global average temperature rise to below 2 °C above pre-industrial levels and to endeavor to remain below 1.5 °C. However, global GHG emissions, the "dominant cause" of observed global warming (IPCC 2014b: 4), to this day "show no

[1]In accordance with the frame analysis approach applied in this thesis, I assume that *every* actor frames in communication. Accordingly, I frame when introducing the present study. I chose a frame highlighting the relevance of the topic of the present thesis by referring to the duality of high stakes related to the impacts of anthropogenic climate change, on one hand, and climate engineering (CE) approaches, on the other. Alternatively, I could have decided, for example, to focus mainly on the impacts of potential CE or on the impacts of climate change.

[2]All accentuations in citations in this thesis (e.g. *italics*) are adopted from the original unless explicitly stated otherwise.

Some ideas discussed in this section have been published in Kreuter (2018) or are currently under review for publication in Kreuter and Lederer (Under review).

J. Kreuter, *Climate Engineering as an Instance of Politicization*,
Springer Climate, https://doi.org/10.1007/978-3-030-60340-3_1

sign of peaking" (UNEP 2018: xv). In this context, a set of technological approaches for "the *deliberate large-scale manipulation of the planetary environment to counteract anthropogenic climate change*"[2] (Shepherd et al. 2009: 1), subsumed under the term 'Climate Engineering' (CE), encounter increasing interest, particularly in academic circles (Matzner and Barben 2018). The role of CE in climate policy is, to date, a peculiar one: Only very few high-level dedicated political decisions have been made determining whether or not these approaches are to be considered in the portfolio of options to counteract climate change impacts. None of the approaches subsumed under the term exist, at this moment, as physically tangible socio-technical systems (Schot and Kanger 2018) capable of delivering a global climate impact. If they did exist in this way, each of these approaches could have significant unintended side effects on both natural and social systems, possibly even some which cannot be foreseen today. At the same time, the ability to use at least some forms of CE (mostly Carbon Dioxide Removal, CDR) can be understood to be implied in the scenarios presented by the Intergovernmental Panel on Climate Change (IPCC) which would allow remaining below the climate targets defined in the UNFCCC process (IPCC 2018a: 19). This peculiar situation has been dubbed by observers "[t]he world's biggest gamble" (Rockström et al. 2016: 465) or "magical thinking" (Rayner 2016: 2). Janos Pasztor, Executive Director at the Carnegie Climate Governance Initiative (C2G), argues: "We are potentially at the dawn of an age of geoengineering.[3] It is time for policymakers to start discussing whether geoengineering is to go forward, and if so, how." (Pasztor 2017a: 420)[4]

It can be assumed that the near future will bring some decisive action with regards to CE, be it the development or even the use of these approaches or decisions to stop their consideration completely. I am interested in assessing what we can learn, at this moment in time, about how this action might look taking into consideration the current state of discussion of CE. In order to (1) analytically and systematically assess how the meaning of CE is constructed in the current mainly academic discussion on CE and (2) normatively, provide decision-makers with tools to take into account the constructedness of this meaning, in the present study I address the following central research question: *How is the meaning of CE approaches constructed in the academic discussion through frames which legitimize or delegitimize the use of these approaches to address anthropogenic climate change*?

The main argument of this study is the following: *I assume that, in a situation of little dedicated political deliberation of CE, the intersubjective meaning of these approaches is defined largely in other fora, in particular in the academic discussion on CE. The empirical analysis conducted for this study indicates that the meaning of CE constructed in the academic discussion through framing is not uniform, but instead composed of a diversity of frames. Two of those frames are particularly frequent in the sample: a security frame legitimizing CE, and a complexity frame*

[3]Instead of the term 'geoengineering', I use the term CE in this thesis to denote the same set of approaches.

[4]The Initiative changed its name in June 2019; up to that point, it was called Carnegie Council Geoengineering Governance Initiative (C2G2) (Pasztor 2019).

delegitimizing these approaches. In this sense, the academic discussion on CE can be understood as an instance of politicization as the diversity of frames may contribute to rendering the issue of CE "the subject of deliberation, decision-making and human agency" (Hay 2007: 81). *This, I claim, provides a tremendous opportunity for decision-makers: My normative claim is that the meaning of CE thus constructed—in particular the diversity of frames of these approaches—should be preserved if and when a high-level decision-making process is initiated.*

1.1 Why Assess Academic Framing of Climate Engineering? Two Wicked Problems, a Lack of Political Decision-Making and a Lively Academic Discussion

Both climate change and CE present wicked problems (Hulme 2014: 117; Zhang and Posch 2014: 392; Rittel and Webber 1973: 155). Wicked problems are problems which do not have a clearly right solution. Instead, every approach to address the problem produces diverse implications, advantageous to some, disadvantageous to others. Accordingly, any way of addressing climate change—be it through intense mitigation or through large-scale CE or a combination of these approaches or a completely different approach—will always have implications that are positive or negative for various groups of people and produce winners and losers. Simultaneously, every way of addressing CE—be it a complete prohibition through a research moratiorium and deployment ban or acceptance of its use in order to counteract inadvertent climate change—will garner implications.

While the object question differs in the two cases—in the first, it is inadvertent anthropogenic climate change, in the second intentional CE—the two wicked problems are intricately connected through the question of whether or not CE should be considered as part of an answer to anthropogenic climate change. The underlying dilemma is whether or not one wicked problem—i.e., climate change—should be addressed by introducing a new wicked problem—i.e., CE. Either decision—to introduce CE as a measure (among others) to address climate change, or not to do so—is fraught with implications: The stakes surrounding this decision are high either way (Linnér and Wibeck 2015).

Political decision-making is aimed at determining the authoritative distribution of material and non-material values in society and between groups in society (see Easton 1953; Lasswell 1958; Pye 2001; Lederer 2017). Empirically, little dedicated political decision-making has been endeavored up to this point concerning the question whether CE should be considered as a measure (among several other) to address anthropogenic climate change (SPP 1689, 2019a: 61), despite various calls for political deliberation of these very complex highly consequential approaches (see, for example, Long et al. 2015: 30; Pasztor 2017a: 419).[5] Several developments make

[5]A few exceptions exist, for example amendments to the London Convention/London Protocol (Ginzky and Frost 2014) and decisions by the parties to the CBD (CBD 2010) concerning CE.

this limitation increasingly relevant: First, the impacts of climate change are now beginning to be felt—not only in remote areas of the world or by small fractions of the world population, but geographically wide-spread and by many people. Secondly, the assumption of being able to use some form of CE is implied in climate models which provide information for climate policy decisions. Thirdly, CE approaches would produce impacts and unintended side-effects. Fourthly, CE approaches can, at this point, not be considered ready for deployment. Meanwhile, not making a decision with regards to these approaches is a decision in its own right: If GHG emissions are allowed to continue unabated—a scenario termed 'business as usual' or BAU—this would incur implications for various groups of global society, thus determining the distribution of non-material and material values just as much as if CE or intensive mitigation or a different approach was chosen intentionally. Not making conscious political decisions about CE implies accepting that these approaches will continue according to their own dynamics. This means that they might disappear from the discussion altogether, for example due to lack of funding, or that they might be developed in a way that does not consider the interests and needs of all those affected, but instead serves certain particular interests and goals.[6] While global political deliberation does not guarantee that the best decision for all concerned will be made, I claim that it is a useful tool at our disposal at this point that can be used to draw near that ideal (SPP 1689, 2019a: 63; Pasztor 2017a: 420).

While little high-level political deliberation is taking place to date concerning the question whether CE should be considered or not, CE approaches are discussed among other actors, for example in academic circles. In these discussions, these actors start to construct the social meaning of CE: These approaches do not yet exist as a physically tangible socio-technical system (Schot and Kanger 2018: 1045; Asayama et al. 2017: 82), i.e. as configurations of technological artifacts, infrastructures, actors and regimes with the capacity fulfill the function of abating climate change. Accordingly, "big questions—about the trajectory of scientific research, the technologies that might be developed and the risks they may entail, and government regulation—are still full of uncertainty and therefore up for discussion." (Carr et al. 2014: 41) In this discussion, the meaning of CE approaches is constructed by highlighting and focusing on specific questions and issues and—inadvertently—sidelining others: As Luokkanen et al. argue,

> interpretative […] flexibility can be assumed to be great in emergent technologies characterized by significant uncertainties, which allows different groups to emphasize different aspects of the technology, like the proverbial blind men examining an elephant. (Luokkanen et al. 2014: 968)

Most recently, Switzerland submitted a draft resolution at the fourth Session of the United Nations Environment Assembly (UNEA) proposing "a global assessment of potential risks and benefits of geoengineering technologies" (IISD Reporting Services 2019: 13). The draft resolution was, however, withdrawn after no agreement could be found on a number of issues pertaining to the resolution.

[6]For example, Reynolds and Wagner (2019) argue that decentralized solar CE "appears at once *technically possible* and *economically feasible*", which would make it available to individual states or even non-state actors.

This early construction of meaning will be relevant for decision-making if and when it does occur, as this meaning can, for example, influence whether CE is considered a legitimate or an illegitimate measure to address climate change (see, for example, Huttunen and Hildén 2014: 5; Hamilton 2015: A29; Bellamy et al. 2012a: 28; Hamilton 2013a: 83). An account of a process in which technology became legitimized is presented in Chap. 4, in which I discuss the role of early academic framing in the legitimization of nuclear technology as an illustration of how academic framing can be relevant for decision-making on emerging technology.

The discussion of CE was started in academic circles—including both social and natural sciences as well as the humanities—and is, to date, arguably liveliest among this group of actors (Matzner and Barben 2018). However, while political decision-making might be influenced by academic considerations of CE, academic research cannot replace political decision-making: While academic research can give answers to a number of important questions surrounding CE in a very precise way—such as what the probable impact of large-scale application of stratospheric aerosol injection (SAI) would be on the photosynthesis rate of plants or whether humans can be expected to uphold mitigation endeavors when they know about the potential impact of CE—it cannot give answers to wicked problems that have no right or wrong answers, no matter how much data is collected and analyzed: "Science may be revealing the intricacies of the climate system, but it is not helping us discover the meaning of climate change." (Hulme 2010: 172) Thus, it is important to assess whether, in this situation of little political decision-making, meanings of CE constructed in the much larger academic discussion might be drawn into this void, thus pre-defining the decision by inadvertently legitimizing or delegitimizing this set of approaches as a measure to address climate change.

1.1.1 Legitimization or Delegitimization of Climate Engineering

Following Maarten A. Hajer, it is assumed here that

> environmental conflict should not be conceptualized as a conflict over a predefined unequivocal environmental problem with competing actors pro and con, but it is to be seen as a continuous struggle over the definition and the meaning of the environmental problem itself. (Hajer 1997: 14)

Thus, the framing of CE—the definition of the central problem, the causal analysis of this problem, the evaluation of measures to address the problem and the proposal of remedies—is highly relevant in constructing the meaning—and the perceived legitimacy—of CE: "[F]raming is believed to be 'sticky,' meaning that after a dominant framing of the issue has settled changes are more difficult […]." (Luokkanen et al. 2014: 967) To acquire the ability to assess whether framing is taking place in the academic discussion which might legitimize or delegitimize CE approaches as measures to address climate change,[7] a theoretical framework was developed for the present thesis on the basis of International Relations (IR) and Science and Technology Studies (STS) theory which allows the identification of elements of the discussion which legitimize or delegitimize the consideration of CE approaches.[8]

By applying this theoretical framework, the present thesis aims to systematically assess how the meaning of CE approaches is constructed in the academic discussion with regards to the legitimacy or illegitimacy of these approaches as measures to address climate change. In order to achieve this, I empirically analyzed the constitution of the academic discussion on CE—operationalized through peer-reviewed academic publications on CE—focusing on its constitution by frames and frame elements. The thesis addresses the following central research question: *How is the meaning of CE approaches constructed in the academic discussion through frames which legitimize or delegitimize the use of these approaches to address anthropogenic climate change?* Assuming that academic actors conduct strict self-regulation through shared standards of objectivity, the intentional shaping of the meaning of CE, e.g. through framing, can be expected to be much less prevalent within this group of actors than with, for example, political or media actors (Lakoff 2014). Nonetheless, inadvertent framing has been found to occur, albeit in very small degrees, in this discussion. If a dominance of a legitimizing or delegitimizing frame was to be found to exist in the academic discussion, I claim, this frame might be picked up if and when high-level political decision-making takes place concerning CE approaches.

The following conclusions are drawn from the theoretical considerations and empirical analysis conducted in order to be able to answer the central research question. The way these conclusions were drawn from the study is elaborated in detail in Chap. 7.

1. Framing of emerging technologies can influence decision-making.
2. In situations of little political deliberation, the academic discussion can predefine the space of political decision-making if and when it occurs.
3. Thematic frames focusing on security, complexity, economy and appropriateness can either legitimize or delegitimize CE as measures to address climate change.
4. Framing in general is rare across the sample.
5. Framing of CE in academic discussion is neither uniform nor clearly polarized, but diverse.
6. This diversity of framing can be traced back, partly, to the presence of texts from different disciplinary perspectives in the sample.
7. This diversity of framing can be traced back, partly, to the presence of texts addressing different CE approaches in the sample.
8. All in all, instances of elements from legitimizing variants of the ideal-type frames are slightly more frequent across the sample than elements from delegitimizing variants.

[7] For a more detailed discussion of the term 'legitimacy', see Sect. 1.1.3 as well as Chap. 3.

[8] For a detailed description of this framework, see Chap. 3.

9. Both the legitimizing and the delegitimizing variants of the ideal-type frames develop erratically over time, but awareness of duality of high stakes seems to increase, in particular with early opponents of CE.
10. Of all ideal-type frames, the legitimizing variant of the security frame and the delegitimizing variant of the complexity frame are most common across the sample.
11. The legitimizing variant of the economy frame, while the first in appearing in the sample, does not dominate over the legitimizing variant of the security frame and the delegitimizing variant of the complexity frame later on.

Thus, I find that the academic discussion of CE does not provide decision-makers with a simple answer to the question of whether CE approaches should be used in order to counter-act inadvertent anthropogenic climate change: I do not find a clearly dominant legitimizing or delegitimizing frame in the discussion. At the same time, neither do I find a clear polarity in the sense of two opposed camps of actors: In contrast, the academic discussion is characterized by a high degree of frame variance and frame contestation, both between texts and inside of texts. At this moment in time, neither legitimizing nor delegitimizing frames clearly dominate this discussion and the participants are found to be conscious of the wickedness of the problem. Decision-makers, at this juncture, still have the opportunity to get involved in the construction of the meaning of CE, which will become much more difficult once a specific meaning has been consolidated.

1.1.2 The Academic Discussion of Climate Engineering as an Instance of Politicization

From the vantage point of political theory, and in particular—but not exclusively—IR theory, the conclusions of this study present an interesting instance of a process of politicization. Politicization, in general terms, has been defined by Zürn as a process of "transporting an issue or an institution into the sphere of politics—making previously unpolitical matters political." (Zürn 2019: 977) Depoliticization is, accordingly, the process in the opposite direction—transporting an issue out of the sphere of politics. Numerous different authors from diverse theoretical backgrounds could, supposedly, agree to this definition. At the same time, it is, in this broad sense, a "chameleon-like concept" (Jessop 2014: 207), which requires clarification if it is to be useful. When clarified, the concept can help to analyze a score of different phenomena assessed in a number of strands of literature concerned with 'politicization' and to define in which way the academic discussion on CE can be perceived as an instance of politicization and what that means for future CE decision-making.

The political as the basic element of the concept of politicization

Firstly, and most importantly, the above definition raises the question: What is this 'political'—and what is it not? Commonly, **the political** is understood as the entirety of processes, actors and systems involved in public authoritative decision-making, defining the distribution of material and non-material values in society and between groups in society (see Easton 1953; Lasswell 1958; Pye 2001).[9] The political, which is, in German, denoted by the single term 'Politik', falls into three different terms in English (Pelinka and Varwick 2010): policy, politics and polity. 'Policy' describes the *content* of political decisions and solutions to political problems and focuses on aims and programs guiding decisions: It highlights the intentional design ["Gestaltung", (Pelinka and Varwick 2010: 21)] of social interaction. 'Politics' describes *processes* of political decision-making, focusing on interests, organization of interests and conflict of interests, procedures, dynamics and relations between actors and processes of policy design, decision and implementation. 'Polity' describes political *structures*, focusing on institutions and organizations, norms and rules and political culture. **Political decision-making** can be understood as a process in which a political problem is identified and defined, a space of possible actions to address the problem opened—by considering various different possible measures—and then closed down by choosing one of these actions—or several in a portfolio—over others. Part of this process is the definition of a central political problem or challenge for which specific measures or policies are to provide solutions. The present thesis is interested in political decision-making concerning one specific central political problem: that of global climate change. I see the definition of a central problem as part of the political decision-making process. Very similarly, Hajer argues:

> In a constructivist approach policy-making is not seen simply as a matter of what I call 'problem closure': defining a set of socially acceptable solutions for well-defined problems. It is, first and foremost, an interpretive activity in which different, often contradictory claims as to what is the case are to be judged, compared, combined, and acted upon. The definition of a policy problem, like for instance acid rain, cannot be taken for granted. (Hajer 1997: 21–22)

Typologies of politicization

When it comes to understanding how an issue can become political or subject of political decision-making, i.e., the process of politicization, Zürn offers two different approaches to analytically differentiate the two states involved in these dynamics, i.e. the political and the non-political: The first approach follows a systems theory approach à la Niklas Luhmann and defines the political as one of various social subsystems, none of which takes precedence over the others and each of which functions according to its very specific logic. The political subsystem is, accordingly, concerned with "the production of collectively binding decisions which are supplied by the application of the central medium of power"[10] (Zürn 2013: 14)—as opposed, for example, to the economic subsystem, which "follows the logic of profit and has the price as central medium"[11] (Zürn 2013: 15). The second approach, in contrast,

[9]For a list of various definitions of the political, see, for example, Pelinka and Varwick (2010: 20).

[10]Translated by JK, original quote: Es geht in diesem System um die Produktion von kollektiv bindenden Entscheidungen, die durch den Einsatz des zentralen Mediums der Macht gespeist werden. (Zürn 2013: 14)

[11]Translated by JK, original quote: [...] Teilsystem Wirtschaft, das der Logik des Gewinns folgt und den Preis als zentrales Medium hat. (Zürn 2013: 15)

is derived from a discourse theoretical perspective and defines the political as the space in which the very decision is made "in which social subsystem and according to which functional logic specific decisions are to be made."[12] (Zürn 2013: 17) In this sense, the political is distinct from other social subsystems, providing a kind of meta-sphere which is, at least analytically, if not chronologically, located upstream of the diverse social subsystems. In short: The political can be defined "either as a process of deliberation or as a mechanism for the provision of collective or public goods" (Hay 2007: 90). The systems theoretical and discursive perspectives on the political, on the one hand, and the three dimensions of the political, on the other hand, are useful in understanding how types of politicization and depoliticization processes can be differentiated.

With regards to the systems theoretical and the discursive theoretical understanding of the political, politicization is both: On the one hand, it can denote a process in which issues "become the subject of deliberation, decision-making and human agency where previously they were not" (Hay 2007: 81) and which analytically precedes any assignment to a particular social subsystem (Zürn 2013: 20). Climate change, in this reading, was politicized when humankind first started to consider measures to address this phenomenon, be it through mitigation of GHGs, adaptation to climate change effects or technologies to modify the climate actively, thus no longer assuming the climate to be 'God-given' or its current configuration a necessary and unchangeable feature of the world. On the other hand, politicization can represent the movement of the issue into the political subsystem as the one functioning according to the logic of producing collectively binding decisions concerning the distribution of material and non-material values in society (Zürn 2013: 20).

A similar differentiation is offered by Hay, who connects the two types of politicization with each other in a model containing three consecutive stages (Hay 2014: 309): discursive, societal and governmental politicization. These three stages denote the promotion of an issue from the "realm of necessity to the private sphere" (discursive p.), "from the private to the public sphere" (societal p.) and "from the public to the governmental sphere" (governmental p.), respectively (Hay 2007: 79). Hay's first stage, I find, bears strong resemblance to Zürn's discourse theoretical politicization, while Hay's second and third stage are more closely related to Zürn's systems theoretical politicization. I am particularly interested in discursive or discourse theoretical politicization as a stage preceding political decision-making in the sense of problem closure, as the issue of CE is, arguably, currently still at this stage of affairs. The politicization and depoliticization of CE revolves around the question: Is the way in which these approaches come into the world predetermined—e.g. by the laws of physics—or is it subject to contingency and human decisions? Politicized CE would highlight the latter contingency, as "a process of discursive politicization [sic] would […] involve the promotion of a topic as a public issue where competing interpretations exist as choices." (Wood and Flinders 2014: 161).

[12]Translated by JK, original quote: Im politischen Raum wird geklärt, in welchem gesellschaftlichen Teilsystem und nach welcher Funktionslogik bestimmte Entscheidungen gefällt werden sollen. (Zürn 2013: 17)

The three dimensions of the political—polity, politics and policy—are useful in further differentiating types of politicization and depoliticization processes. For instance, *polity* as a "spatial concept [...] contrasts the sphere of society in which political activities occur with other, non-political spheres, such as religion, the economy, law, education, or science." (Jessop 2014: 208) Accordingly, Jessop identifies discursive depoliticization processes of "sacralisation, marketisation, juridification, scientisation" (Jessop 2014: 212). Similarly, Methmann et al. identify three types of depoliticization processes at stake in climate governance:

> If we were to isolate a general conclusion from the richness and plurality of the previous chapters [of the edited volume], it would be this: global climate governance is being depoliticized through processes of technocratization, economization and securitization. (Methmann et al. 2013c: 249)

Such thematic types of depoliticization—the moving of an issue from the political sphere to the economic, religious, judicial or scientific, for example—or, conversely, politicization—the moving of an issue into the political sphere—are identified by Zürn, from a systems theoretical perspective, as processes of horizontal politicization: Horizontal p. denotes the process of a "shift between functionally differentiated spheres"[13] (Zürn 2013: 21).[14] The polity dimension, thus, highlights the difference of thematic foci on the issue in question:

> For the polity, then, a key aspect of re- and depoliticisation [sic] is the redrawing of the 'lines of difference' between the political and one or more ostensibly non-political spheres – whether the latter are defined as an unmarked residuum in terms of their location outside the political sphere (for example, state vs society, public vs private) or as marked spheres with their own institutional order, operational logics, subjects, and practices (for example, the religious, economic, legal, educational, or scientific fields. (Jessop 2014: 212)

With regards to the dimension of *policy*, this redrawing can take place in the shape of specific policies or political contents: "In particular, many of the modes for removing issues from open political contention can also be found in the formation of specific policies, policy making, policy taking, and policy implementation." (Jessop 2014: 216)

In contrast, *politics*-politicization involves moving an issue into an arena of contestation between diverse interests, potentially "reshaping political practices [...] [,] for example, changing the balance of forces and stimulating new political claims and movements" (Jessop 2014: 208). For example, Zürn differentiates between the individual actors involved in these politicization processes: "Is it states, nongovernmental organizations, media institutions or individual experts?"[15] (Zürn 2013: 21)

---

[13]Translated by JK; original quote: "Verschiebung zwischen funktional differenzierten Sphären" (Zürn 2013: 21)

[14]Zürn uses the term 'horizontal' to contrast vertical p., which indicates the shift of an issue inside one specific subsystem from one geographical level to another, e.g. from the national to the international level. (Zürn 2013: 21)

[15]Translated by JK; original quote: "Sind es Staaten, Nichtregierungsorganisationen, Medien oder individuelle Experten?" (Zürn 2013: 21).

Further, one may ask whether politicization has to be intended or can be an unintended outcome. Jessop, for example, poses the question "whether these processes are intended outcomes of deliberate action or unintended, possibly unacknowledged, effects of societal trends, other processes, or practices with other goals." (Jessop 2014: 207) For Methmann et al., politicization and depoliticization moves are intentional:

> A politicizing treatment of a particular problem, in this sense, is one that brings the underlying antagonisms to the fore, opens up discursive contestation and aims for the transformation of sedimented social structures. Depoliticization, by contrast, involves all counter-strategies, which seek to conceal the contingency of social reality, disclose discursive struggle by silencing alternative views or channel dislocations in a way that fundamental social structures remain untouched. (Methmann et al. 2013b: 13)

I claim that either process does not have to be intended by the actors involved, i.e. politicization and depoliticization moves are not necessarily strategies. Instead, they can conglomerate from the interaction of various discursive processes instigated by various actors with a multitude of objectives in mind. This makes the endeavor all the more relevant to assess how the meaning of CE is established in this early phase of its existence, before the meaning is settled. The present thesis touches upon each of these dimensions of the political: In focusing on decision-making in the context of the construction of meaning of CE, I am interested in how the framing of CE can influence the *content* of climate policy—e.g., by representing CE as a useful or relevant part of climate policy or by representing it as an impossible measure to address climate change. The interactions of various members of the academic community as studied in the empirical part of this thesis describes the *process* in which frames of CE are introduced, reinforced and contested. By defining the meaning of CE, these frames add to the political *structure* of CE, determining or contesting whether CE approaches are to be considered legitimate or illegitimate tools in addressing climate change.

Beyond these various ways of analytically differentiating types of politicization and depoliticization processes according to the system theoretical and the discourse theoretical perspective or the three dimensions of the political, various authors have addressed the issue of normativity: Is politicization to be understood as a normative concept, or not? And if it is, is politicization as a process desirable, or is it undesirable? Zürn argues that various types of literature on politicization differ in their evaluation of the phenomenon:

> Whereas European Union (EU) studies in general terms mainly ask about the disintegrative effects of politicization […], the study of de-politicization in the national context and of politicization in international institutions often emphasizes the normatively positive aspects of increased mobilization. (Zürn 2019: 978)

Other authors, in particular from discourse theoretical perspectives, make explicit their own normative position with regards to these questions. For instance, Hay argues:

> Though I acknowledge this to be a personal view I, for one, would wish staunchly to defend the normative stance of much of the (first wave) literature on depoliticisation, just as I would wish equally staunchly to defend the idea that politicized governance (which is visible deliberative and accountable) is normatively preferable to depoliticised governance (in which

> decisions are made ostensibly in the public good but in private and without the capacity for public scrutiny). (Hay 2014: 297)

Similarly, with a view to the specific issue of climate change, Pepermans and Maeseele claim that

> [t]he dominant consensus-building perspective problematizes the politicization of climate change as an important barrier to climate action [...]. In contrast to this, the critical debate perspective problematizes climate change's capture in a depoliticized consensus and calls for its repoliticization to revive democratic debate and citizenship. (Pepermans and Maeseele 2016: 478)

Zürn argues that politicization is both a precondition for and, if excessive, destruction of democratic and liberal society:

> On the one hand, we can see that politicization is both precondition of democratization of a society and feature and best protective mechanism of a democratic society. [...] On the other hand, excessive politicization destroys the liberal and civil society foundation of a democratic society. (Zürn 2019: 24–25)

The typologies of politicization and depoliticization outlined above, in particular the question of normativity, provide dividing lines for a number of different strands of literature applying the concept of politicization to empirical analysis. A short overview over this very rich literature allows me to situate my own research endeavor more clearly with regards to this "chameleon-like concept" (Jessop 2014: 207).

Politicization as a concept in research use—Different strands of literature[16]

Zürn offers a classification of politicization studies in the literature along "different political levels" (Zürn 2019: 978), namely the domestic or national, the regional (specifically the European) and the global level. For the national level, i.e. in "comparative politics, the study of politicization was always part of the field—without always using the term itself." (Zürn 2019: 979) Zürn considers the decline of classic forms of political participation—such as voter turnout as well as party membership—as well a recent rises of right-wing populist parties as processes of de- or repoliticization, respectively. On the European level, "the broadening of the integration process into non-economic issue areas led a rise in identity politics, which entangled European integration in a growing conflict between elites and masses: from permissive consensus to constraining dissensus." (Zürn 2019: 982) Thus, a formerly non-politicized EU became the arena of political struggle. On the global level, as a consequence of an increasing authority shift from the national to the international level, Zürn finds politicization to be a concept in particular in the study of "inter- and transnational institutions" (Zürn 2019: 985).

Zürn criticizes the level-specific foci of these three strands of literature: He argues that "[a] focus on only one political level in the study of politicization runs the danger of overlooking counter-trends on the other political levels" (Zürn 2019: 987) such as the depoliticization on the national and a concurrent politicization on the regional or

---

[16]This is not an attempt to offer a comprehensive overview over this very rich strand of literature, but rather to provide some systematic guidance in this field in order to be able, as a next step, to situate the present thesis in it.

global level. I claim that the level is not the most useful criterion to classify politicization studies for the use of this thesis, as I am not interested in any specific political level, but rather on a discourse than spans various geographical levels. Instead, and inspired by the discussion offered by Jessop (2014), I propose to represent the literature on politicization according to the three dimensions of the political: The differentiation between politicization in a politics, a polity and a policy dimension allows best to understand the specific mechanisms of politicization in question in each individual study. The notion of 'dimension' highlights that the three classes of studies are not mutually exclusive: Many studies are relevant for more than one of the three dimensions. In contrast, I find that the classification according to political level veils the relations between studies addressing different political levels. For example, I find that there is a number of studies on *global* politicization processes which do not share a focus on institutional change, as the global-level studies identified by Zürn, but instead argue from critical theory perspectives about discursive processes of, for example, economization of political issues (Jessop 2014; Wood and Flinders 2014; Foster et al. 2014; Burnham 2014; see, for example, Burnham 2001; Donmez 2019; Warner 2019; Flinders and Wood 2014).[17] While Zürn does not take these into account in his classification in 2019, he notes critical perspectives on the *domestic* level, for example in Hay (2007).[18]

A number of studies in the literature directs attention to politicization in the *politics* dimension—i.e. on processes through which an issue becomes subject to "formally instituted, organized or informal practices that are directly oriented to, or otherwise shape, the exercise of […] power" (Jessop 2014: 209) according to potentially conflicting sets of interests expressed by various actors, when it was not subject to these practices before. Prominently, I understand studies on the politicization of climate science, in particular in recent domestic US-American politics, to belong to this strand (see, for example, Bolsen and Druckman 2018; Bolsen et al. 2019; Schmid-Petri 2017; Akerlof et al. 2012; McCright and Dunlap 2011; Pepermans and Maeseele 2014; Colvin Under review). Recent discussion of the discursive constitution of the Brexit also belong to this set (Zappettini and Krzyzanowski 2019; Dennison and Geddes 2018). In this strand of literature, politicization can be evaluated as a positive phenomenon which can, for example, contribute to increased political mobilization "as a response to a period of de-politicization in liberal democracies" (Zürn 2019: 977) or which, from a critical theory perspective, can help "revive democratic debate and citizenship" (Pepermans and Maeseele 2016: 478). It can, however, also be evaluated as a problem, for example when "partisan group identity and the politicization of science" undermines "the impact of a scientific consensus message about human-induced climate change" (Bolsen and Druckman 2018: 389). These examples indicate how texts that I subsume under the politics dimension also

[17]This set of studies often applies the British notation 'politicisation' instead of the US-American spelling 'politicization'.

[18]Interestingly, Zürn in 2008 discusses the critical claim that many issues of international political action have become economized—shifted to a sphere of market logic—and offers the interpretation that "as the market gains grounds on politics, the market becomes politicized." (Zürn 2008: 293)

touch upon the other dimensions of politicization. For example, the politicization of climate change in the domestic US-American discussion relates not only to politics as processes of negotiating interests and applying power, but also to polity as the spatial sphere and the established structures "in which political activities occur" (Jessop 2014: 208): The politicization of climate change can be understood as an attempt to 'de-scientisate' climate change, i.e., to move the issue out of a sphere which follows the "logic of truth and uses reputation as central medium"[19] (Zürn 2013: 15) into the political sphere.

Politicization in the *polity* dimension has been addressed by studies, for instance, on the moving away of the EU from being a "regulatory state focus[ing] neither on distributive nor on salient issues but manag[ing] efficiency and coordination problems" (Zürn 2019: 982). Similarly, literature on public opposition contesting the authority of inter- and transnational institutions, thus moving international relations out of the field of technocratic execution (see, for example, Zürn 2014; Ecker-Ehrhardt 2018) can be systematized through this focus. Such studies can also, conversely, relate to the politics and policy dimensions of politicization. EU studies touch upon politics dimension of politicization as they address, inter alia, "a *polarisation of opinion*, and [...] an *expansion of actors and audiences* engaged in monitoring EU affairs" (de Wilde et al. 2016: 4) or to the policy dimension when they touch upon the content of collectively binding decisions, for example when "the EU [...] might be turning more divided about the appropriateness of the norms that underpin EU external relations." (Costa 2019: 790).

For the purpose of the present thesis, the third or *policy* dimension of politicization is of most interest, i.e. the assessment of moving the content of specific collectively binding decisions into the realm of contingency—or out of it. Literature from critical perspectives have been highly insightful with regards to this dimension, for example regarding environmental or climate policy (see, e.g., Pepermans and Maeseele 2014; Methmann et al. 2013a; Kvidal-Rovik 2018; Remling 2018; Weisser and Müller-Mahn 2017; Goeminne 2012),[20] economic policy (see, e.g., Jessop 2014; Foster et al. 2014; Burnham 2014; Warner 2019; Donmez 2019) or technology policy (see, e.g., Bates et al. 2014). In this strand of literature, depoliticization on various levels and across levels is criticized. Depoliticizing instances of economization, technocratization and securitization are seen to narrow down the field of possible action in a very specific way, for example with regards to issues such as climate change or financial crisis:

> The contributors of this book have engaged with three [...] stories: the technocratization, the economization and the securitization of climate change. And they found that most of the plots of these stories are highly depoliticized; that is, these stories picture a superficial and highly contingent picture of climate change; a picture that downplays the structural entanglement of climate change with the basic structures of our lives [...] and narrows it

[19] Translated by JK; original quote: "Wissenschaft unterliegt der Logik der Wahrheit und nutzt Reputation als zentrales Medium." (Zürn 2013: 15)

[20] An early account of a depoliticizing tendency in environmental politics is offered by Hajer (1997: 54).

down to a technical problem, an economic risk, or a security threat. (Methmann et al. 2013c: 249)

Arguably, many of these studies concentrate on the politics dimension of politicization, for example by assessing the role of various actors and groups of actors shaping, in some cases intentionally, processes of depoliticization and repoliticization (Anshelm et al. 2018: 14; Methmann et al. 2013b: 13). Furthermore, many of the studies in this strand are united in dedication to the normativity of the concept of politicization, and a strongly positive understanding of the respective processes, and a negative evaluation of de-politicization processes (Pepermans and Maeseele 2016; Methmann et al. 2013a; Weisser and Müller-Mahn 2017).

The present study as a contribution to the politicization debate

In contrast to these studies providing insight into the policy dimension of politicization, the present thesis is not based on a critical theory perspective such as Neo-Gramscianism or Critical Political Economy (Jessop 2014; Warner 2019). Also, and related to this point, it does not start from a dichotomous differentiation of what, for example, Pepermans and Maeseele, call the 'foundational' and 'post-foundational' outlooks on climate change: According to Pepermans and Maeseele, the foundational outlook "understands climate change first and foremost as a physical phenomenon that can be observed, discerned, quantified, and managed with some degree of scientific objectivity" (Pepermans and Maeseele 2016: 480). The authors contrast this position to the post-foundational outlook, which "implies that climate change is only observable through representations, which are always the provisional, contingent, and changeable result of power relations and struggles" (Pepermans and Maeseele 2016: 480). While the former, epistemologically and ontologically, "entails a strict distinction between, on the one hand, indisputable, dispassionate facts, and ideological, passionate, interest-driven politics, on the other", the latter "rejects strict dichotomies between science and politics, knowledge and power, or normative judgement and empirical analysis" (Pepermans and Maeseele 2016: 480).

The modernist social constructivist perspective espoused in this thesis is in agreement with the rejection of *strict* dichotomies between science and politics or knowledge and power.[21] However, from this perspective, rejection of this strict dichotomy is not in contradiction with understanding climate change *both* as a "physical phenomenon that can be observed, discerned, quantified, and managed with *some degree* of scientific objectivity" (Pepermans and Maeseele 2016: 480, italics mine) *and* a phenomenon that, by humans, is "only observable through representations, which are always" *to some degree* "the provisional, contingent, and changeable result of power relations and struggles" (Pepermans and Maeseele 2016: 480). Social reality, in this view, is made up both of material objects with given physical characteristics and of social meanings of these material objects. This means that I assume that science and politics or knowledge and power are not entirely dichotomous or separate, but at the same time, they are also not completely identical: Neither are scientific actors only engaged in power struggles, nor are they exclusively engaged in the production of objective knowledge—but both. Giving up this strong dichotomy

[21] See Chap. 3.

allows me to take seriously both the contributions of science—or, more broadly speaking, academic research—to our knowledge about phenomena in the world, and their contribution to constructing the social meaning of these phenomena,[22] for example through framing.[23] Accordingly, I do not agree with Pepermans' and Maeeseele's claim that "only the assumption of the critical debate perspective allow for the recognition that society is contingent, that it can be understood and organized differently, and that there should be scope for questioning currently existing social arrangements." (Pepermans and Maeseele 2016: 483) The modernist social constructivist perspective takes these issues into account as well, without, however, rejecting academic knowledge about phenomena such as climate change wholesale.

The contribution of the present thesis to the debate on politicization, then, is the modernist social constructivist description and analysis of a specific type of politicization and depoliticization processes: *In the academic discussion on CE, various frame types legitimizing and delegitimizing these diverse emerging approaches compete with each other, contributing to constructing the meaning of CE approaches as either political or non-political issues and, accordingly, open up or narrow down the scope of possible action to handle CE.* I am particularly interested in the discourse theoretical understanding, in the processes through which issues "become the subject of deliberation, decision making and human agency" (Hay 2007: 81). I do not assume intent to be a defining feature of politicization or de-politicization processes: I think that these processes can be either "intended outcomes of deliberate action or unintended, possibly unacknowledged, effects of societal trends, other processes, or practices with other goals." (Jessop 2014: 207) Accordingly, the way in which various academic actors frame CE does not necessarily have to be understood as serving these actors' intentions in rendering CE approaches political or non-political. These processes are, I argue further, not confined to a specific political level, but take place—due to the transnational nature of academic interactions—beyond clearly defined national or regional boundaries (Zürn 2019).

The academic discussion on CE would represent an instance of de-politicization if one particular frame could be found to be dominant in the discussion, for example clearly representing CE as a technological challenge or a security problem. Then, CE would not be understood as subject of deliberation, but consensus would be assumed concerning the nature of the issue, as well as the proper way of addressing it—for example, according to the logic of security through action 'necessary' to prevent a threat, or according to a technocratic logic through providing fixes to clearly defined technological problems. Instead, in the research conducted for the present thesis, I find that various legitimizing and delegitimizing frames are present in the discussion, often even within one and the same text. This leads me to suppose that the academic discussion on CE is an instance of politicization of the issue of CE, as the presence of various, partly contradictory, frames opens up the discussion and creates a space for deliberation and dispute.

---

[22] See Chap. 3.

[23] See Chap. 3.

From a systems theoretical perspective on politicization, the frames present in the academic discussion on CE can be understood to spark movements of the issue from various functionally differentiated thematic spheres—for example from the scientific or the economic sphere—to the political sphere, for instance when calls are voiced that CE should be considered by elected political decision-makers and decisions be made to define conditions of CE research, development and use (see, e.g., Pasztor 2017a); and vice versa, movements of the issue from the political sphere to the economic or ethical sphere, for example when actors consider only economic criteria in assessing these approaches. The ideal-type frames making up the theoretical construct applied in the present study correspond with the depoliticization processes of securitization, technocratization and economization identified in the politicization literature. I add a fourth theme of appropriateness which might spark a process of normativization-depoliticization, i.e. the movement of an issue into a sphere in which clearly defined norms are the central medium.

This thesis, then, contributes to the political science debate on politicization and depoliticization by (1) indicating a possible middle ground between what Pepermans and Maeseele (2016: 478) call "consensus-building" and "critical" positions on the politicization of climate change, (2) offering an example of systematically assessing how transnational actors other than protesters can play a role in de-/politicization processes, (3) adding a fourth type of depoliticizing processes to the three types identified by Methmann et al. (2013c).

Concerning (1), a middle ground can be provided to two divergent "perspectives in the social sciences regarding the desirability of (dis-)agreement about climate change" (Pepermans and Maeseele 2016: 478). This middle ground consists in reconciling the post-foundational and the foundational outlook on climate change, thus allowing me to take into account both the contributions of academic research to our knowledge about phenomena in the world, and their contribution to constructing the social meaning of these phenomena. This reconciliation might help, in the literature of politicization of climate change, to move beyond the entrenchment of positions understanding the process as normatively positive—for example as a way to "revive democratic debate and citizenship" (Pepermans and Maeseele 2016: 478)—and those regarding the process as normatively negative—for example when "the inherent uncertainty of science is emphasized to cast doubt on scientific consensus" (Bolsen et al. 2019: 147). This, in turn, would allow these two different strands of literature to find common ground from which to exchange insights and further understand the phenomena of politicization of climate change.

Regarding (2), the study of de-/politicization in IR has, up to this point, addressed the roles of various actor groups such as "individual citizens, non-governmental organizations (NGOs), parties, lobby groups, and governmental bodies" (Zürn 2019: 985). I contribute to the literature by assessing the role of a different group of actors, namely academic actors involved in the CE discussion. In doing so, I connect insights from STS to the politicization debate, such as the conceptualization of academic actors through the notion of Epistemic Communities (EC). Regarding (3), the depoliticization process connected with appropriateness frames may increase in relevance in the aftermath of the IPCC 1.5 °C report, which arguably applies an appropriateness frame by stressing in its executive summary that "[e]thical considerations, and the principle of equity in particular, are central to this report" (Allen et al. 2018: 51). Considering an issue 'normatively imperative' can lead to depoliticization as the issue in question is then no more one of deliberation, but one of normative certainty.

1.2 Important Terms

While the score of terms relevant to this thesis will be defined in the course of the discussion, some terms relevant to the research design warrant clarification at this point. This is done chiefly in order to disclose the position adopted here with regards to questions of definition and in order to enable the reader to pinpoint the location of the present thesis on the map drawn by the relevant literatures. Further clarification of the terms defined below is provided in the following chapters.

Climate Engineering and climate change

In the influential Royal Society Report of 2009, 'Climate Engineering' (CE) has been defined as "the *deliberate large-scale manipulation of the planetary environment to counteract anthropogenic climate change*" (Shepherd et al. 2009: 1; Rickels et al. 2011b; NAS 2015b). Following the authors of the IPCC AR5, 'climate change' is defined as "a change in the state of the climate that can be identified (e.g., by using statistical tests) by changes in the mean and/or the variability of its properties, and that persists for an extended period, typically decades or longer." (IPCC 2013a: 1450) In line with the UNFCCC, anthropogenic climate change is understood as "a change of climate which is attributed directly or indirectly to human activity that alters the composition of the global atmosphere and which is in addition to natural climate variability observed over comparable time periods." (UN 1992: Art. 1)

Various definitions of CE have been discussed critically in the literature (Bellamy et al. 2012b: 599; Cairns 2013; Markusson et al. 2016). In contrast to Cairns and Stirling (2014), I decided to adopt the Royal Society definition in order to be able to identify discussions of CE approaches even if they were not designated with the term 'climate engineering'. Further decisions were made for this study with regard to three ambiguities connected to the concept of CE: First, various different terms exist in literature denoting roughly the same set of approaches, among them 'climate engineering', 'geoengineering', less commonly 'climate modification' and 'climate remediation' (SPP 1689 2019a: 24). While 'geoengineering' is more commonly used in publications from the Anglo-American language area and has the advantage of precedence—the term has been coined in 1977 (Marchetti 1977)—'climate engineering' has been shown in empirical studies to function more efficiently for people previously unacquainted with the concept in differentiating it from, for example, geotechnical engineering (Mercer et al. 2011: 8).[24] The term is more specific in

[24] Geotechnical engineering is a sub-discipline of civil engineering and refers to "the manipulation or study of components of the earth" (Mercer et al. 2011: 4). Geotechnical engineering projects aim at, for example, protecting populations against the impacts of earth quakes or designing earthworks.

denoting the field that might be subjected to engineering by specifying the climate instead of applying the broader concept of Earth represented by the prefix 'geo'.[25]

Secondly, different positions exist in the community with regards to the question whether an umbrella term is useful at all, or whether approaches should rather be discussed individually or under more specific categories. The US National Academy of Sciences (NAS) chose to publish two individual reports, one on Carbon Dioxide Removal and one on Solar Radiation Management approaches (NAS 2015a, c). Conversely, a recent report by Priority Program 1689 (SPP 1689) for the assessment of CE founded by the German Research Foundation (*Deutsche Forschungsgemeinschaft*, DFG) discusses both sets of approaches under the umbrella term (SPP 1689 2019a). The two sets of approaches differ in their point of influence on the climate system—*Carbon Dioxide Removal (CDR)* influencing the carbon cycle by removing carbon that has already been emitted, Solar Radiation Management (SRM) affecting the global radiation budget by reducing the amount of "sunlight absorbed by Earth" (NAS 2015b)—as well as many of their physical, biological, economic, social, political and ethical prerequisites and consequences. CDR approaches "reduce the levels of carbon dioxide (CO_2) in the atmosphere, allowing outgoing long-wave (thermal infra-red) heat radiation to escape more easily" (Shepherd, Cox et al. 2009, 1). A similar, albeit broader, term is that of negative emissions technologies (NETs) (see Minx et al. 2018; Fuss et al. 2016). The objective of CDR methods, then, is to increase the capacity of natural carbon sinks (terrestrial or oceanic biological systems) or to create artificial new sinks. Examples are ideas such as the fertilization of the world's oceans with iron in order to enhance the uptake of CO_2 by phytoplankton (ocean fertilization, OF), or the extraction of CO_2 from the air through chemicals to create carbon solids (direct air capture, DAC). SRM approaches, in contrast, "reduce the net incoming short-wave (ultra-violet and visible) solar radiation received, by deflecting sunlight, or by increasing the reflectivity of the atmosphere, clouds or the Earth's surface" (Shepherd et al. 2009: 1). The objective of these methods is the reduction of the global temperature by influencing the radiation budget (the ratio between incoming and outgoing radiation) of the atmosphere. SRM approaches do not aim at influencing the carbon cycle, thus do not intentionally affect other detrimental effects of CO_2 increases, such as ocean acidification. Examples for SRM methods are the brightening of marine clouds to increase the albedo—the fraction of radiation reflected back into space—over the ocean (cloud modification, CM) or the creation of an artificial shield against solar radiation by spraying sulfate aerosols into the stratosphere (stratospheric aerosol injection, SAI). However, both sets of approaches pursue the same higher level goal, namely the large-scale intervention of the climate system. As such, both sets are highly relevant with regards to climate decision-making as one or several approaches from either category could, theoretically, be chosen as a measure to address climate change (Image 1.1).

Thirdly, positions differ with regards to the question how CE approaches should be categorized in relation to other climate measures such as mitigation and adaptation

[25] Both terms 'climate engineering' and 'geoengineering' were applied in the document search to generate the corpus of data as the latter is highly frequent in particular in Anglo-American literature.

Image 1.1 Definition of CE applied in this study

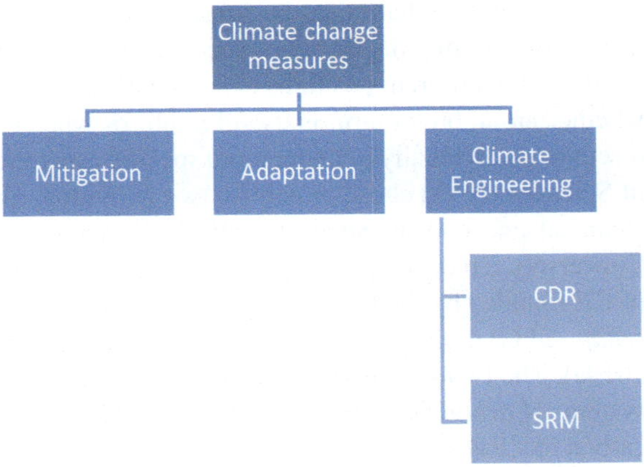

(Heyward 2013). With regards to the question of the position of CE relative to measures such as mitigation and adaptation, CE will be understood, here, as a third category of options (Bellamy 2016: 136).[26] The three decisions made here with regards to the three ambiguities serve to place CE approaches on the map of climate policy: The first defines the scope of the term, the second its relation to other climate measures, the third its relation to further classifications of approaches.

Discourse, practice and meaning

'*Discourse*' is understood, here, following Maarten Hajer, as "an ensemble of ideas, concepts and categories through which meaning is given to social and physical phenomena, and which is produced and reproduced through an identifiable set of practices." (Hajer 2006: 67) In discourse, *meaning* is created, enacted and changed which can, in turn, influence and shape political phenomena, actions and institutions (Anshelm and Hultman 2015: 11): "Meanings are not just representations of people's beliefs and sentiments about political phenomena; they fashion these phenomena." (Wagenaar 2011: 3) I consider the process of constructing the meaning of CE not to be determined exclusively either by social structures or political actors: Humans both capture social meanings which objects have, and change or reproduce these meanings (Giddens 1984). I apply a very broad sense to the term discourse here: I subsume under the term both linguistic and non-linguistic elements of meaning-construction, i.e. both speech and practice. '*Practices*' are defined as "meaningful patterns of action, which, in being performed more or less competently, simultaneously embody, act out, and possibly reify background knowledge in and around the material world." (Adler and Pouliot 2011: 4) Both meaningful action—or practice—and meaningful talk—or, in the narrower sense, discourse—can construct or deconstruct the meaning of an object: "[D]iscourses are a matter of practice as well as words, for actions in the social realm are always accompanied by language that establishes the meaning of action." (Dryzek 2006: 3)

[26] A convincing alternative differentiation is proposed by Heyward (2013), who differentiates five types of responses to climate change, two of which are CDR and SRM, respectively.

Frames and the framing effect
'*Frames*' are understood here as micro level elements of the macro level phenomenon of discourse (Markusson 2013: 6). As such, they represent one way among others of assessing the constitution of a discourse. Other elements of discourse are, for example, metaphors, storylines, arguments, narratives, or imaginaries (see Uther 2014: 68). These elements are not mutually exclusive, but instead offer different perspectives on the discourse. Metaphors, for example, are often used in frames. The frame, according to Entman, is structured in such a way as to offer a proposal for action as one of four frame elements expressed in their functions: Frames

> *define problems* – determine what a causal agent is doing with what costs and benefits, usually measured in terms of common cultural values; [...] *diagnose causes* – identify the forces creating the problem; [...] *make moral judgements* – evaluate causal agents and their effects; [...] and *suggest remedies* – offer and justify treatments for the problems and predict their likely effects. (Entman 1993: 52)

Thus, the concept of frame is highly relevant for the objective of this study as it connects meaning with action, providing insight into how meaning can become causally relevant: "A frame is a perspective from which an amorphous, ill-defined, problematic situation can be made sense of and acted on." (Rein and Schön 1993: 146) Frames do not only present interpretations of the central problem at hand, but also connect them with specific modes of action: "The problem-setting stories [...] link causal accounts of policy problems to particular proposals for action and facilitate the normative leap from 'is' to 'ought'." (Rein and Schön 1993: 148) This means that the frame connects the two elements—identification of problem and choice of measure to address the problem—of political decision-making outlined above. This *framing effect* is famously discussed by Tversky and Kahneman, who find that "[i]ndividuals who face a decision problem and have a definite preference [...] might have a different preference in a different framing of the same problem [...]" (Tversky and Kahneman 1981: 457; see also Kahneman and Tversky 1984).
Construction and constitution
The present thesis aims to systematically assess how the meaning of CE approaches is constructed in the academic discussion with regards to the legitimacy or illegitimacy of these approaches as measures to address climate change. To do so, I empirically analyze the constitution of the academic discussion on CE, focusing on its constitution by frames and frame elements. '*Construction*' denotes, here, the process through which the social meaning of objects is developed in discourse, for example through frames. The construction of the meaning of CE lies in the focus of this thesis. '*Constitution*', in contrast, describes the status of an object as a system made up of various parts with specific properties. In assessing the constitution of an object, we ask: "'how are things in the world put together so that they have the properties that they do?'" (Wendt 1998: 103) I am interested, in this thesis, in how the meaning of CE is constituted in order to have the property to either legitimize or delegitimize this set of approaches as measures to address climate change.
Legitimacy
'Legitimacy' is considered, in this thesis, as an element of the meaning of an object. I do not apply a prescriptive understanding of legitimacy as a "normative quality

that is attributed by theorists to certain political systems" and defining under "which conditions governance deserves the predicate legitimate" (Steffek 2003: 253). The notion of legitimacy that is at issue here is decidedly descriptive: the study aims to show which frames can be traced that legitimize a technology as a governance measure, and which frames exist which delegitimize said technology. The analysis of legitimacy in its descriptive conception was founded by sociologist Max Weber, who defines the term as the prestige of being considered binding ("das Prestige der Verbindlichkeit") (Weber 1980 [1922]: 16). This shifts the focus towards the "social functions of legitimacy rather than in the normative justification of governance" (Steffek 2003: 253). In the case of the present thesis, the question is whether CE as a measure to address anthropogenic climate change is considered something that should be done, or must not be done.

Technology

When I write about CE technologies, I use the term 'technology' not in the narrow sense, denoting individual technical devices or artefacts (Winner 1977; Bijker 2006)—for example a blimp or airplane for carrying aerosol particles into the stratosphere in order to conduct SAI. Instead, when I discuss technology, I apply the term in a broader sense, referring to a complete socio-technical system in the sense of a "configuration[…] of actors, technologies [in the narrower sense] and institutions for the fulfilment of societal functions" (Schot and Kanger 2018: 1045). The adoption of this broader sense of the term is relevant as only CE approaches developed in this sense could be expected to be influential on the global climate: as socio-technical systems spanning all three dimensions of this concept. These dimensions are

> (1) tangible technologies, for example, artefacts, devices, and infrastructures; (2) actors and social groups who develop, use, maintain, and regulate technologies; and (3) rules (understood as regimes) that guide perceptions and activities of actors and social groups (Geels 2007: 127).[27]

An individual DAC plant cannot be expected to make an impact on the global climate—a whole infrastructure would need to be in place, as well as a regime regulating processes and actors with specific interests and knowledge with regards to the plants and infrastructures. As the defining characteristic of a socio-technical system is the "fulfilment of societal functions" (Schot and Kanger 2018: 1045), a socio-technical system of CE would be defined by providing the societal function of "*counteract*[ing] *anthropogenic climate change*" (Shepherd et al. 2009: 1). The approaches subsumed under the term CE are considered, at this point, to be 'emerging' technologies as none of them, as of yet, present a complete socio-technical system.

[27]In this, I follow the understanding of the term 'socio-technical system' as encompassing all three dimensions as espoused in Geels (2007: 127); Schot and Kanger (2018: 1045), rather than conceive of the 'system' as comprising only the material dimension (devices and infrastructures) and differentiating from it the 'regime' (rules and procedures) and the actors involved (Geels 2004: 898, 903).

1.3 The Interplay of Theory and Data

The design of a research project is a "plan that defines how the research project is to be conducted and, in particular, how empirical evidence is to be used in order to receive answers to the research questions" (Gschwend and Schimmelfennig 2007: 13).[28] Thus, academic research can be understood as a "dialogue between theory and data" (Gschwend and Schimmelfennig 2007: 14)[29] or an "interplay between theory and data" (Behnke et al. 2012: 27).[30] The research design is a model of the project in the sense of a "simplified description [...] of a system or process" (OUP 2019). It provides an overview of the specific way in which the two elements of research—empirical data and theoretical statements—are brought into contact with each other in the concrete case of the individual research project[31]: The research design, then, shall incorporate the following elements: the identification of the empirical starting point which motivates the project, the definition of the central research question, the identification of central theoretical considerations, the determination of the methodical approach, and the choice of cases and observations, the interactions between theory and data (see Gschwend and Schimmelfennig 2007: 15).

1.3.1 An Imbalance Between Relevance and Political Consideration—The Empirical Starting Point

The present thesis is motivated by the *observation* that a peculiar imbalance exists regarding CE technologies concerning, on the one hand, their *potential relevance for climate governance and impact on our climate*, and, on the other hand, the current *lack of political deliberation and decision-making* on the question whether and how we want these technologies to enter our lives (Image 1.2). On the basis of the first aspect, interest in CE outside of academia can be expected to rise. Accordingly, it is important to assess what the understanding of CE would be that this interest would

[28]Translated by JK; Original: "Ein Forschungsdesign ist ein Plan, der festlegt, wie das Forschungsprojekt ausgeführt werden soll, und insbesondere, wie empirische Evidenz dafür verwendet werden soll, um Antworten auf die Forschungsfragen zu erhalten." (Gschwend and Schimmelfennig 2007: 13)

[29]Translated by JK; Original: "Grob gesagt kann man wissenschaftliche Forschung als Dialog zwischen Theorie und Daten verstehen." (Gschwend and Schimmelfennig 2007: 14)

[30]Translated by JK; Original: "Dabei gibt es ein Wechselspiel zwischen Theorien und Daten, das in beide Richtungen verläuft." (Behnke et al. 2012: 27)

[31]A theory is, for this purpose, defined as a set of general statements about reality (Behnke et al. 2012: 21; Spindler and Schieder 2010: 20). Theories of international relations are defined following Spindler and Schieder (2010: 20): "*Theorien* der Internationalen Beziehungen treffen [...] allgemeine Aussagen über [das] [...] Beziehungsgeflecht grenzüberschreitender Interaktionen und das darin politisch relevante, auf Wertzuweisungen ausgerichtete Handeln von staatlichen und nichtstaatlichen Akteuren."

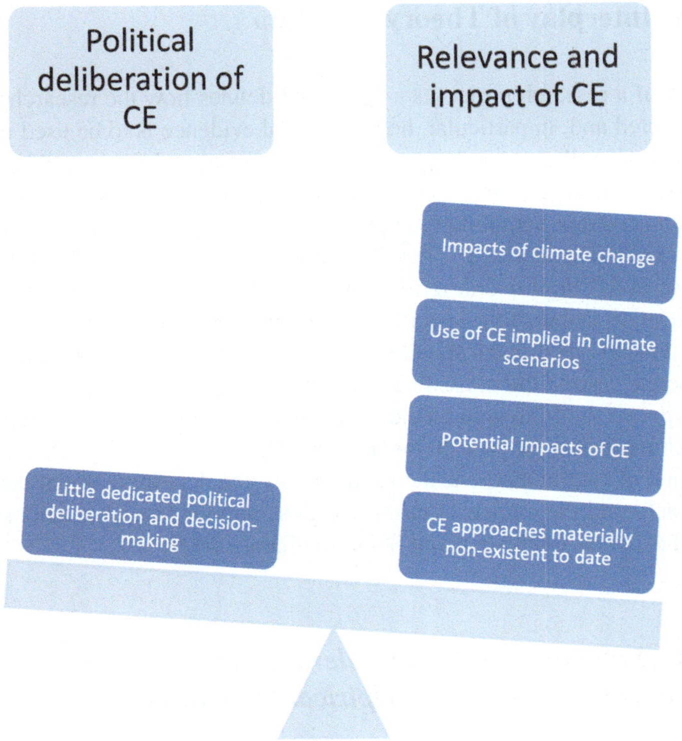

Image 1.2 Empirical observation motivating the present study

start from. Before I present the theoretical and methodical approach espoused here to do so, I present the imbalance in more detail.

On the one hand, a number of issues feed into the potential relevance of CE approaches for climate governance: Firstly, the *real-world impact* of inadvertent anthropogenic climate change already at present has been identified in scientific study and experienced in everyday life. Atmospheric CO_2 concentrations reached 405.5 parts per million (ppm) in 2017 (WMO 2019: 9). Since industrialization began, humankind has emitted approximately 2200 billion tons of CO_2 into the atmosphere (Rogelj et al. 2018: 107). Concerning average temperatures, the years between 2015 and 2018 have been found to be the "four warmest years on record" (WMO 2019: 2). Anthropogenic activities are estimated to "have caused approximately 1.0 °C of global warming above pre-industrial levels, with a likely range of 0.8–1.2 °C" (IPCC 2018a: 6). In Germany, for example, the average annual temperature even rose by 1.3 °C between 1881 and 2014 (SPP 1689 2019a: 9). The IPCC expresses high confidence that global warming will "reach 1.5 °C between 2030 and 2052 if it continues to increase at the current rate" (IPCC 2018a: 6). Climatic changes such as changes in average temperature and precipitation patterns already have impacts on physical and social systems such as ice coverage, drought and flood patterns, coastal erosion, terrestrial and marine ecosystems, wildfire, food production, livelihoods, health and economics (IPCC 2014b: 50; Hoegh-Guldberg et al. 2018: 212ff.). For example, Europe experienced the socio-economic impacts of climate change recently in the summer of 2018, when agriculture, energy production, infrastructures for the transport of goods and other social and physical systems suffered from extreme temperatures and low precipitation rates (dpa 2018; Tagesschau.de 2018; DWD 2019). Similar temperature and drought patterns are expected to be 30 times more likely due to anthropogenic climate change than they would have been without human interference (Met Office 2018; DWD 2019).

Secondly, the ability to use some approaches of CE can be understood to be *assumed in many climate scenarios* which allow adherence to the climate targets set in multi-lateral political decision-making. Concerted international efforts have been made, and are currently being made, to modify inadvertent human interaction with the global climate, in particular to mitigate the emission of GHGs for example on the international level through the negotiations under the UNFCCC. The 2015 Paris Agreement (PA) of the Parties to the UNFCCC emphasizes the aim of "holding the increase in the global average temperature to well below 2 °C above pre-industrial levels and pursuing efforts to limit the temperature increase to 1.5 °C above pre-industrial levels." (UNFCCC 2015: Art 2) However, significant gaps remain between climate action that has been implemented—or pledged to be implemented in the future—by the Parties of the UNFCCC on the one hand and the amount of GHG mitigation required in order to remain inside of the Paris limits (UNEP 2018: xiv) on the other. In spite of the pledges, "[g]lobal greenhouse gas emissions show no signs of peaking" in reality (UNEP 2018: xv): Instead, emissions from sectors such as energy production continue to increase globally (IEA 2019).

The authors of the 2018 IPCC Special Report 'Global Warming of 1.5 °C' (1.5 °C Report) highlight that the question whether it is feasible to remain below the 1.5 °C threshold is a complex one, involving "geophysical, environmental-ecological, technological, economic, socio-cultural and institutional" dimensions (Allen et al. 2018: 52). If a remaining carbon budget—an amount of 'allowed' emissions to remain below the threshold—of 580 $GtCO_2$ is assumed,[32] at current emission rates (around 42 $GtCO_2$ in 2017), "net zero global CO_2 emissions [would need to be met] in about 30 years" (Rogelj et al. 2018: 107) in order to remain below 1.5 °C of warming. As CO_2 emissions are produced through numerous and various human activities—from food or energy production over mobility, housing and heating to manufacturing—the goal of net zero CO_2 emissions requires a shift to renewable energy production, the electrification of machinery, cars and heating, the adoption of GHG-free fuel for cars, trains and planes, the significant reduction of meat consumption and various other measures. A number of observers argue that the PA's objective will very probably require consideration of, at the very least, some forms of CDR (C. H. Greene et al. 2017: 279; Horton et al. 2016: 3; Minx et al. 2017: 2; Goldberg and Lackner 2015: 238; Jones et al. 2016: 2; Gasser et al. 2015: 5; Smith et al. 2016: 1400; Williamson 2016: 155; Rockström et al. 2016: 468; Kriegler et al. 2018: 2; Fuss et al. 2016: 2;

[32] This number is subject to a range of uncertainties elaborated in Rogelj et al. (2018: 107).

Van Vuuren et al. 2018: 391) in order to counteract residual emissions even if all these measures are successfully put in place. The majority of scenarios in line with remaining below 2 °C or even 1.5 °C of warming above pre-industrial levels in the IPCC AR5 assume the large-scale application of at least some form of CDR such as afforestation or bioenergy production with carbon capture and storage (BECCS) (IPCC 2014a: 57; Fuss et al. 2016: 2). All four illustrative model pathways allowing to remain below global warming of 1.5 °C outlined in the IPCC 1.5 °C Report make use of carbon dioxide removal approaches (IPCC 2018a: 16): "All pathways that limit global warming to 1.5 °C with limited or no overshoot project the use of carbon dioxide removal (CDR) on the order to 100–1000 $GtCO_2$ over the twenty-first century." (IPCC 2018a: 19) Van Vuuren et al. assess alternative, deep mitigation pathways to remain below 1.5 °C of warming and find that approaches such as "lifestyle change, additional reduction of non-CO_2 greenhouse gases and more rapid electrification of energy demand based on renewable energy" can "significantly reduce the need for CDR, but not fully eliminate it." (Van Vuuren et al. 2018: 391)

Thirdly, each of the various CE approaches would, however, have significant *impacts*—not only by counteracting anthropogenic global warming, but also through unintended side effects—*both on natural and on social systems*. While the various approaches differ vastly with regards to which specific side-effects they produce and to the degree in which they would impact various social and physical systems, none of them is free of unintended consequences if applied on a scale large enough to counteract inadvertent climate change (IPCC 2014b: 89; Rickels et al. 2011b: 33ff.). The specific side effects of individual approaches have been the focus of numerous research endeavors (see, for example, Fuss et al. 2018; Keller et al. 2018; Parker and Irvine 2018; Eastham et al. 2018; Cairns 2014; Biermann and Möller 2019; Kreidenweis et al. 2016) discussing potential impacts on such diverse systems as precipitation and water cycles, ozone concentrations in the atmosphere, extreme weather events, biodiversity, agricultural productivity, economic factors and forms of government. As only a very limited number of field tests have been conducted, most knowledge about side-effects of CE approaches is derived from climate models or analogical reasoning—e.g. by drawing conclusions from effects of natural aerosol injections through volcanoe eruptions to potential effects of intentional aerosol injection through SAI. Beyond the side-effects which have been identified through these means, a number of potential effects are, at this point, unknown. It is apparent, accordingly, that no CE approach can provide a 'simple fix' to the challenge of anthropogenic climate change.

Fourthly, existing technologies have been deemed "*insufficiently mature* to rely on them for climate stabilization" (Fuss et al. 2016: 1; see also Rayner 2016; Lawrence et al. 2018)[33] even if it was decided that they should be considered. CE approaches do not exist in the material world as "physically-tangible [sic] technological object[s]" (Asayama et al. 2017: 82), much less as a complete socio-technical system—including not only technological objects, but also infrastructures, regimes and actors needed in order to allow the technology to fulfill its purpose (Schot and

[33] Italics are mine.

Kanger 2018: 1045). To date, CE approaches have been the object of numerous modelling studies in order to assess their possible uses, impacts and side-effects. Beyond this, they have entered the material world through a limited number of field experiments and,[34] in case of DAC, a small number of commercial plants. This means that research and development would need to be conducted *if* the decision was made that these approaches should be considered as part of a response to climate change. This also means that a number of questions remain unanswered to date concerning the impact and side-effects of these approaches.

These four interrelated issues, (1) the increasing impacts of climate change on both social and physical systems, (2) the assumption of being able to use at least some form of CE implied in climate models, (3) the side effects of CE approaches IF they were used to stabilize climate, and (4) the non-existence of technologies able to ensure climate stabilization, outline the peculiar situation, which Rayner calls "magical thinking" (Rayner 2016) and Rockström et al. discuss as "[t]he world's biggest gamble" (Rockström et al. 2016: 465): In order to produce scenarios which allow us to prevent climate impacts, we assume to be able to do something that we cannot, yet, do, and without deciding whether or not, and how, we would do it. If the decision was that CE should be considered as a measure (among others) to address climate change, then many questions would remain to be answered before these approaches were ready for deployment, so that research, development and governance would need to be undertaken. If the decision was made that CE should not be considered, then scenarios would need to be developed which do not assume the use of these approaches, but instead inform decision-makers on other ways in which the climate targets can be achieved.

The academic community studying CE from a diversity of disciplinary perspectives are highly aware of the high stakes inherent both in the challenge of climate change and in the set of approaches called CE (Linnér and Wibeck 2015). In particular in order to alleviate "substantial uncertainties and risks" and to consider "ethical and governance dilemmas" (Lawrence et al. 2018: 1) inherent in various CE approaches, the need for deliberation and decision-making concerning CE approaches has been pointed out (see, for example, Long et al. 2015; Jinnah 2018; Pasztor 2017a; Pasztor et al. 2017; Parson and Keith 2013; Kreuter 2018). However, to date, political deliberation and decision-making concerning CE approaches is limited (Morrow 2017: 3; Pasztor et al. 2017: 231; SPP 1689 2019a).[35] The imbalance between potential relevance and impact of CE approaches on the one hand and the limits of deliberation and decision-making on the other hand recently has seen various actors aiming to enlarge the discussion on CE into the public and political sphere: The DFG Priority Program 'Climate Engineering' (SPP-CE) picked up the necessity to start a political debate in their recent report "Climate Engineering und unsere Klimaziele—eine

[34] Such as LOHAFEX, SOIREE or E-PEACE.

[35] Instances of political decisions on CE approaches are presented, for example, by amendments to the London Convention/London Protocol (Ginzky and Frost 2014) and decisions by the parties to the CBD (2010).

überfällige Debatte" (SPP 1689 2019a).[36] The Carnegie Climate Geoengineering Governance Initiative (C2G2) at the Carnegie Council for Ethics in International Affairs was founded in order to "expand the conversation from the scientific and research community to the global policy-making arena, and to encourage a society-wide discussion about the risks, potential benefits, ethical and governance challenges raised by these approaches." (C2G2 2019)[37] The challenges inherent in the presence of imbalance between the relevance of CE approaches and the lack of political deliberations has been pointed out by Talberg et al. (2018) and Gupta and Möller (2018) who argue that, due to a lack of dedicated governance frameworks, informal governance patterns emerge which structure the field of CE, a finding which "draws attention to the need for political oversight" (Gupta and Möller 2018: 497).

Starting from the observation of the imbalance outlined above, research can be conducted in two main directions: Backwards in time, one could assess how this imbalance came about and which factors can explain its existence. Looking to the future, one can attempt to provide insights into how this current imbalance might influence future decision-making concerning CE: The lack of current political deliberation means that the social meaning of these approaches is developed elsewhere. The present thesis aims to give an account of how the current, mainly academic discussion of CE might construct the meaning of these approaches as legitimate or illegitimate to address the challenge of climate change, and indicate how this can be taken into account in climate decision-making.

The decision was made to follow the latter direction in order to be able to provide decision-makers with tools to reflect carefully on the construction of the meaning of CE approaches and its role in the decision-making process. Controversial technologies have, in the past, in some instances, been accepted as a means to address challenges to society without comprehensive political deliberation—such as, for example, nuclear weapons technology (Lederer and Kreuter 2018). However, a complete moratorium of CE is not inconceivable. The potentially high relevance of CE to achieve the objectives of the PA, on one hand, and the potentially large unintended impacts of these approaches, on the other, render both wholesale acceptance and complete rejection without effective climate protection measures problematic. Mike Hulme outlines the danger of being drawn into use of CE: "As we move from one rhetorical climate policy deadline to another, there is a real danger that a hyper-ventilating condition of despair and panic will lead society into making either hubristic and authoritarian responses to climate change." (Hulme 2010: 173) Other authors have argued that a moratorium on CE could deprive society of important knowledge for climate action (Keith 2013: 13, 88; Long et al. 2015: 30), even if it was only the knowledge that specific CE approaches are not feasible and thus not available in a portfolio of measures to address the challenge of climate change. The case of CE provides an opportunity for an inclusive, reflexive and comprehensive deliberative

[36]Report in German; title translated by JK: "Climate Engineering and our climate objectives—a debate that is past due".

[37]C2G2 recently changed its name to Carnegie Climate Governance Initiative (C2G), see Pasztor (2019).

process of decision-making as the academic discussion of these approaches does not present unqualified support or rejection of these approaches, but instead a strong consciousness of the duality of high stakes: the impacts of unmitigated climate change on one hand and the impacts of CE, on the other (Linnér and Wibeck 2015; Kreuter 2018).

1.3.2 Central Research Question and Interactions of Theory and Data

The following central research question is addressed in the present thesis: *How is the meaning of CE constructed in the academic discussion through frames which legitimize or delegitimize the use of these approaches to address anthropogenic climate change?* The research question defines the research interest and serves as the central criterion for decision whenever, during the research process, various different steps are possible. The step chosen is to be the one deemed most useful in order to answer the research question. Through this question, the present thesis pursues a two-fold *research objective*: (1) descriptively and analytically, to assess systematically how the meaning of CE is constructed through frames in the academic discussion and whether one of two extremes—legitimization or delegitimization of CE—is dominant; (2) normatively, to provide decision-makers with tools to assess and take into account the constructedness of this legitimacy if and when, in the near future, decisions about CE will move onto the political agenda, without denying the accuracy or relevance of findings from disciplinary and interdisciplinary research on CE (Image 1.3).

The theoretical perspective from which this research question is posed is modernist social constructivism. The theoretical perspective serves the purpose of defining the epistemological position of the study ("Definitionsleistung", Spindler and Schieder (2010: 23). Espousing this perspective, I assume that social reality is made up of both ideational and material objects. The latter have given physical characteristics but their meaning is socially constructed and can, in turn, through human action, be causally effective. The construction of this meaning, then, becomes highly interesting; one can ask: "'how are things in the world put together so that they have

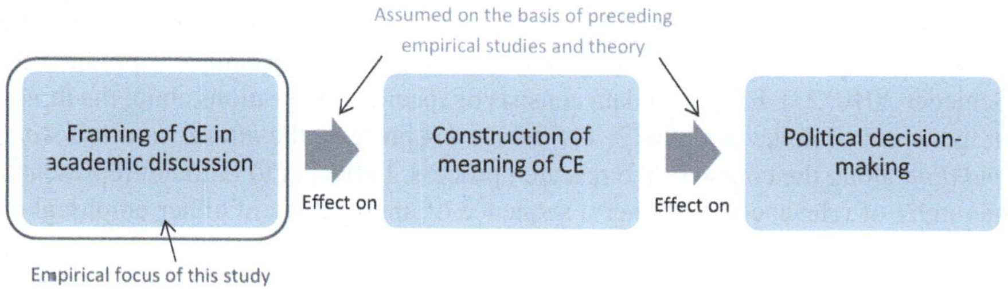

Image 1.3 Empirical focus of the present study

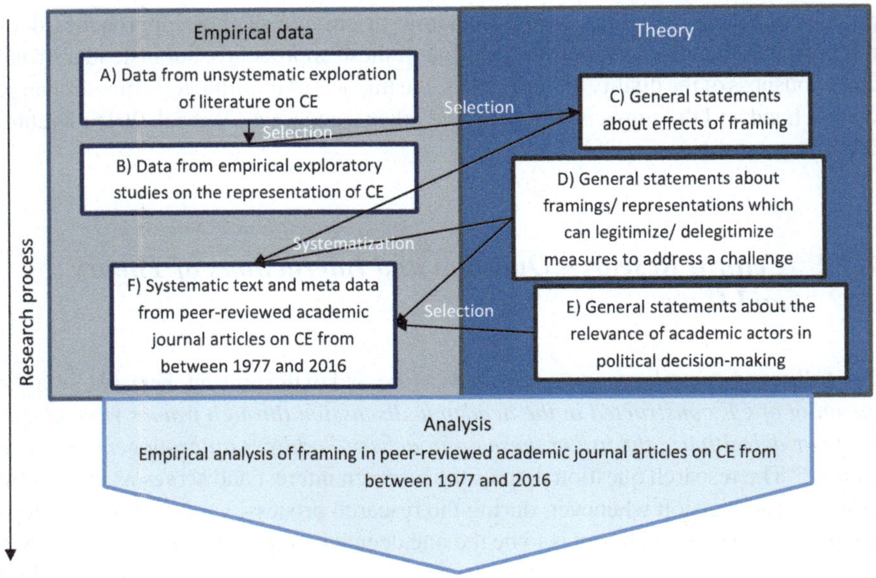

Image 1.4 Interplay of data and theory in the present study

the properties that they do?'" (Wendt 1998: 103), e.g., how is the meaning of CE constructed so that it can influence decision-making?[38] In studying the *construction* of the meaning of CE (instead, for example, the causes for its consideration or the effects of its deployment in discussion), this thesis is, in its empirical analysis, dedicated to providing a constitutive account instead of a causal one (Wendt 1998: 103). However, it is informed by and builds upon theoretical statements about the potential *causal relevance* of the meaning of CE, derived in particular from frame theory, for example the assumption that the specific meaning constructed for CE can influence political decision-making.

This means that empirical data and theoretical statements interact with each other in several ways and at several points during the process in the design of this research project. The central research question serves as the criterion for deciding whether a specific set of information—be it empirical or theoretical—is relevant for the present study or not. Theoretical information consists of general statements about specific relationships. As such, theory is used in this study always for some specific purpose providing a specific benefit in the pursuit of the research objectives (Spindler and Schieder 2010: 23). Empirical data consists of specific observations about the things in the world (Behnke et al. 2012: 21). Image 1.4 presents the *interactions of theory and data* along the course of the research process. Letters A to F do not represent a hierarchy of relevance, but rather a sequence of applications of either empirical or

[38]In the present thesis, the property of interest is the effect of the meaning of CE on decision-making. This is, however, not based on observation—as CE is not yet, commonly, discussed in decision-making arenas—but derived from theory (e.g., frame theory, practice theory).

theoretical information in the present research project in order to answer the research question.

A first overview over the subject matter—i.e. CE—was gained through (A) an unsystematic exploration of literature on CE from various fields of research as well as media and policy consultancy. From this information an assumption was inferred about the general relevance of the representation of CE in discourse for the construction of its meaning in society, for example with regards to climate decision-making. Accordingly, a literature review was conducted identifying (B) studies which assess the representation of CE in various discursive settings. The findings of these studies provided an overview over the role of various elements of discourse for the construction of the meaning of CE—such as narratives, imaginaries, frames, metaphors etc. The concept of frame (C) was identified to define an element of discourse most useful to answer the research question, as theoretical statements about frames have been made and tested not only on their constitution, but also on their causal effects on decision-making and the perception of legitimacy. Thus, the empirical findings from earlier studies (B) indicate the usefulness of the theoretical concept of framing for answering the central research question of this project. Frame theory (C), in turn, provides a framework to systematize empirical information gathered to answer the research question ("Systematisierungsleistung", Spindler and Schieder (2010: 23). Theoretical statements about processes which can legitimize or delegitimize actions (e.g. from securitization theory and the concept of technological fix) (D) further differentiate this systematization. Further, theoretical statements from frame theory (C) as well as from practice theory, in particular from the EC concept (E), inform the selection of relevant empirical information ("Selektionsleistung", Spindler and Schieder (2010: 23)). The relevant information, in the case of the present study, is the corpus of all documents pertaining to the academic discussion of CE. This was narrowed down further to (F) a sample of peer-reviewed academic journal articles published between 1977 and 2016 for methodical reasons of data availability and accessibility as well as limits in methodical possibilities.

The research project, thus, follows a partly *inductive*, partly *deductive* research design (Behnke et al. 2012: 28): Inductive and deductive approaches differ in whether inferences are drawn from data to theory (inductive) or from theory to data (deductive). Purely deductive and inductive approaches can clearly be delineated epistemologically, but these lines become blurred in practice (Behnke et al. 2012: 32) (Table 1.1).

1.3.3 The Tools Applied to the Quest for Answers—Theoretical and Methodical Approach

The *theoretical framework* guiding the present thesis spans three levels of abstraction: (1) the theoretical perspective, which provides a set of assumptions about the world and about the nature of knowledge, defines the epistemological and ontological

Table 1.1 Overview of the research project

Element of research design	Present study
Empirical starting point	Empirical observation of imbalance between relevance for climate policy and potential impact on climate of CE on the one hand and little dedicated political decision-making on the other
Central research question	How is the meaning of CE approaches constructed in the academic discussion through frames which legitimize or delegitimize the use of these appraoches to address anthropogenic climate change?
Objective	*Systematically describe* the construction of the meaning of CE approaches in the academic discussion, motivated by its relevance for climate decision-making
Position in process of academic research (Gschwend and Schimmelfennig 2007: 14)	*Deductive* analysis of academic discussion on CE, *systematizing* findings of preceding studies and *generating hypotheses* for future studies
Theoretical perspective with regards to epistemology and ontology	*Definition of ontological and epistemological position*: Modernist social constructivism
Theoretical construct for analysis	*Selection of relevant information*: Epistemic communities approach *Systematization of empirical information*: Frame theory; four ideal-type frame themes: Security, Complexity, Economy, Appropriateness
Method of empirical analysis	Qualitative and quantitative content analysis
Choice of empirical case	Academic discussion of CE approaches; choice informed by empirical starting point, findings from preceding empirical studies and practice theory
Choice of data	Representative sample generated from the corpus of peer-reviewed journal articles on CE or specific approaches published between 1977 and 2016

'playing field' in which the analysis is situated; (2) a set of more specific theoretical assumptions which serve to narrow down the scope of the study and justify its relevance; (3) the theoretical construct which serves to systematize data analysis according to theoretical statements from frame theory as well as several strands of IR theory. The theoretical perspective (1) is, as discussed before, that of modernist social constructivism (Wendt 1998; Balzacq 2010). The more specific theoretical assumptions (2) to define the scope and the relevance of the study are derived from politicization theory (Hay 2007; Wood and Flinders 2014), practice theory (Adler and Pouliot 2011; Haas 1992), Science and Technology Studies (Bijker 2006; Schot and Kanger 2018; Geels and Schot 2007) and frame theory (Tversky and Kahneman 1987; Entman 1993). The theoretical construct for analysis (3) is developed from frame theory and based on Max Weber's differentiation of causes determining social action (Weber 1980 [1922]) as well as the notions of securitization (Balzacq 2010), technological fix (Weinberg 1966) and economization (Hajer 1997). For each of these elements of the theoretical approach to the present study, specific questions guided the process.

In order to illustrate the use of the theoretical construct to provide a systematic account of the discursive legitimization or delegitimization of emerging technology through academic actors and its relevance for political decision-making, Chap. 4 turns to the case of the development of nuclear technology in the 1930s and 1940s. In that case, the discussion of an emerging technology was started in academic circles in the USA as well as several European countries, representing nuclear technology in particular through legitimizing security frames such as the representation of the potential technology as a safeguard against the Nazi German threat. And in that case, decisions were made to develop and even use the technology, against the opposition of some of the researchers who had originally pushed for its political consideration. This account cannot provide the basis for a deterministic argument about the power of legitimizing framings in causing specific decisions. But it can provide an example of a situation in which academic framing legitimized the consideration of a specific emerging technology and in which said technology was, eventually, deployed, despite known negative repercussions for hundreds of thousands of people.

The *research method* serves to operationalize the theoretical construct in such a way as to be able to recognize and identify empirical observations corresponding to the construct (Behnke et al. 2012: 23): It allows for the transformation of the theoretical construct into a data model, i.e. the empirical data structured according to the theoretical construct (Früh 2011: 23). Thus, a methodical approach is what allows the researcher to connect theory and data in the research process. The research method applied in this thesis in order to generate and analyze empirical data in accordance with the theoretical approach is qualitative and quantitative content analysis (Mayring 2014; Früh 2011; Rössler 2010). This method provides a systematic procedure comprised of a number of individual steps to be taken in order to ensure that the findings from the study, i.e. the results of the interaction between theory and data, comply with quality criteria of scientific enquiry such as "objectivity, reliability and validity" (Mayring 2014: 42). The method of qualitative and quantitative content analysis is described in detail and put to use for the present study in Chap. 5.

The *empirical case* is the central research object on which the study is focused. It is a phenomenon that can be defined and delineated in time and space. From this case, the empirical starting point is derived. The empirical case in focus in the present study is the early academic discussion of CE approaches as possible measures to address climate change. The reasons for focusing on the academic discussion, instead of the political or public, for example, are provided by theoretical considerations concerning the special role of academics in cases of decision-making in conditions of uncertainty as well as by empirical information about the special role of academics in the discussion of CE. The corpus of data to be analyzed is narrowed down to peer-reviewed academic journal articles from between 1977 and 2016. Peer-reviewed journal articles were chosen as their impact on the community is measured

Table 1.2 Questions guiding the individual chapters (subordinated to main research question)

Questions	Resources applied to answer questions
Which are the general epistemological and ontological assumptions of the research perspective espoused here?	Modernist social constructivism
How is climate change constructed as a political problem?	Politicization theory
In what way are academic actors relevant for decision-making concerning CE?	Practice theory, in particular epistemic communities-approach Science and technology studies
How can technology be conceptualized for this study?	Science and technology studies
How can framing influence decision-making?	Frame theory
Which types of frames can legitimize or delegitimize a technological approach to address climate change and how can these be conceptualized?	Causes for determination of social action; securitization theory, concepts of ecological modernization, technological fix
Which role have the ideal-type frames in academic discussion played for decision-making on emerging technology in the past?	Secondary empirical information about early stages of development of nuclear weapons
How can frames in the academic discussion assessed systematically? How can data be selected?	Method of qualitative and quantitative content analysis
Which frames can be identified in the academic discussion on CE?	Empirical analysis of sample of peer-reviewed academic publications on CE from between 1977 and 2016

(e.g. through the number of citations), which does not apply for other types of oral or written communication (such as conference presentations, speeches, books, or newspaper commentaries). English-language articles were chosen as a large portion of the discussion takes place in English, as well as due to language limitations on the part of the author. 1977 was chosen as the starting point as this is the year of publication of the first article found through a systematic query in the database Web of Science.[39] A representative sample from this corpus is applied to qualitative and quantitative content analysis in the present study (Table 1.2).

References

Adler E, Pouliot V (2011) International practices. Int Theor 3(1):1–36. https://doi.org/10.1017/S175297191000031X

[39] www.webofknowledge.com, Clarivate.

Akerlof K, Rowan KE, Fitzgerald D, Cedeno AY (2012) Communication of climate projections in US media amid politicization of model science. Nature Clim Change 2(9):648–654. https://doi.org/10.1038/NCLIMATE1542

Allen, M. R., Dube, O. P., Solecki, W., Aragón-Durand, F., Cramer, W., Humphreys, S., et al. (2018). Framing and context. In: Masson-Delmotte V, Zhai P, Pörtner HO, Roberts D, Skea J, Shukla PR et al (eds) Global warming of 1.5 °C. An IPCC special report on the impacts of global warming of 1.5 °C above pre-industrial levels and related global greenhouse gas emission pathways, in the context of strengthening the global response to the threat of climate change, sustainable development, and efforts to eradicate poverty. Geneva, IPCC, pp 49–92

Anshelm J, Hultman M (2015) discourses of global climate change: apocalyptic framing and political antagonisms. Routledge, London

Anshelm J, Haikola S, Wallsten B (2018) Politicizing environmental governance—a case study of heterogeneous alliances and juridical struggles around Ojnare Forest, Sweden. Geoforum 91:206–215. https://doi.org/10.1016/j.geoforum.2018.03.003

Asayama S, Sugiyama M, Ishii A (2017) Ambivalent climate of opinions: tensions and dilemmas in understanding geoengineering experimentation. Geoforum 80:82–92

Balzacq T (2010) Constructivism and securitization studies. In: Cavelty MD, Mauer V (eds) The Routledge handbook of security studies. Routledge, London, pp 56–72

Bates S, Jenkins L, Amery F (2014) (De)politicisation and the father's clause parliamentary debates. Pol Polit 42(2):243–258

Behnke J, Baur N, Behnke N (2012) Empirische Methoden der Politikwissenschaft (2. aktualisierte, Auflage edn. Ferninand Schöningh, Paderborn

Bellamy R (2016) A sociotechnical framework for governing climate engineering. Sci Technol Human Values 41(2):135–162. https://doi.org/10.1177/0162243915591855

Bellamy R, Chilvers J, Vaughan NE, Lenton TM (2012b) A review of climate geoengineering appraisals. WIREs Clim Change 3(6):597–615

Bellamy R, Chilvers J, Vaughan NE, Lenton TM (2012a) Appraising geoengineering. Tyndall Working Paper Series, 153

Biermann F, Möller I (2019) Rich man's solution? Climate engineering discourses and the marginalization of the South. International Environmental Agreements, Online first, 1–17

Bijker WE (2006) Why and how technology matters. In: Goodin RE, Tilly C (eds) The Oxford handbook of contextual political analysis. Oxford University Press, New York, pp 681–706

Bolsen T, Druckman JN (2018) Do partisanship and politicization undermine the impact of a scientific consensus message about climate change? Group Process Intergr Relat 21(3):389–402. https://doi.org/10.1177/1368430217737855

Bolsen T, Palm R, Kingsland JT (2019) Counteracting climate science politicization with effective frames and imagery. Sci Commun 41(2):147–171. https://doi.org/10.1177/1075547019834565

Burnham P (2001) New Labour and the politics of depoliticisation. Brit J Polit Int Relat 3(2):127–149

Burnham P (2014) Depoliticisation: economic crisis and political management. Pol Polit 42(2):189–206

C2G2 (2019) C2G2 mission. https://www.c2g2.net/c2g2-mission/. Accessed 22 Mar 2019

Cairns R (2013) Examining framings of geoengineering using Q methodology. Climate Geoengineering Governance Working Paper Series, 002

Cairns R (2014) Climate geoengineering: issues of path-dependence and socio-technical lock-in. WIREs Clim Change 5:649–661. https://doi.org/10.1002/wcc.296

Cairns R, Stirling A (2014) 'Maintaining planetary systems' or 'concentrating global power'? High stakes in contending framings of climate geoengineering. Glob Environ Change 28:25–38

Carr W, Yung L, Preston CJ (2014) Swimming upstream: engaging the American public early on climate engineering. Bull Atom Scient 70(3):38–48. https://doi.org/10.1177/0096340214531180

CBD (2010) Decision adopted by the conference of the parties to the convention on biological diversity at its tenth meeting: X/33. Biodiversity and Climate Change. Nagoya: Convention on Biological Diversity

Costa O (2019) The politicization of EU external relations. Journal of European Public Policy 26(5):790–802. https://doi.org/10.1080/13501763.2018.1478878

Dennison J, Geddes A (2018) Brexit and the perils of 'Europeanised' migration. J Europ Publ Pol 25(8):1137–1153. https://doi.org/10.1080/13501763.2018.1467953

de Wilde P, Leupold A, Schmidtke H (2016) Introduction: the differentiated politicisation of European governance. West Europ Polit 39(1):3–22. https://doi.org/10.1080/01402382.2015.1081505

Dryzek JS (2006) Deliberative global politics. Discourse and democracy in a divided world. Cambridge et al, Polity

DWD (2019) Pressemitteilung: Klima-Pressekonferenz 2019 des Deutschen Wetterdienstes. Neue Langfristvorhersagen ermöglichen Prognose von Dürren in Deutschland. *Pressemitteilungen des DWD*. https://www.dwd.de/DE/presse/pressekonferenzen/DE/2019/PK_26_03_2019/pressekonferenz.html

Eastham SD, Weisenstein DK, Keith DW, Barrett SRH (2018) Quantifying the impact of sulfate geoenginering on mortality from air quality and UV-B exposure. Atmos Environ 187:424–434. https://doi.org/10.1016/j.atmosenv.2018.05.047

Easton D (1953) Political system. an inquiry into the state of political science. David A. Knopf, New York

Ecker-Ehrhardt M (2018) Self-legitimation in the face of politicization: Why international organizations centralized public communication. Rev Int Organ 13(4):519–546. https://doi.org/10.1007/s11558-017-9287-y

Entman RM (1993) Framing: toward clarification of a fractured paradigm. J Commun 43(4):51–58

Flinders M, Wood M (2014) Depoliticisation, governance and the state. Pol Polit 42(2):139–149

Foster EA, Kerr P, Byrne C (2014) Rolling back to roll forward: depoliticisation and the extension of government. Pol Polit 42(2):225–241

Früh W (2011) Inhaltsanalyse: theorie und praxis. UKV, Konstanz

Fuss S, Jones CD, Kraxner F, Peters GP, Smith P, Tavoni M et al (2016) Research priorities for negative emissions. Environ Res Lett 11:115007. https://doi.org/10.1088/1748-9326/11/11/115007

Fuss S, Lamb WF, Callaghan MW, Hilaire J, Creutzig F, Amann T et al (2018) Negative emissions—Part 2: costs, potentials and side effects. Environ Res Lett 13. https://doi.org/10.1088/1748-9326/aabf9f

Gasser T, Guivarch C, Tachiiri K, Jones CD, Ciais P (2015) Negative emissions physically needed to keep global warming below 2 & C. Nat Commun 6:7958. https://doi.org/10.1038/ncomms8958

Geels FW (2004) From sectoral systems of innovation to socio-technical systems. Insights about dynamics and change from sociology and institutional theory. Res Pol 33:897–920

Geels FW (2007) Transformations of large technical systems: a multilevel analysis of the dutch highway system (1950–2000). Sci Technol Human Values 32(2):123–149

Geels FW, Schot J (2007) Typology of sociotechnical transition pathways. Res Pol 36:399–417. https://doi.org/10.1016/j.respol.2007.01.003

Giddens A (1984) The constitution of society: outline of the theory of structuration. Polity Press, Cambridge

Ginzky H, Frost R (2014) Marine geo-engineering: legally binding regulation under the london protocol. Carbon Clim Law Rev 8(2):82–96

Goeminne G (2012) Lost in translation: climate denial and the return of the political. Global Environ Politics 12(2):1–8. https://doi.org/10.1162/GLEP_a_00104

Goldberg D, Lackner KS (2015) Creating negative emissions at remote CO_2 sequestration sites. Greenhouse Gases Sci Technol 5:238–240. https://doi.org/10.1002/ghg

Greene CH, Huntley ME, Archibald I, Gerber LN, Sills DL, Granados J et al (2017) Geoengineering, marine microalgae, and climate stabilization in the 21st century. Earth's Future 5:278–284. https://doi.org/10.1002/2016EF000486

Gschwend T, Schimmelfennig F (2007) Forschungsdesign in der Politikwissenschaft: ein Dialog zwischen Theorie und Daten. In: Gschwend T, Schimmelfennig F (eds) Forschungsdesign in der

Politikwissenschaft: Probleme —Strategien—Anwendungen. Campus, Frankfurt am Main, pp 13–35

Gupta A, Möller I (2018) De facto governance: how authoritative assessments construct climate engineering as an object of governance. Environ Polit. https://doi.org/10.1080/09644016.2018.1452373

Haas PM (1992) Introduction: epistemic communities and international policy coordination. Int Org 46(1):1–35

Hajer M (1997) The politics of environmental discourse. Ecological modernization and the policy process. Clarendon Press, Oxford

Hajer M (2006) Doing discourse analysis: coalitions, practices, meaning. In: Van den Brink M, Metze T (eds) Words matter in policy and planning. Discourse theory and method in the social sciences. Utrecht, KNAG, pp 65–74

Hamilton C (2013) Earthmasters. the dawn of the age of climate engineering. Yale University Press, New Haven, CT

Hay C (2014) Depoliticisation as process, governance as practice: what did the 'first wave' get wrong and do we need a 'second wave' to put it right? Pol Polit 42(2):293–311. https://doi.org/10.1332/030557314X13959960668217

Hay C (2007) Why we hate politics. Cambridge, MA, Polity

Heyward C (2013) Situating and abandoning geoengineering: a typology of five responses to dangerous climate change. Polit Sci Polit, PS, pp 23–27

Hoegh-Guldberg O, Jacob D, Taylor M, Bindi M, Brown S, Camilloni I et al (2018) Impacts of 1.5 °C global warming on natural and human systems. In: Masson-Delmotte V, Zhai P, Pörtner HO, Roberts D, Skea J, Shukla PR et al (eds) Global Warming of 1.5 °C. An IPCC special report on the impacts of global warming of 1.5 °C above pre-industrial levels and related global greenhouse gas emission pathways in the context of strengthening the global response to the threat of climate change, sustainable development, and efforts to eradicate poverty. Geneva, IPCC, pp 175–312

Horton JB, Keith DW, Honegger M (2016) Implications of the Paris agreement for carbon dioxide removal and solar geoengineering. Harvard project on climate agreements—viewpoints, July 2016

Hulme M (2014) Can science fix climate change? A case against climate engineering. Polity Press, Cambridge, UK

Hulme M (2010) The idea of climate change: exploring complexity, plurality and opportunity. Gaia 19(3):171–174

Huttunen S, Hildén M (2014) Framing the controversial: geoengineering in academic literature. Sci Commun 36(3):3–29

IEA (2019) Global energy and CO_2 status report. The latest trends in energy and emissions in 2018. In: Agency IE (ed), Paris, France

IISD Reporting Services (2019) Summary of the fourth session of the united nations environment assembly. Earth Negotiations Bulletin 16(153)

IPCC (2013a) Climate change 2013: the physical science basis. Contribution of working group I to the fifth assessment report of the intergovernmental panel on climate change. Cambridge UK, IPCC

IPCC (2014a) Climate Change 2014: Mitigation of Climate Change. Working Group III Contribution to the Fifth Assessment Report of the Intergovernmental Panel on Climate Change. Cambridge UK: IPCC

IPCC (2014b) Climate change 2014: synthesis report. Contribution of working groups I, II and III to the fifth assessment report of the intergovernmental panel on climate change. In: Pachauri RK, Meyer LA (eds) Geneva: IPCC

IPCC (2018a) Global Warming of 1.5 °C. An IPCC Special Report on the impacts of global warming of 1.5 °C above pre-industrial levels and related global greenhouse gas emission pathways, in the context of strengthening the global response to the threat of climate change, sustainable development, and efforts to eradicate poverty. Summary for Policymakers. In IPCC (ed). Geneva, IPCC

Jessop B (2014) Repoliticising depoliticisation: theoretical preliminaries on some responses to the American fiscal and Eurozone debt crises. Pol Polit 42(2):207–223
Jinnah S (2018) Why govern climate engineering? A preliminary framework for demand-based governance. Int Stud Rev 20(2):272–282. https://doi.org/10.1093/isr/viy022
Jones CD, Ciais P, Davis SJ, Friedlingstein P, Gasser T, Peters GP et al (2016) Simulating the Earth system response to negative emissions. Environ Res Lett 11. https://doi.org/10.1088/1748-9326/11/9/095012
Kahneman D, Tversky A (1984) Choices, values, and frames. Am Psychol 39(4):341–350
Keith DW (2013) A case for climate engineering. Boston Review, Cambridge/London
Keller DP, Lenton A, Littleton EW, Oschlies A, Scott V, Vaughan NE (2018) The effects of carbon dioxide removal on the carbon cycle. Curr Clim Change Rep 4:250–265. https://doi.org/10.1007/s40641-018-0104-3
Kreidenweis U, Humpenöder F, Stevanovic M, Bodirsky BL, Kriegler E, Lotze-Campen E et al (2016) Afforestation to mitigate climate change: impacts on food prices under consideration of albedo effects. Environ Res Lett 11:085001. https://doi.org/10.1088/1748-9326/11/8/085001
Kreuter J (2018) Climate engineering and political decision-making: the importance of polarity in the academic debate. INoGOV Policy Brief
Kreuter J, Lederer M (Under review) The geopolitics of negative emissions technologies—learning lessons from REDD+ and Renewable Energies for Afforestation, BECCS and Direct Air Capture
Kriegler E, Luderer G, Bauer N, Baumstark L, Popp A, Rogelj J, et al (2018) Pathways to limiting warming to 1.5 °C: a tale of turning around in no time? Philos Trans Royal Soc A 376(20160457). https://doi.org/10.1098/rsta.2016.0457
Kvidal-Rovik T (2018) Nature articulations in norwegian advertising discourse: a depolicitized discourse of climate change. Environ Commun J Nat Cult 12(6):794–806. https://doi.org/10.1080/17524032.2018.1482835
Lakoff G (2014) The all new don't think of an elephant!: Know your values and frame the debate, 2nd edn. Chelsea Green Publishing, Chelsea, VT
Lasswell HD (1958) Politics: who gets what, when, how. Meridian Books, New York
Lawrence MG, Schäfer S, Muri H, Scott V, Oschlies A, Vaughan NE et al (2018) Evaluation climate geoengineering proposals in the context of the Paris agreement temperature goals. Nat Commun 9(3734). https://doi.org/10.1038/s41467-018-05938-3
Lederer M (2017) Carbon trading: who gets what, when, and how? Global Environ Polit 17(3):134–140. https://doi.org/10.1162/GLEP_a_00419
Lederer M, Kreuter J (2018) Organising the unthinkable in times of crises: will climate engineering become the weapon of last resort in the Anthropocene? Organization 25(4). https://doi.org/10.1177/1350508418759186
Linnér B-O, Wibeck V (2015) Dual high-stake emerging technologies: a review of the climate engineering research literature. WIREs Clim Change 6:255–268
Long J, Loy F, Morgan MG (2015) Policy: start research on climate engineering. Nature 518(7537)
Luokkanen M, Huttunen S, Hildén M (2014) Geoengineering, news media and metaphors: framing the controversial. Public Understand Sci 23(8):966–981. https://doi.org/10.1177/0963662513475966
Marchetti C (1977) On geoengineering and the CO_2 problem. Clim Change 1:59–68
Markusson N (2013) Tensions in framings of geoengineering: constitutive diversity and ambivalence. Climate Geoengineering Governance Working Paper Series, *003*
Markusson N, Venturini T, Laniado D, Kaltenbrunner A (2016) Contrasting medium and genre von Wikipedia to open up the dominating definition and classification of geoengineering. Big Data and Society, July–December 2016, pp 1–17
Matzner N, Barben D (2018) Verantwortungsvoll das Klima manipulieren? Unsicherheit und Verantwortung im Diskurs um climate engineering. In: Janich N, Rhein L (eds) Unsicherheit als Herauforderung für die Wissenschaft. Peter Lang, Berlin et al., pp 143–179
Mayring P (2014) Qualitative content analysis: theoretical foundation, basic procedures and software solution. Gesis, Klagenfurth

McCright AM, Dunlap RE (2011) The politicization of climate change and polarization in the American public's view of global warming, 2001–2010. Sociol Q 52(2):155–194. https://doi.org/10.1111/j.1533-8525.2011.01198.x
Mercer AM, Keith DW, Sharp JD (2011) Public understanding of solar radiation management. Environ Res Lett 6:1–9. https://doi.org/10.1088/1748-9326/6/4/044006
Met Office (2018) Chance of summer heatwaves now thirty times more likely
Methmann C, Rothe D, Stephan B (2013a) Interpretive approaches to global climate governance. (De-)constructing the greenhouse. London, Routledge
Methmann C, Rothe D, Stephan B (2013b) Introduction. How and why to deconstruct the greenhouse. In: Methmann C, Rothe D, Stephan B (eds) Interpretive approaches to global climate governance. (De)constructing the greenhouse. London, Routledge, pp 1–22
Methmann C, Rothe D, Stephan B (2013c) Reflections. In: Methmann C, Rothe D, Stephan B (eds) Interpretive approaches to global climate governance. In: (De)constructing the greenhouse. London, Routledge, pp 248–256
Minx JC, Lamb WF, Callaghan MW, Bornmann L, Fuss S (2017) Fast growing research on negative emissions. Environ Res Lett 12. https://doi.org/10.1088/1748-9326/aa5ee5
Minx JC, Lamb WF, Callaghan MW, Fuss S, Hilaire J, Creutzig F et al (2018) Negative emissions—part 1: research landscape and synthesis. Environ Res Lett 13. https://doi.org/10.1088/1748-9326/aabf9b
Morrow DR (2017) International governance of climate engineering. A survey of Reports on climate engineering, 2009–2015. FCEA Working Paper Series
NAS (2015a) Climate Intervention: Reflecting Sunlight to Cool Earth. The National Academy of Sciences, Washington, D.C
NAS (2015b) Climate intervention: carbon dioxide removal and reliable sequestration. Reflecting sunlight to cool earth. Washington, DC, National Academy of Sciences
NAS (2015b) Climate intervention: carbon dioxide removal and reliable sequestration. The National Academy of Sciences, Washington, D.C
OUP (2019 Definition of model in English. https://en.oxforddictionaries.com/definition/model. Accessed 28 Mar 2019
Parker A, Irvine JP (2018) The risk of termination shock from solar geoengineering. Earth's Future 6:456–467. https://doi.org/10.1002/2017EF000735
Parson EA, Keith DW (2013) End the deadlock on governance of geoengineering research. Science 339:1278–1279. https://doi.org/10.1126/science.1232527
Pasztor J (2017) The need for governance of climate engineering. Ethics Int Affairs 41(4):419–430. https://doi.org/10.1017/S0892679417000405
Pasztor J (2019) What's in a name? Why we became C2G. C2G Blog
Pasztor J, Scharf C, Schmidt K-U (2017) How to govern geoengineering? Science 357(6348):231
Pelinka A, Varwick J (2010) Grundzüge der Politikwissenschaft, 2nd edn. UTB, Böhlau
Pepermans Y, Maeseele P (2014) Democratic debate and mediated discourses on climate change: from consensus to de/politicization. Environ Commun J Nature Cult 8(2):216–232. https://doi.org/10.1080/17524032.2014.906482
Pepermans Y, Maeseele P (2016) The politicization of climate change: problem or solution? WIREs Clim Change 7:478–485. https://doi.org/10.1002/wcc.405
Pye LW (2001) Political science. In: Krieger J, Crahan ME, Jacbos LR, Joseph WA, Nzogola-Ntalaja G, Paul JA (eds) The oxford companion to politics of the world, 2nd edn. Oxford University Press, Oxford, pp 674–676
Rayner S (2016) What might evans-pritchard have made of two degrees? Anthropol Today 32(4):1–2
Rein M, Schön D (1993) Reframing policy discourse. In: Fischer F, Forester J (eds) The argumentative turn in policy analysis and planning. Duke University Press, Durham, pp 145–166
Remling E (2018) Depoliticization adaptation: a critical analysis of EU climate adaptation policy. Environ Polit 27(3):477–497. https://doi.org/10.1080/09644016.2018.1429207
Reynolds J, Wagner G (2019) Highly decentralized solar geoengineering. Environmental Politics, Online First, 1–17. https://doi.org/10.1080/09644016.2019.1648169

Rickels W, Klepper G, Dovern J, Betz G, Brachatzek N, Cacean S et al (2011b) Large-scale intentional interventions into the climate system? assessing the climate engineering debate. Scoping report conducted on behalf of the german federal ministry of education and research (BMBF). Kiel, Kiel Earth Institute

Rittel HWJ, Webber MM (1973) Dilemmas in a general theory of planning. Policy Sci 4:155–169

Rockström J, Schellnhuber HJ, Hoskins B, Ramanathan V, Schlosser P, Brasseur GP et al (2016) The world's biggest gamble. *earth's*. Future 4(10):465–470. https://doi.org/2016EF000392

Rogelj J, Shindell D, Jiang K, Fifita S, Ginzburg V, Handa C et al (2018) Mitigation pathways compatible with 1.5 °C in the context of sustainable development. In: Masson-Delmotte V, Zhai P, Pörtner HO, Roberts D, Skea J, Shukla PR et al (eds) Global warming of 1.5 °C. An IPCC Special Report on the impacts of global warming of 1.5 °C above pre-industrial levels and related global greenhouse gas emission pathways, in the context of strengthening the global response to the threat of climate change, sustainable development, and efforts to eradicate poverty. Geneva, IPCC, pp 93–174

Rössler P (2010) Inhaltsanalyse, 2nd edn. UKV, Konstanz

Schmid-Petri H (2017) Politicization of science: how climate change skeptics use experts and scientific evidence in their online communication. Clim Change 145:523–537. https://doi.org/10.1007/s10584-017-2112-z

Schot J, Kanger L (2018) Deep transitions: emergence, acceleration, stabilization and directionality. Res Policy 47:1045–1059

UN Secretary-General (2018) Secretary-general's remarks on climate change [as delivered]

Shepherd J, Cox P, Haigh D, Keith DW, Launder B, Mace G et al (2009) Geoengineering the climate: science, governance and uncertainty. The Royal Society, London

Smith P, Haszeldine RS, Smith SM (2016) Preliminary assessment of the potential for, and limitiations to, terrestrial emission technologies in the UK. Environ Sci Process Impact 18:1400–1405

Spindler M, Schieder S (2010) Theorien in der Lehre von den internationalen Beziehungen. In: Schieder S, Spindler M (eds) Theorien der Internationalen Beziehungen, Auflage edn. Barbara Budrich, Opladen, pp 9–38

SPP 1689 (2019a) Climate Engineering und unsere Klimaziele—eine überfällige Debatte. In C. Schwerpunktprogramm 1689 der Deutschen Forschungsgemeinschaft „Climate Engineering: Risks, Opportunities?" (SPP 1689) (Ed.). Kiel: Kiel Earth Institute

Steffek J (2003) The legitimation of international governance: a discourse approach. Europ J Int Relat 9(2):249–275

Tagesschau.de (2018) Schiffahrt auf Rhein beeinträchtigt. Bundesregierung gibt Ölreserven frei. *Tagesschau.de*

Talberg A, Christoff P, Thomas S, Karoly D (2018) Geoengineering governance-by-default: an earth system governance perspective. Int Environ Agreem 18:229–253. https://doi.org/10.1007/s10784-017-9374-9

Tversky A, Kahneman D (1987) Rational choice and the framing of decisions. In: Hogarth RM, Reder MW (eds) Rational choice: the contrast between economics and psychology. University of Chicago Press, Chicago, pp 67–94

Tversky A, Kahneman D (1981) The framing of decisions and the psychology of choice. Sci New Ser 211(4481):453–458

UN (1992) United nations framework convention on climate change

UNEP (2018) Emissions gap report 2018. In UNEP (Ed.). Nairobi, Kenya: UNEP

UNFCCC (2015) Adoption of the Paris agreement. In U. N. F. C. o. C. Change (ed) (Vol. FCCC/CP/2015/L.9/Rev.1). Paris: United Nations

Uther S (2014) Diskurse des Climate Engineering. Argumente, Akteure und Koalitionen in Deutschland und Großbritannien. Wiesbaden, Springer Fachmedien

Van Vuuren D, Stehfest E, Gernaat DEHJ, Van den Berg M, Bijl DL, de Boer HS et al (2018) Alternative pathways to the 1.5 °C target reduce the need for negative emissions. Nat Clim Change 8:391–397

Wagenaar H (2011) Meaning in action: interpretation and dialogue in policy analysis. M. E. Sharp, New York

Warner S (2019) (Re)politicising 'the governmental': resisting the Industrial Relations Act 1971. Brit J Polit Int Relat Online First. https://doi.org/10.1177/1369148119845336

Weber M (1980) Wirtschaft und Gesellschaft. Grundriss der verstehenden Soziologie (5. Revidierte, Studienausgabe edn. Mohr, Tübingen

Weinberg AM (1966) Can technology replace social engineering? Bull Atom Sci 22(10):4–8

Weisser F, Müller-Mahn D (2017) No place for the political: micro-geographies of the Paris climate conference 2015. Antipode 49(3):802–820. https://doi.org/10.1111/anti.12290

Wendt A (1998) On constitution and causation in international relations. Rev Int Stud 24(Special Issue), 101–117

Williamson P (2016) Scrutinize CO_2 removal methods. Nature 530:153–155

Winner L (1977) Autonomous technology: technics-out-of-control as a theme in political thought. MIT Press, Cambridge, MA

WMO (2019) WMO statement on the state of the climate in 2018. WMO, Geneva

Wood M, Flinders M (2014) Rethinking depoliticisation: beyond the governmental. Pol Polit 42(2):151–170

Zappettini F, Krzyzanowski M (2019) The critical juncture of Brexit in media & political discourses: from national-populist imaginary to cross-national social and political crisis. Crit Discourse Stud 16(4):381–388. https://doi.org/10.1080/17405904.2019.1592767

Zhang Y, Posch A (2014) The wickedness and complexity of decision making in geoengineering. Challenges 5:390–408. https://doi.org/10.3390/challe5020390

Zürn M (2008) The politicization of economization? On the current relationship between politics and economics. In: Scherer AG, Palazzo G (eds) Handbook of research on global corporate citizenship. Edward Elgar, Cheltenham, pp 293–311

Zürn M (2013) Politisierung als Konzept der Internationalen Beziehungen. In: Zürn M, Ecker-Ehrhardt M (eds) Die Politisierung der Weltpolitik. Umkämpfte internationale Institutionen. Berlin, Suhrkamp, pp 7–35

Zürn M (2014) The politicization of world politics and its effects: eight propositions. Eur Polit Sci Rev 6(1):47–71. https://doi.org/10.1017/S1755773912000276

Zürn M (2019) Policization compared: at national, European, and global levels. J Eur Publ Pol 26(7):977–995. https://doi.org/10.1080/13501763.2019.1619188

Chapter 2
An Idea—And More Than an Idea: Climate Engineering in Research and Decision-Making

Abstract This chapter offers a concise summary of the current state of Climate Engineering, introducing these emerging technologies as existing, to a large degree, so far only in the minds of people and the calculations for climate models. As well as outlining the ideas on which these approaches are based which date back decades, in some cases centuries, the chapter outlines the currently few political discussions of CE at the margins of climate policy. Kreuter discusses this peculiar state of affairs and explores the relevance of social construction in this context based on the state of the art of social science analysis of CE.

2.1 Past and Current Developments

This study cannot and is not intended to give a comprehensive account of the history of CE,[1] or of the current state of affairs with regards to recent research and decision-making. Instead, I outline three central observations which provide a background to the assessment of the academic discussion of CE and its relevance for decision-making as conducted in this study: (1) Concerning their history, CE approaches share in common two seemingly contradictory features: On the one hand, they are commonly considered 'emerging technologies'. On the other hand, the ideas behind these technologies date back decades, in some cases centuries. (2) Concerning the arenas in which CE is discussed, various CE approaches are recently addressed not only in academic circles, but also in political arenas, albeit often indirectly and at the margins. (3) Concerning their current state of existence, no CE approaches exist to date materially at such a scale as to actually influence the climate globally,[2] and few field experiments have been conducted. However, an increasing number of research projects in various parts of the world have assessed preconditions and implications of various CE approaches from a number of disciplinary, interdisciplinary and transdisciplinary perspectives. Beyond this, specific approaches such as DAC have recently

---

[1] For insightful overviews, see Fleming (2010); Keith (2000); Bonnheim (2010); Bowden (2010); Fleming (2012, 2013).

[2] To the knowledge of the author.

J. Kreuter, *Climate Engineering as an Instance of Politicization*,
Springer Climate, https://doi.org/10.1007/978-3-030-60340-3_2

become the basis of the business model of various companies and have most recently sparked interest of state actors such as the UK government (HM Treasury 2020).

2.1.1 Climate Engineering in History—Technologies Both New and Old

CE approaches can be described as both new and old, each in a different respect: They can be considered new or 'emerging technologies' as none of them has been implemented to date in the sense of a fully developed socio-technical system as would be required in order to actually effectually influence the climate on a global scale (Asayama et al. 2017: 82; Rayner 2016; Carr et al. 2014: 41). Aside from material devices and infrastructures, which have, so far, been developed only on small scale, "political cultures and coalitions, social values, world-views, and paradigms" make up such a system (Smit 2006: 734). These are not in place for any CE approach as of the date of this writing. At the same time, CE is "not a new idea" (Shepherd et al. 2009: 4), as it is based on research findings and objectives which date back decades, if not centuries: Modifying climate and weather conditions in order to make them more convenient for human habitation and well-being arguably has been the intention of a number of societies (Fleming 2010). Technological advancement has allowed for significant weather modification projects in the nineteenth and twentieth century. Early proposals to influence weather and climate in the 19th and early twentieth century were conducted in the context of various aims ranging from increasing agricultural yield (Shepherd et al. 2009; Keith 2000) to using weather as a weapon. For instance, in the late 1940s, an idea of unleashing the energy of storm clouds around microscopic chemicals was developed by Irving Langmuir, which would, in his estimates, release at least as much energy as an atomic bomb (Fleming 2010: 170). Proposals to use weather or climate as a weapon reached their summit during the Cold War (Fleming 2010: 182). In the course of the growing environmental movement in the 1960s and 1970s, 48 members of the United Nations, among them the USA and the Russian Federation, signed the 'Convention on the Prohibition of Military and any other Hostile Use of Environmental Modification Techniques' (ENMOD) in 1977 (UN 1977; Fleming 2010: 184).

The notion to counteract inadvertent anthropogenic climate impacts through CE measures can be traced back at least to the 1960s (Rasch et al. 2008): The possibilities of weather and climate modification through cloud seeding as well as its possible social, legal, biological and internationally political implications were discussed in a dedicated Report by the National Science Foundation Special Commission on Weather Modification (1965). The idea of combatting adverse effects of human activity on the climate—e.g. of air and heat pollution and growing populations—through diverse CE techniques was outlined in the 1960s and 1970s both by the Soviet geoscientist Mikhail Budyko and Joseph O. Fletcher of the US RAND Corporation (Fletcher 1968: 23, 1969: 21; Budyko 1977: 246; Fleming 2010). Arguably, CE could be described as the original attempt to confront the challenge of anthropogenic climate change, before mitigation and adaptation were even discussed (Bellamy et al. 2012a: 4). In this perspective, CE research in the 2000s can be understood as a continuation of early climate modification schemes (Keith 2000: 250).

While some authors observe a degree of hesitance on the part of researchers to delve further into CE after the end of the Cold War (N. Klein 2014: 261; Granger Morgan et al. 2013: 40), the topic was not completely off the academic table: In 1992, the U.S. National Academy of Sciences published a report on *Policy Implications of Greenhouse Warming*, which contained a full chapter on CE options (NAS 1992: 433–462). Individual researchers continued to publish on the issue after the end of the Cold War, for example in a special issue in *Climatic Change* in 1996 (see, for example, Teller et al. 1997; Teller et al. 2002; Bodansky 1996; Schneider 1996; Dickinson 1996; Marland 1996; Schelling 1996). In the late 2000s, academic interest in CE—if measured by the number of publications on the issue, for example—increased intensively (Matzner and Barben 2018: 150; Oldham et al. 2014: 5; Linnér and Wibeck 2015: 257; Belter and Seidel 2013: 420; Lawrence and Crutzen 2017: 137). This increase coincides with the publication of Nobel Prize laureate Paul Crutzen's lead essay 'Albedo enhancement by stratospheric sulfur injections: A contribution to resolve a policy dilemma' (Crutzen 2006) in a special section of *Climatic Change*, accompanied by a number of articles addressing questions around whether or not CE should be researched (Bengtsson 2006; Cicerone 2006; Kiehl 2006; Lawrence 2006). It has been argued that Crutzen's contribution opened up CE for deliberation and research in a larger academic community (Kerr 2006: 401; Keith 2013: 93; Preston 2012b: 2; Larson 2016: 173; Lawrence and Crutzen 2017: 136).

2.1.2 *Climate Engineering in Climate Policy—Discussions and Decisions at the Margins*

While CE research never took place in a space devoid of political considerations, in recent years these approaches have increasingly entered the sphere of climate policy, but only at the margins. This is the case in several respects: Firstly, climate policy and the perception of its outcomes can be argued to be relevant to motivate researchers to address CE as a research object: Interest in CE by academics increased together with the perception of a failure of multilateral climate negotiations to achieve a binding agreement for mitigation measures (see, for example, Launder and Thompson 2008). Klein argues that "Crutzen created some space for preliminary research to take place, but geoengineering's real breakthrough came after the Copenhagen summit flopped in 2009 […]" (Klein 2014: 262).

Secondly, through various channels such as dedicated reports by ad-hoc academic consortia and established organizations, the discussion of CE approaches moved from academic to political arenas: A number of reports were published in the late 2000s and the 2010s, authored by academic authors (Morrow 2017), but carrying the discussions outside of the academic sphere. For example, the Royal Society published a report on CE in 2009 (Shepherd et al. 2009), and a report by the NAS in 2015 explicitly recommends the research into SRM techniques (NAS 2015c: 10). Arguably, "[t]his report could open the floodgates for significant U.S. government funding for research into stratospheric aerosol injections" (Larson 2016: 178). Further, established organizations at the boundary between climate science and climate policy (Hoppe 2005) picked up CE issues, for example the IPCC in its most recent fifth Assessment Report (AR5), as ideas that have been brought forth as potential solutions to some climate change problems (IPCC 2014b: 89), albeit not supporting these approaches.

Thirdly, CE surfaces now and again on the agenda of political bodies involved in climate policy. For example, in 2010 the parties to the Convention on Biological Diversity (CBD) decided to invite all members to.

> [e]nsure […] in the absence of science based, global, transparent and effective control and regulatory mechanisms for geo-engineering […] that no climate-related geo-engineering activities that may affect biodiversity take place, until there is an adequate scientific basis on which to justify such activities and appropriate consideration of the associated risks […] (CBD 2010: Art. 8).

In contrast, in 2019, at the 4th session of the United Nations Environment Assembly (UNEA), a resolution to conduct "a global assessment of potential risks and benefits of geoengineering technologies" (IISD Reporting Services 2019: 13) proposed by Switzerland was withdrawn after no agreement could be found. At the COPs of the UNFCCC, CE approaches, in particular CDR, find increasing interest in side events, but are not subject of high-level decision-making (IISD Reporting Services 2018). Interestingly, funding for research and development of DAC is part of a UK government plan for economic consolidation after the Covid-19 spring 2020 pandemic (HM Treasury 2020). Further, the overall 2020 budget of the U.S. National Oceanic and Atmospheric Administration (NOAA) includes a comparatively small appropriation ($ 4 million) to research potential effects of SAI (Fialka 2020).

Considered as an object of political interest, CE approaches can be assessed either as the object of governance—as something that can be regulated—or as a measure of governance, or as a policy, i.e. as something that can help to regulate and govern some other issue. Concerning the governance of CE approaches, authors assess the existing landscape of environmental, technology and similar governance in order to identify which rules and regulations pertain to CE approaches (Larson 2016; Burns and Nicholson 2016; Scott 2013; Reynolds 2018), study new governance approaches (Huttunen et al. 2015; Armeni 2015; Böttcher et al. 2016), enquire into the specific governance requirements of CE or offer governance frameworks dedicated to this 'new' category of climate policy (Markusson and Wong 2015; Foley et al. 2015; Bellamy 2016; Hubert and Reichwein 2015; Pasztor 2017b; Bodansky 2013; Turkaly et al. 2017; Burns 2016; Gardiner and Fragnière 2018; Nicholson et al. 2018). Some authors argue that the current mainly academic engagement with the topic in itself constitutes a form of governance, described, for example, as "de facto governance" (Gupta and Möller 2018: 480) with a focus on the structuring of the field through authoritative assessment, or "governance by default" (Talberg et al.

2018: 229). Lederer and Kreuter (2018) make the argument that CE governance, instead of being decided in legalistic channels—e.g. through international treaties—might "pursue a logic of technical feasibility, political acceptance and bureaucratic momentum" (Lederer and Kreuter 2018: 1). This notion of hidden momentum has been picked up in the discussion of CE as a policy to address the challenge of climate change. Whether various CE approaches could play a role in the climate policy portfolio, which role this could be and how this might come about has been assessed in the literature (Bahn et al. 2015; Geden 2019; Geden et al. 2018; Peters and Geden 2017; Buck 2016; MacMartin et al. 2017).

Thus, while CE approaches are not the main focus of climate policy actors, to some degree climate policy decisions rely, arguably, on the assumption of CE. This reliance on a non-existing technology in order to achieve the 1.5 °C climate target (Lawrence et al. 2018) could also mean, however, that the current, mainly academic discussion on CE might define the "conditions of possibility" (Wendt 1998: 106) of a later, political discussion. To prevent such a narrowing-down, initiatives such as the C2G2 aim to "catalyze the creation of effective governance of [...] Geoengineering" and to "expand the conversation from the scientific and research community to the global policy-making arena" (C2G2 2019).

2.1.3 *The Current State of Affairs—Models and a Few Experiments*

No CE approach exists today in the sense of a socio-technical system sufficient to influence the global average temperature in a meaningful way (Lawrence et al. 2018; SPP 1689 2019a). Also, only a small number of natural science field experiments have been conducted (Bellamy 2014: 5; Doughty 2015). For many members of the CE research community the question still remains whether field research of CE is necessary and/or feasible (see, for example, Stilgoe 2015: 2). However, there are a few exceptions. A number of carbon dioxide removal experiments have been conducted in the past, for example the OF experiments SOIREE in 1999 (Boyd and Law 2001; Boyd and Abraham 2001), LOHAFEX in 2009 (Thiele et al. 2012: 8803) and the 2012 Haida Salmon Restoration Corporation's iron fertilization (Bellamy 2014: 4; Tollefson 2012).[3] Beyond this, CE approaches such as DAC have become the centerpiece of the business model of a small but growing number of companies[4] as well as the focus of new political constituencies[5] (see, for example, Speicher 2017). Concerning SRM, the Eastern Pacific Emitted Aerosol Cloud Experiment (E-PEACE) conducted in 2011 (Russell et al. 2013: 709) "while not identifying as

[3] An overview over the history of ocean iron fertilization research and experimentation can be found in Strong et al. (2009).

[4] Such as Climeworks (https://www.climeworks.com/) or Carbon Engineering (https://carbonengineering.com/).

[5] Such as the NGO Carbon180 (https://carbon180.org/work).

SRM" (Doughty 2015: 2), produced results relevant for Marine Cloud Brightening methods of CE (Russell 2012: 12). An early SAI experiment was conducted in Russia, which brought aerosol generators aboard helicopters into the lower troposphere and measured the refractive index of the generated aerosols (Izrael et al. 2009: 635). The SPICE experiment, part of the UK Stratospheric Particle Injection for Climate Engineering (SPICE) project, was conceived of in 2010, but never took place (Macnaghten and Owen 2011: 293; Pidgeon et al. 2013: 452) due to a "lack of rules governing such geoengineering experiments" (Cressey 2012). Further experiments of different CE methods have been discussed (Latham et al. 2012: 4249; Parson and Keith 2013: 1278; Keith et al. 2014), while some scientists argue that SAI "cannot be tested in the atmosphere without full-scale implementation" (Robock et al. 2010: 530; see also Klein 2014: 269).

Most knowledge about these approaches stems from modelling studies and laboratory research as well social science field research—conducted, for example, through interviews and surveys. Further, CE approaches are the topic of a number of past or ongoing research programs, many of which are characterized by a high level of interdisciplinarity. A number of national and international research projects, publicly and/or privately funded, have recently analyzed or are currently analyzing CE from a number of perspectives in order to discover political, social, legal, and ethical, as well as physical, chemical, biological and engineering issues posed by these methods (Kreuter et al. 2020). As early as 2008, the presidents of 13 large national academies argued in a common statement that "[t]here is also an opportunity to promote research on approaches which may contribute to towards maintaining a stable climate (including so-called geoengineering technologies [...])" (Joint Science Academies 2008: 2). Examples of research programs are the UK projects IAGP (Integrated Assessment of Geoengineering Proposals) and CGG (Climate Geoengineering Governance), the French project REAGIR (Reflections on Environmental Geoengineering), the German program "Climate engineering: risks, challenges, and opportunities" (DFG-SPP 1689), the EU project EuTRACE (European Trans-Disciplinary Assessment of Climate Engineering), the US-American "Ethics of Geoengineering" project (Szerszynski and Galarraga 2013: 2819), and Harvard's Solar Geoengineering Research Programme.

2.2 Climate Engineering as a Topic of Social Science Scrutiny[6]

CE has, as early as the special issue on the topic in *Climatic Change* in 1996, been studied not only from natural science and engineering perspectives, but also social sciences and humanities (Bodansky 1996; Jamieson 1996; Schelling 1996). In recent years, the numbers of social science and humanities research articles on the matter have increased (Linnér and Wibeck 2015: 258). Particularly the wide, in some cases

[6]Some ideas expressed in this section have been published in Kreuter (2015a, b).

global, scope of effects of CE technologies as well as their deep implications for human society make these technologies interesting for analysis and evaluation in social sciences as well as humanities. While the objects of interest, the research questions, the scope, the underlying theories, the methodologies and the data used in social science studies on CE differ widely, the importance of how CE is discursively represented and which impacts these representations can have is widely acknowledged (see, for example, Rickels et al. 2011a: 23; Wiertz 2012: 53; Cairns and Stirling 2014: 36). In media coverage, scientific publication and political debate alike, CE is always presented in a certain way which alleviates the importance of some aspects of it and neglects others—in some types of communication more so than in others. As Mike Hulme (2014: 5) points out: "In controversial public debates it matters how terms are defined and understood."

Accordingly, the focus of this study lies on the framing of CE approaches in the academic discussion, following the assumption that this framing can influence decision-making. The following literature review provides an overview of studies from social science and humanities which touch upon the question of representing CE discursively, and upon the issue of decision-making concerning the phenomenon of CE. I differentiate roughly three sets of studies, based on their respective research foci. The first set is united in studying *public perceptions* of these approaches. The second set, covering *empirical analyses of discourse*, targets the ways in which CE is presented in various discursive arenas and who speaks in these arenas. The third set focuses on the discussion of *specific frames* or, more broadly speaking, of discursive representation of CE.

The three sets of literature taken together map out the research gap which this study fills: Exploratory discourse analyses have provided remarkable insight into various arenas of CE discourse and identified a number of representations of CE. However, the exploratory approach—letting the material 'speak for itself' without any preconceived idea of the structure of the discourse—cannot provide information on what impact these representations might have on the meaning of CE. This is because exploratory studies explicitly do not start from deductively developed categories or types of representation. Conversely, perception studies have found empirical indications and discussions of specific representations have made theoretical arguments that specific ways of representing CE might increase or decrease the acceptance of these approaches. However, these studies do not assess how prevalent these ways of representation are in the discourse. Furthermore, and more specifically, none of these studies make a systematic argument about how these representations—or the meaning constructed through them—might influence future decision-making concerning CE.

2.2.1 *Understanding Opinions of the Unknown—Studying Public Perception*

The first set of studies focuses on how different groups of people perceive CE under various circumstances. These studies are relevant to the present research endeavor as they provide an overview over the perceptions of CE present in various groups, i.e., what the grounds are like on which various frames of CE will fall or already have fallen. While the studies on public perception of CE do not cover more than a fraction of the world's population, they share a number of important findings concerning CE as a public issue: Firstly, public awareness of CE can, still, be considered to be low (Burns et al. 2016; Pidgeon et al. 2012; Wibeck et al. 2015; Scheer and Renn 2014: 4; Mercer et al. 2011: 1; Sugiyama et al. 2016: 6)—with variations between regions and social groups. While early studies focused, primarily, on publics in the UK (NERC 2010), Germany (Rickels et al. 2011a) or North America (Mercer et al. 2011), more recently, individuals from other states (Sugiyama et al. 2016; Asayama et al. 2017; Carr and Yung 2018; Carr and Preston 2017) as well as from specific local communities (Gannon and Hulme 2018; Buck 2018) have moved more into focus.

Secondly, existing opinions on the topic are very diverse and depend on a number of factors such as context, framing and access to information (Bellamy et al. 2017). Cautious positions are common but all in all, positions are not grouped into two clear, opposing camps. Instead, they are found to be diverse (Hiller and Renn 2012: 217; Cairns and Stirling 2014: 36; Bellamy et al. 2016: 282). The type of technology proposed is found to play a role (Braun et al. 2018: 471), CDR measures apparently enjoying more acceptance than SRM (Pidgeon et al. 2012), as well as the pre-existing degree of concern for climate issues (Wibeck et al. 2015: 28). The degree of activity has been found to be relevant: While model or laboratory research is generally considered acceptable, field research and actual deployment enjoy much less acceptance (Asayama et al. 2017: 82; Scheer and Renn 2014: 5; Mercer et al. 2011: 1; Merk et al. 2015: 299) Further, the geographic location—i.e. if someone lives in Europe or in the USA—seems to make a difference concerning individuals' perceptions of CE (Mahajan et al. 2018: 536; Carr and Preston 2017).

Thirdly, the way that the issue is framed in the media, in politics and by other actors can be expected to have a strong influence on how it is perceived by the public. E. T. Burns et al. (2016: 537) argue that "[u]nfamiliarity with solar geoengineering brings a key methodological challenge, since the mere introduction of the topic introduces a potential bias". Studies found influence of such ideas as 'messing with nature' (Corner et al. 2013: 938), moral hazard (Corner and Pidgeon 2014: 11), natural analogies—such as presenting stratospheric aerosol injections as "no different to a volcano" or comparing carbon capturing chemical vents with "artificial trees" (Corner and Pidgeon 2015: 5) and the'climate emergency' framing (Asayama et al. 2017: 82). In an experiment with respondents from the USA, Raimi et al. (2019: 300) find that "[w]hen geoengineering is framed as a major solution, people worry less about climate change, leading to reduced mitigation support." More generally, a number of researchers, on the basis of currently low rates of awareness of and knowledge about CE, assume that perceptions and positions concerning CE can be expected to fluctuate, dependent, inter alia, on external stimuli (Scheer and Renn 2014: 1) such as "framing, future information on risks and benefits, and changes to context" (Mercer et al. 2011: 1) as well as ongoing debate (Merk et al. 2015: 299). Many questions remain unanswered to date concerning public perception and opinions of CE, and calls for more public engagement are common in the debate (see, for example, Carr et al. 2014; Carr 2013; Bellamy and Lezaun 2015; Buck 2010).

2.2.2 *How and by Whom is Climate Engineering Represented?—Exploring the Discourse*

The analysis of public perception of CE stresses the relevance of assessing the way in which these approaches are discursively represented by various actors in various communication situations. A number of empirical studies have assessed CE discourse—for example, metaphors, imaginaries, arguments or storylines (see Uther 2014: 68)—or assessed frames but applying a different definition to the term—and studied the voices heard in this discourse.

The special *relevance of expert* or academic representations[7] of CE is pointed out by Bowden, who argues that CE was marginalized early on due to issues of "cost, feasibility, uncertainty, and risk" (Bowden 2010: 70), but then re-embraced in the 2000s and introduced into public and political discourse. The academic discussion is the focus of a number of studies (see, for example, Bellamy et al. 2012b; Huttunen and Hildén 2014; Markusson 2013; Linnér and Wibeck 2015; Wiertz 2016; Low 2017; Uther 2014; Himmelsbach 2018). Low argues that "[r]esearchers from across the natural, applied, and social sciences—at least for now—play constitutive roles in framing issues and setting agendas at the boundary spaces between science, society and policy" (Low 2017: 69). The special role of experts is identified also in empirical assessments of media coverage of CE, as members of this group enjoy large representation in media content on CE (Buck 2013: 173; Uther 2014: 107, 158).[8]

Certain *themes* surface again and again in the discourse on CE, and in various arenas such as the academic discussion, media coverage or advocacy debate, not least the four themes of most interest for this study: security, complexity, economics and appropriateness. The renaissance of the notion of CE in the 2000s, Bowden argues, was due to two new features in the discourse:

[7] I use the terms ‚expert ' or ‚academic ' here instead of ‚scientific ' in order to highlight that this discursive field is not limited to natural scientists, but includes also engineers, social scientists and scholars of the humanities. In this, I follow the definition presented in the Epistemic Communities approach (Adler and Haas 1992; Haas 1992).

[8] Other assessments of the relative representation of speakers from specific social groups in the discussion about CE differentiate, for example, between global regions (Biermann and Möller 2019; Buck 2013) or membership in indigenous groups (Whyte 2019).

> First, the science became more certain and the urgency of the problem more immediate (both because of the possibility of tipping points and because the Kyoto Protocol turned out to be largely ineffective and global emissions continued to rise). Second, the optimism that a negotiated political solution could be found dissipated as emissions continued to rise [...]. (Bowden 2010: 75).

Both features, the notion of urgency and impending crisis as well as the notion of a political failure, are found in later studies of the academic discussion of CE, namely in frames of security (based on the logic of necessity to protect against an existential threat) and of technological simplicity (based on the idea that complex social problems might be reducible to simple technological challenges). Inductive empirical studies found discursive representations concerning issues such as climate emergency, risk and benefit, governance, knowledge, complexity, control, society and justice (Bellamy et al. 2012b; Huttunen and Hildén 2014; Bellamy 2013, 2014; Himmelsbach 2018). Thus, these findings present frames other than the security and the complexity frame: economic and political considerations are also identified in framing CE. But not only linguistic patterns such as frames constitute the meaning of CE: Heyen et al. (2015) draw attention to how the decisions made in modelling with regards to assumptions about preferences can change the concordant view in political and social implications of SRM. Wiertz (2016) focuses on the practice of climate simulations in order to assess how the meaning of SRM specifically is constructed. Low (2017) scrutinizes the role of both framing and modelling conducted by academic experts, and argues that "[s]cholarship must recognize that in assessing future-oriented technologies, experts are implicated in claiming and making the future—and help render these dynamics transparent [...]" (Low 2017: 70).

The discussion and representation of CE in the media identifies a number of similar frames in newspaper articles and other media outlets as present in the academic discussion, represented by notions of urgency and security, complexity, justice and economics. They include frames of "Innovation, Risk, Governance and Accountability, Economics, Morality, Security and Justice" (Porter and Hulme 2013: 342), catastrophic, cautionary, spatiotemporal struggle, managerial and Bildungsroman framings (Buck 2013: 170), metaphors of fight and war, control, analogies with natural phenomena and man-made mechanisms (Luokkanen et al. 2014: 972) orand metaphors such as "'The planet is a body', 'The planet is a machine,' and 'The planet is a patient/addict'" (Nerlich and Jaspal 2012: 131). Scholte et al. find that there is "strong support for an opening of the debate [...] given the decline of overly deterministic frames, the emergence of frames related to sociopolitical issues and an overall more balanced distribution of the various frames" (Scholte et al. 2013: 1). Anshelm and Hansson (2014a) find that only 8% of the articles analyzed were critical of CE, and isolate the concerns raised: the fear of a "technological gamble with the planet", the "inability to handle structural dysfunction", the "geoclique" furthering the "Trojan horse" of CE to promote their personal interests and the fear of a "democratic deficit" (Anshelm and Hansson 2014a: 137–140). Anshelm and Hansson (2014b) find "four coherent storylines that represent most of the geoengineering advocacy in the public discourse in mass media" (Anshelm and Hansson 2014b: 101). These four storylines, they call "the scientist's double fear", "the failure of politics and cynical industrial fatalism", "pure technology: a bridge to a sustainable future" and "just mimicking nature" (Anshelm and Hansson 2014b: 107, 110, 112, 113). These storylines capture some of the notions from the academic discussion, in particular the notion of high stakes related to both climate change and CE, the notion of political failure and the notion of the simplicity of technology in the face of complex social challenges.

Political CE advocacy is the focus of Sikka's study of frames used in "speeches, articles, research, media submissions and policy reports given by a variety of organizations and individuals supportive of geoengineering" (Sikka 2012a: 167). Here, again, notions of security as well as economic considerations are present: Sikka finds that two frames in particular are salient. One is what she calls 'exceptionalism': the idea that climate change is such a catastrophic problem that CE is the "only option" (Sikka 2012a: 167) to deal with it. The other framing connects CE with economic prosperity, "positioning opposition to geoengineering as opposition to the economy" (Sikka 2012a: 170). Two more frames that Sikka isolates consider CE in terms of a technology: that of "a claim to scientific neutrality" and that of "technological determinism" (Sikka 2012a: 167).

The discourse is, however, characterized not only by some specific frames of CE, but also by *tensions between these frames*, present in particular in the academic discussion. (Markusson 2013: 31; Cairns and Stirling 2014; Matzner and Barben 2018). Linnér and Wibeck (2015) argue that there existed a peculiar polarity in the academic discussion of CE which differentiated this from similar discourses on comparable emerging technologies:

> Most peer-reviewed climate engineering literature does not weigh the risks and new, additional, benefits of the various technologies, but emphasizes either the potential dangers of climate engineering or the climate change consequences of refraining from considering the research, development, demonstration and/or deployment of climate engineering technologies. (Linnér and Wibeck 2015: 255).

A gap remains concerning whether there is a pattern behind the polarity identified by these authors. The present thesis assesses the role of three factors—disciplinary focus, time of publication and specific CE approach of academic publications on CE—in order to identify whether any of these factors can help understand the polarity.[9]

The findings of these mainly inductive and exploratory studies indicate that (1) framing, in its narrow sense, and discursive representation, in the wider understanding, of CE is very important; that (2) the themes of Security, Economy, Complexity and Appropriateness are uncovered in the discussion regularly; and that (3) ambivalence is a common feature in the discussion. Also, experts play a special role in the discourse on CE (Tables 2.1 and 2.2).

[9]See Chaps. 5 and 6.

Table 2.1 Overview over analyses of CE discourse

Level of discourse analyzed/discursive arena analyzed	Macro (entire discourse)	Micro (specific elements of discourse, e.g. arguments or frames or imaginaries)
Several	Uther (2014)	Matzner and Barben (2018), Harnisch et al. (2015), Neuber (2018), Flegal (2018)
Academic	Low (2017), Szerszynski and Galarraga (2013), Gupta and Möller (2018)	**This study** Bellamy et al. (2012a), Markusson (2013), Huttunen and Hildén (2014), Bellamy (2014, 2013), Wiertz (2016)
Media/public	Macnaghten and Szerszynski (2013), Tingley and Wagner (2017)	Buck (2013), Porter and Hulme (2013), Scholte et al. (2013), Luokkanen et al. (2014), Anshelm and Hansson (2014b), Nerlich and Jaspal (2012)
Political		Sikka (2012a, 2019)

Table 2.2 Overview over frame analyses on CE (detail of column 'micro' above)

Research design/ Discursive arena analyzed	Inductive/exploratory	Deductive
Academic	Bellamy et al. (2012a), Markusson (2013), Linnér and Wibeck (2015), Bellamy (2014), Huttunen and Hildén (2014), Bellamy (2013)	**This study**
Media	Buck (2013), Porter and Hulme (2013), Scholte et al. (2013)	Scholte et al. (2013)

2.2.3 *What Does It Mean?—Examination of Specific Discursive Elements*

A number of studies focus on specific frames, arguments or representations and assess these, either theoretically or through empirical study, and to uncover potential assumptions and implications. This literature is relevant for the present study as it allows for the critical reflection of the frames which are assessed here, and more specifically, for the development of a pattern for the mechanisms through which they influence legitimacy of CE approaches. The studies are subdivided below into the four themes of interest to the present study: security, economy, complexity and appropriateness.

Various frames and arguments pertaining to these themes have been examined in the literature. With regards to the *security* theme, discursive representations can be divided into presenting CE either as a source of danger, on one hand, or as a shield against (potential) threats, on the other. The notion of climate emergency, the buying time-argument, the idea of CE as the lesser of two evils and the notion of

CE as a Plan B are all united in presenting CE approaches as a potential (part of a) measure to address a looming threat. The notion of moral hazard or the slippery slope and the role of CE with regards to international conflict are representations of CE approaches as threats to valued referent objects. Hamilton, among others (see, for example, Gardiner 2013a: 28; Burns 2013: 218; Horton 2015), discusses the notion of 'climate emergency' as an argument for the development and deployment of CE technologies:

> The Earth's climate is a 'non-linear system, that is, changes in one variable do not lead to simple proportional changes in related ones […] In non-linear systems, a small change in one state may initially have only small effects but at some point a threshold may be crossed where the system, driven by amplifying feedbacks, flips suddenly into a new state. (Hamilton 2013a: 12).

This danger of "abrupt climate change" (Hamilton 2013a: 13; see also Hulme 2003; Bellamy and Hulme 2011) is based on the identification of certain tipping points or boundaries in the global climate system (Lenton et al. 2008: 1786; Lenton 2011; Steffen et al. 2018) and is, according to Hamilton, in turn the basis for the idea "that within the next few decades, we may face a 'climate emergency'" (Hamilton 2013a: 13). Various authors have argued for the relevance of the presence of the notion of climate emergency in the discourse on CE (see, for example, Hamilton 2013a: 13; Gardiner 2013b: 11; Victor et al. 2009: 64). The notion is relevant in particular as a basis for the conclusion that CE might be inevitable (see, for example, Michaelson 2013: 83; Davies 2013: 64; Güssow et al. 2013: 243; Burns 2013: 200; Heyward and Rayner 2013: 28).

The assumptions going into and the implications deriving from framing CE with regards to a climate emergency have been elaborated upon in several studies: Various authors criticize the idea that a climate emergency might be objectively identified (see, for example, Sillmann et al. 2015; Horton 2015): "Most obvious is the problem of defining, detecting and announcing a climate emergency. What exactly *is*, or what *could* be, a climate emergency, and who is authorized to define one?" (Hulme 2014: 23) Horton argues that "an emergency (climate or otherwise) is ultimately a shared idea, rather than objective reality" (Horton 2015: 2), which means that there is no objective threshold after which the use of CE was objectively necessary. Asayama argues that the emergency argument is part of what he calls "apocalyptic catastrophism" (Asayama 2014: 89):

> The fear of future impacts of a changed climate rather than a changed climate *itself* legitimizes the option of geoengineering. As such, what underlies the idea of a 'techno-fix' through geoengineering is catastrophism that imagines the future climate as the apocalypse. (Asayama 2014: 90).

Concerning implications, various authors argue that the climate emergency might provide CE with a feature of necessity, thereby closing down spaces for deliberation and starting off a momentum that could turn out hard to control (Corner et al. 2011: 13–14; Markusson et al. 2013: 281; Horton 2015: 3).

An argument closely linked to this notion is the idea of 'buying time' through CE (Rayner et al. 2013: 501; Schäfer et al. 2014: 243; Neuber 2018: 127): "Unlike the idea of entering an immediate crisis, the 'buying time' frame presents CE as bridging the gap until emissions can be reduced to the amount necessary to achieve climate goals" (Lederer and Kreuter 2018: 6). Other notions related to the framing of CE as a protective mechanism is the representation of CE as the 'lesser of two evils' (Svoboda 2015; Hulme 2017), in which the potentially threatening character of CE approaches is accepted, but assumed to be lesser in degree than the impact of 'runaway' climate change, or as a 'Plan B' (Corry 2014, 2017; Anshelm and Hansson 2015; Schäfer et al. 2014), i.e. a second way to go forward if climate policy is not successful.

The discussion of dangers potentially emanating from CE research or deployment presents the mirror image of the set of ideas presented above: Instead of protection against a danger, CE approaches themselves are considered potentially dangerous. 'Moral hazard' or 'risk compensation' describes the common concern that the increasing investment of effort into CE might have detrimental effects on the effort undertaken to further mitigation and adaptation measures (see, for example, Heyen 2012: 43; Reynolds 2015: 2; Bunzl 2009: 2; Shepherd et al. 2009: 37; Rayner 2014: 6; Corner and Pidgeon 2010: 30; W. C. G. Burns 2013: 209; Michaelson 2013: 100; Corner and Pidgeon 2014: 2; Robock 2008: 15). The representation of CE approaches as moral hazard has been criticized from many sides. Various researchers, among them the authors of a report commissioned in 2011 by the German Federal Ministry of Education and Research, point out the weakness of its empirical basis for the case of CE:

> While the *moral hazard* phenomenon can be statistically proven in the case of other technological innovations, for example for the introduction of safety belts in cars, in the context of Climate Engineering, there is no empirical material which supports the *moral hazard* argument.[10] (Rickels et al. 2011a: 86).

The authors of the Royal Society turn the argument around, suggesting that "it is possible that geoengineering could galvanize people into demanding more effective mitigation action" (Shepherd et al. 2009: 39). A number of studies have, subsequently, gone in to assess the empirical validity of the moral hazard claim with outcomes diverging based on the specific technology considered, the political preferences of respondents, as well as other factors: (Reynolds 2015: 174; Kahan et al. 2015: 203–204; Merk et al. 2016: 1; 2018; Fairbrother 2016: 1; Campbell-Arvai et al. 2017: 321). Thus, it seems that further empirical study is warranted in order to uncover under which circumstances the moral hazard claim might hold true: "What seems almost certain is that some actors—potentially including some states—will experience an incentive to reduce or defer mitigation in the presence of SRM, while others do not" (McLaren 2016: 600). Further criticisms are based, not on empirical assessment of the moral hazard claim, but on theoretical considerations of the foundation of the claim (Hale 2012: 115; Michaelson 2013: 101; Morrow 2014a: 1; Reynolds 2015:

[10]Translated by JK; original: „Während das *moral-hazard*-Phänomen bei anderen technischen Neuerungen, wie zum Beispiel dem Einführen von Sicherheitsgurten in Autos, statistisch belegt werden kann, gibt es im Kontext von Climate Engineering kein empirisches Material, das die *moral-hazard*-These unterstützt." (Rickels et al. 2011a: 86).

174) or based on highlighting the way it is used instead of the argument itself (Bunzl 2009: 2; Wiertz 2012: 52; Lin 2013: 673).

The political configurations of potential CE development or use are the focus of studies considering the role of these approaches as a political arena or as a factor in transboundary conflicts: Assessing CE as a political arena, researchers have made arguments about how CE approaches might highlight and decision-making concerning CE be influenced by vested interests of political parties (Nicholson and Thompson 2015: 1; Hamilton 2013a: 90; Horton and Reynolds 2016: 438–439), economic entities (Sikka 2012a: 173; Hamilton 2013a: 80) or even researchers (Egede-Nissen 2010: 1; Oldham et al. 2014: 17; Kintisch 2010: 8). On the international levels, arguments are made about the potential impact of CE on transboundary conflicts, for example as a factor aggravating existing and introducing new conflicts or as a potential source of conflict reduction (Brzoska et al. 2012; Maas and Scheffran 2012), about the difficulty of attributing specific climate effects to specific sources (Nightingale and Cairns 2014), and the role of CE for future geopolitics (Dalby 2015; Boyd 2016; Yusoff 2013). One commonly discussed notion in this context is the idea of unilateral climate action through CE: The observation that CE technologies, other than mitigation, do not need concerted action from a number of states to be effective, and the subsequent concern that they might be developed or even deployed by a single country unilaterally, are common in the literature (see, for example, Victor 2008: 324; Shepherd et al. 2009: 40; Rickels et al. 2011a: 69; Brzoska et al. 2012: 191; Edney and Symons 2014: 5). Horton, however, challenges the argument of unilateralism and argues "that this fear [...] is largely misplaced, grounded more in unexamined policy assumptions than in reasoned analysis of the strategic situation faced by states" (Horton 2013: 168). Rabitz, similarly, argues that "only great powers could unequivocally resort to unilateralism" (Rabitz 2016: 106). Accordingly, this danger, similar to the notion of the moral hazard, does not seem to be inherent to the approaches, but rather depend, inter alia, on the political context of their use (Lloyd and Oppenheimer 2014: 52; Moore et al. 2016: 588; Singer et al. 2014: 1; Bluemling et al. 2019: 1; Weili and Ying 2014: 1). Various scenario and game experiments have isolated factors which might, in an international setting, have an impact on international interaction concerning CDR or SRM. While these findings can never be prescriptive, they can "illustrate possible future conflicts" (Matzner and Herrenbrück 2016). Findings indicate that the expected climate effects hitting the respective actor (Singer and Matchett 2015: 202), power struggles between '"big players"' (Matzner and Herrenbrück 2016: 15), the specific approach in question (Sandler 2018: 105) and the presence or absence of cooperation (Emmerling and Tavoni 2018: 244; Ricke et al. 2013: 7) can influence collective action concerning CE. Further studies will be required in order to stabilize findings in this regard.

Concerning the theme of *complexity*, two notions discussed in the literature are of particular interest for the present study: the idea of a 'technological fix' on one hand, and the notion of the 'Anthropocene' on the other. The notion of the 'technological fix' or 'technofix' is an expression of the belief that technologies, as tools to serve a specific purpose, can help fix any problem, potentially by reducing the complexity of a social challenge to the simplicity of a technological one (Weinberg 1966: 5).

It is not specific to the CE discourse, but surfaces here regularly (see, for example, Hulme 2014: 1; Corner and Pidgeon 2010: 30; D. Scott 2012: 151; Borgmann 2012: 189). As Dane Scott elaborates, those notions are comprised in the philosophy of 'technological progressivism'. This view on the relationship between humans and technology maintains that "the consistent application of science and technology is humanity's greatest hope for improving human life" (Scott 2013: 2).[11] The representation of CE as technofixes, however, also come with their own set of problems, which have been highlighted in the literature. For example, technofixes are criticized for not getting at the root cause of the problem (Scott 2013: 7; Sikka 2012b: 112; Szerszynski 2010: 22; Markusson et al. 2017: 1).

Another important notion in the context of the theme of complexity is that of the 'Anthropocene' (Crutzen and Stoermer 2000: 17): It suggests.

> (i) that the Earth is now moving out of its current geological epoch, called the Holocene and (ii) that human activity is largely responsible for this exit from the Holocene, that is, that humankind has become a global geological force in its own right. (Steffen et al. 2011a: 843).

CE is considered as an indicator that humanity stands at the doorstep of the Anthropocene, a new epoch in Earth history, or that it has already entered into it (see, for example, Steffen et al. 2011a: 858; Schäfer et al. 2014: 239; Galaz 2012: 25; Lederer and Kreuter 2018: 2) and for the way in which society might deal with "the complex set of questions that emerges with the ability to fundamentally intervene in natural processes on global scale [...], and how this interacts with social and political order" (Schäfer et al. 2014: 243). The Antropocene, then, might introduce complexity on both sides of the equation, as both inadvertent impacts of humankind on the climate system and the intentional modification of the climate through CE are fraught with uncertainties and challenges. The notion is strongly criticized, for example from geological viewpoints with regards to how this shift can be identified (Steffen et al. 2015) and from a critical perspective for being a "fetishism" (Cunha 2015: 65) originating from capitalist ideology. Other authors argue that the notion implies 'hubris' (Kiehl 2006; Szerszynski 2014) and a "determination to shape the world around us to suit our desires, desires that have no limit" (Hamilton 2013b: 56). Dalby puts the notion into the context of geopolitics, arguing that "[w]hile geopolitics used to be about the context of global politics, now in the Anthropocene, it has become a matter of remaking that context rather than taking it as a given" (Dalby 2015: 190).

Concerning the theme of *economy*, numerous authors have weighed in on the discussion of whether—and if so, which—CE approaches can be considered of low cost, how this 'cheapness' might and should be defined and identified and what

[11] Scott (2013) differentiates three 'philosophies of technology': technological progressivism, technological pessimism and technological pragmatism (D. Scott 2013). According to Scott, the main idea of technological pessimism can be described in the statement that "human history is now on a dangerous trajectory where technology is destined to become the 'determining and controlling influence on society and culture'" (D. Scott 2013: 5; see also White 1967: 1206) Technological pragmatism espouses a more nuanced take on technology: The assumption is that technology is not inherently good or bad (see Borgmann 1987: 40ff.).

implications findings from this quarter may have. Cost–benefit-analysis or risk–risk-analysis is discussed or applied to one or more approaches in a number of publications (see, for example, Shepherd et al. 2009: 44ff.; McClellan et al. 2012; Fernow 2012; Klepper 2012; Barrett 2008; Schneider 2008; Doda 2014; Charles Greene 2010; Chuck Greene et al. 2010; Bickel and Agrawal 2013; Anthony Harding and Moreno-Cruz 2019; NAS 2015a, c; Dietz et al. 2018; Roshan et al. 2019; Moreno-Cruz et al. 2015). The direct costs of some CE strategies such as sulfate aerosol dispersal in the Stratosphere or ocean iron fertilization are considered comparatively low by most authors (see, for example, McClellan et al. 2012: 6; Lane 2013: 116; Güssow et al. 2013: 251; Bickel 2010: 25; Bickel and Agrawal 2013: 993; Moreno-Cruz and Keith 2013: 443)—or, as Barrett puts it, "incredible" (Barrett 2008: 49).[12] However, there exists a wide range of opinions and calculations, and accordingly, some substantial disagreement, concerning indirect and hidden costs of these technologies (see, for example, Klepper 2012: 211; Doda 2014: 19; Fernow 2012: 17; Bickel and Agrawal 2013: 993; Goes et al. 2011: 720; Klepper and Rickels 2014; Amelung et al. 2012; Heyen 2012; Barrett et al. 2014). Additionally, the incapability of SRM measures to counter other climate change consequences such as ocean acidification or the build-up of greenhouse gases is pointed out (McClellan et al. 2012: 7; Moreno-Cruz and Smulders 2010: 25). Including factors such as uncertainty about climate sensitivity and CE side effects (Heutel et al. 2018) or intergenerational considerations (Goeschl et al. 2013) have influence on the findings of economic modelling concerning CE. A. Harding and Moreno-Cruz (2016: 575) give a concise overview over the development of economic analysis of SRM, arguing that "[e]conomists have been at the forefront of the research on solar geoengineering, but a more careful and detailed analysis needs to be put forward."

With regards to *appropriateness*, a number of studies assess normative implications and preconditions of CE—i.e. issues of how the world should be. Questions of right and wrong (moral questions) and considerations of how to systematize and define rules of how to decide between right and wrong (ethical questions) are commonly invoked in the discussion of CE (Hamilton 2013a: 158; Shepherd et al. 2009: 39). Gardiner describes CE as a 'perfect moral storm', which is defined as "the unusual intersection of a number of serious, and mutually reinforcing, problems, which [...] are all obstacles to our ability to behave ethically" (Gardiner 2011a: 7). Further studies either apply a specific set of values—such as Christian ethics (Clingerman and O'Brien 2014) or Confucian ethics (Wong 2015), human rights norms (Svoboda et al. 2018), equity considerations (Burns 2013), justice conceptions (Svoboda et al. 2019; McLaren 2015) or a critical theory perspective (Buck 2012)—to evaluate CE approaches and their preconditions and implications, assess arguments made in support or opposition of CE from various perspectives (Morrow 2014b; Preston 2012a; Whyte 2012; Gardiner 2013a, b, 2011b) or propose-specific guidelines for CE research and consideration (Morrow et al. 2009; Preston and Carr 2019; Hourdequin 2018, 2019).

[12] This dictum has recently been qualified in Barrett et al. (2014).

In summary, arguments and frames with regards to the themes of security, complexity, economy and appropriateness are critically assessed in a number of studies from the social science and humanities literature on CE. On this basis, many of the arguments, frames and notions present in the discussion are based on specific presuppositions and imply certain consequences, not all of which are intended by the respective speakers. The present study aims to add to this rich literature by providing a systematic theoretical framework to describe how, i.e., through which processes, these four themes can provide CE approaches with legitimacy or illegitimacy in decision-making.

This review of the literature indicates that we have learned a lot, in recent years, about which factors influence perception of CE, which themes and representations are present in the discussion and how some of these representations function in influencing the perception or meaning or decision-making on CE. However, a gap remains which can be filled by drawing together the findings of this rich literature and using them as a starting point for empirical analysis. The research gap is twofold: Firstly, and more generally, it consists of a lack of systematic aggregation of research findings concerning the implications of specific discursive representations of CE for the construction of the meaning of these approaches. Secondly, and more specifically, it consists of a lack of assessing the role of this meaning for future decision-making concerning CE. In order to fill this gap, a more theoretically grounded approach needs to be espoused to study the CE discussion than has been done in previous, exploratory assessments. This approach allows for the identification of frame types legitimizing or delegitimizing CE and subsequently assessing the empirical material through this categorization of frame types. In this way, the material can be structured concerning the comparative prevalence of different frame types. Furthermore, we can assess how specific factors such as diverging prevalence of frame types in different disciplinary publications, at different chronological stages of the discussion or in the discussion of different CE approaches, can help understand this structure.

References

Adler E, Haas PM (1992) Conclusion: epistemic communities, world order, and the creation of a reflective research program. Int Organ 46(1):367–390

Amelung D, Dietz W, Fernow H, Heyen D, Reichwein D, Wiertz T (2012) Beyond calculation: climate engineering risks from a social sciences perspective (Forum Marsilius-Kolleg). Marsilius-Kolleg, Heidelberg

Anshelm J, Hansson A (2014a) Battling promethean dreams and trojan horses: revealing the critical discourses on geoengineering. Energy Res Soc Sci 2:135–144. https://doi.org/10.1016/j.erss.2014.04.001

Anshelm J, Hansson A (2014b) The last chance to save the planet? An analysis of the geoengineering advocacy discourse in the public debate. Environ Human 5:101–123

Anshelm J, Hansson A (2015) Has the grand idea of geoengineering as Plan B run out of steam? Anthrop Rev 1–11. https://doi.org/10.1177/2053019615614592

Armeni C (2015) Global experimentalist governance, international law and climate change technologies. Int Compar Law Q 64(4):875–904

Asayama S (2014) Catastrophism toward 'opening up' or 'closing down'? Going beyond the apocalyptic future and geoengineering. Cur Sociol 63(1):89–93. https://doi.org/10.1177/0011392114559849

Asayama S, Sugiyama M, Ishii A (2017) Ambivalent climate of opinions: Tensions and dilemmas in understanding geoengineering experimentation. Geoforum 80:82–92

Bahn O, Chesney M, Gheyssens J, Knutti R, Pana AC (2015) Is there room for geoengineering in the optimal climate policy mix? Environ Sci Pol 48:67–75. https://doi.org/10.1016/j.envsci.2014.12.014

Barrett S (2008) The incredible economics of geoengineering. Environ Resource Econ 39:45–54

Barrett S, Lenton TM, Millner A, Tavoni A, Carpenter SR, Anderies JM et al (2014) Climate engineering reconsidered. . Nature Clim Change 4:527–529. https://doi.org/10.1038/nclimate2278

Bellamy R (2013) Framing geoengineering assessment. Geoengineering Our Climate Working Paper and Opinion Article Series

Bellamy R (2014) Safety first! framing and governing climate geoengineering experimentation. Climate Geoengineering Governance Working Paper Series, 014

Bellamy R (2016) A sociotechnical framework for governing climate engineering. Sci Technol Human Values 41(2):135–162. https://doi.org/10.1177/0162243915591855

Bellamy R, Hulme M (2011) Beyond the tipping point: understanding perceptions of abrupt climate change and their implications. Weather, Clim Soc 3:48–60. https://doi.org/10.1175/2011WCAS1081.1

Bellamy R, Lezaun J (2015) A review of deliberative public engagements with climate engineering. Climate Geoengineering Governance Working Paper Series, 019

Bellamy R, Chilvers J, Vaughan NE, Lenton TM (2012) A review of climate geoengineering appraisals. WIREs Clim Change 3(6):597–615

Bellamy R, Chilvers J, Vaughan NE, Lenton TM (2012a) Appraising geoengineering. Tyndall Working Paper Series, 153

Bellamy R, Chilvers J, Vaughan NE (2016) Deliberative mapping of options for tackling climate change: citizens and specialists 'open up' appraisal of geoengineering. Publ Understand Sci 25(3):269–286. https://doi.org/10.1177/0963662514548628

Bellamy R, Lezaun J, Palmer J (2017) Public perceptions of geoengineering research governance: an experimental deliberative approach. Global Environ Change 45:194–202

Belter CW, Seidel DJ (2013) A bibliometric analysis of climate engineering research. WIREs Climate Change 4(5):417–427. https://doi.org/10.1002/wcc.229

Bengtsson T (2006) Geo-engineering to confine climate change: Is it at all feasible? Climatic Change 77(3–4):229–234

Bickel JE (2010) The climate engineering option: economic and policy implications. Prepared for the AEI

Bickel JE, Agrawal S (2013) Reexamining the economics of aerosol geoengineering. Clim Change 119(3–4):993–1006. https://doi.org/10.1007/s10584-012-0619-x

Biermann F, Möller I (2019) Rich man's solution? Climate engineering discourses and the marginalization of the South. International Environmental Agreements, Online first, 1–17

Bluemling B, Kim RE, Biermann F (2019) Seeding the clouds to reach the sky: will China's weather modification practices support the legitimization of climate engineering? AMBIO. https://doi.org/10.1007/s13280-019-01180-3

Bodansky D (1996) May We Engineer the Climate? Climatic Change 33(3):309–321

Bodansky D (2013) The who, what, and wherefore of geoengineering governance. Climatic Change 121:539–551. https://doi.org/10.1007/s10584-013-0759-7

Bornheim NB (2010) History of climate engineering. WIREs Clim Change 1(6):891–897. https://doi.org/10.1002/wcc.82

Borgmann A (1987) Technology and the character of contemporary life: a philosophical inquiry, reprint. The University of Chicago Press, Chicago, IL

Borgmann A (2012) The setting of the scene: technological fixes and the design of the good life. In: Preston CJ (ed) Engineering the climate: the ethics of solar radiation management. Lexington, Plymouth, UK, pp 189–200
Böttcher M, Gabriel J, Low S (2016) Solar radiation management: foresight for governance. IASS Working Paper Series
Bowden G (2010) Rock, paper, scissors: uncertainty, innovation, and the evolving ecoloy of climate change discourse. Proc Media Ecol Assoc 11:67–78
Boyd PW (2016) Development of geopolitically-relevant ranking criteria for geoengineering methods. Earth's Future 4(11):523–531. eft2.2016EF000447
Boyd PW, Abraham ER (2001) Iron-mediated changes in phytoplankton photosynthetic competence during SOIREE. Deep-Sea Res II 48:2529–2550
Boyd PW, Law CS (2001) The southern ocean iron release experiment (SOIREE)—introduction and summary. Deep-Sea Res II 48:2425–2438
Braun C, Merk C, Pönitzsch G, Rehdanz K, Schmidt U (2018) Public perception of climate engineering and carbon capture and storage in Germany: survey evidence. Clim Pol 18(4):471–484. https://doi.org/10.1080/14693062.2017.1304888
Brzoska M, Link PM, Neuneck G (2012) Geoengineering—Möglichkeiten und Risiken. Sicherheit & Frieden 30(4):185–193
Buck HJ (2010) What can geoengineering do for us? Public participation and the new media landscape. Working Paper
Buck HJ (2012) Geoengineering: re-making climate for profit or humanitarian intervention? Develop Change 43(1):253–270. https://doi.org/10.1111/j.1467-7660.2011.01744.x
Buck HJ (2013) Climate engineering: spectacle, tragedy or solution? A content analysis of news media framing. In: Methmann C, Rothe D, Stephan B (eds) Interpretive approaches to global climate governance. (De)constructing the greenhouse. London, Routledge, pp 166–180
Buck HJ (2016) Rapid scale-up of negative emissions technologies: social barriers and social implications. Clim Change https://doi.org/10.1007/s10584-016-1770-6
Buck HJ (2018) Perspectives on solar geoengineering from Finnish Lapland: local insights on the global imaginary of Arctic geoengineering. Geoforum 91:78–86. https://doi.org/10.1016/j.geoforum.2018.02.020
Budyko MI (1977) Climatic changes. American Geophysical Union, Washington, DC
Bunzl M (2009) Researching geoengineering: should not or could not? Environ Res Lett 4(4):1–3
Burns WCG (2013) Climate geoengineering: solar radiation management and its implications for intergenerational equity. In: Burns WCG, Strauss AL (eds) Climate change geoengineering: philosophical perspectives, legal issues, and governance frameworks. Cambridge University Press, New York, pp 200–220
Burns WCG (2016) The Paris agreement and climate geoengineering governance: the need for a human rights-based component. CIGI Papers
Burns WCG, Nicholson S (2016) Governing climate engineering. In: Nicholson S, Jinnah S (eds) New earth politics. Essays from the Anthropocene. Cambridge (MA), MIT Press, pp 343–366
Burns ET, Flegal J, Keith DW, Mahajan A, Tingley D, Wagner G (2016) What do people think when they about solar geoengineering? A review of empirical social science literature, and prospects for future research. Earth's Future 4:536–542
C2G2 (2019) C2G2 mission. https://www.c2g2.net/c2g2-mission/. Accessed 22 Mar 2019
Cairns R (2013) Examining framings of geoengineering using Q methodology. Climate Geoengineering Governance Working Paper Series, 002
Cairns R, Stirling A (2014) Maintaining planetary systems' or 'concentrating global power'? High stakes in contending framings of climate geoengineering. Global Environ Change 28:25–38
Campbell-Arvai V, Hart PS, Raimi KT, Wolske KS (2017) The influence of learning about carbon dioxide removal (CDR) on support for mitigation policies. Clim Change 143:321–336. https://doi.org/10.1007/s10584-017-2005-1
Carr W (2013) Public engagement on solar radiation management and why it needs to happen now. Clim Change 121(3):567–577

Carr W, Preston CJ (2017) Skewed vulnerabilities and moral corruption in global perspectives on climate engineering. Environ Values 26:757–777. https://doi.org/10.3197/096327117X15046905490371

Carr W, Yung L (2018) Perceptions of climate engineering in the South Pacific, Sub-Saharan Africa, and North American Arctic. Clim Change 147:119–132. https://doi.org/10.1007/s10584-018-2138-x

Carr W, Yung L, Preston CJ (2014) Swimming upstream: engaging the American public early on climate engineering. Bulletin of the Atomic Scientists 70(3):38–48. https://doi.org/10.1177/0096340214531180

CBD (2010) Decision adopted by the conference of the parties to the convention on biological diversity at its tenth meeting: X/33. Biodiversity and Climate Change. Nagoya: Convention on Biological Diversity

Cicerone RJ (2006) Geoengineering: encouraging research and overseeing implementation. Climatic Change 77(3–4):221–226. https://doi.org/10.1007/s10584-006-9102-x

Clingerman F, O'Brien KJ (2014) Playing God: Why religion belongs in the climate engineering debate. Bull Atom Scient 70(3):27–37. https://doi.org/10.1177/0096340214531181

Corner A, Pidgeon N (2015) Like artificial trees? The effect of framing by natural analogy on public perceptions of geoengineering. Clim Change 130:425–438. https://doi.org/10.1007/s10584-014-1148-6

Corner A, Pidgeon N (2010) Geoengineering the climate: the social and ethical implications. Environ Sci Pol Sustain Dev 52(1):24–37

Corner A, Parkhill K, Pidgeon N (2011) Experiment earth: reflections on a public dialogue on geoengineering. Working Paper

Corner A, Parkhill K, Pidgeon N, Vaughan NE (2013) Messing with nature? Exploring public perceptions of geoengineering in the UK. Global Environ Change 23(5):938–947

Corner A, Pidgeon N (2014) Geoengineering, climate change scepticism and the 'moral hazard' argument: an experimental study of UK public perceptions. Philos Trans Royal Soc A 372(2031):1–14. https://doi.org/10.1098/rsta.2014.0063

Corry O (2014) Questioning the plan-B framing of climate engineering: political feasibility and emergency measures. In: Paper presented at the climate engineering conference "critical global discussions". Berlin, 18–21 Aug 2014

Corry O (2017) The international politics of geoengineering: the feasibility of Plan B for tackling climate change. Secur Dial 48(8):297–315. https://doi.org/10.1177/0967010617704142

Cressey D (2012) Geoengineering experiment cancelled amid patent row. Nat News 485:429

Crutzen PJ (2006) Albedo enhancement by stratospheric sulfur injections: a contribution to resolve a policy dilemma. Climatic Change 77:211–219

Crutzen PJ, Stoermer EF (2000) The "Anthropocene" . Global Change Newsletter 41:17–18

Cunha D (2015) The anthropocene as fetishism. Mediations—J Marxist Lit Group 28(2):65–77

Dalby S (2015) Geoengineering: the next era of geopolitics? Geogr Compass 9(4):190–201. https://doi.org/10.1111/gec3.12195

Davies G (2013) The psychological costs of geoengineering: why it may be hard to accept even if it works. In: Burns WCG, Strauss AL (eds) Climate change geoenegineering: philosophical perspectives, legal issues, and governance frameworks. Cambridge University Press, New York, USA, pp 59–80

Dickinson RE (1996) Climate engineering. A review of aerosol approaches to changing the global energy balance. Climatic Change 33(3):279–290

Dietz S, Bowen A, Doda B, Gambhir A, Warren R (2018) The economics of 1.5 °C climate change. Ann Rev Environ Resour 43:455–480

Doda B (2014) Why is geoengineering so tempting? Centre for Climate Change Economics and Policy Working Paper Series

Doughty J (2015) Past forays into srm field research and implications for future governance. Geoengineering Our Climate Working Paper and Opinion Article Series

Edney K, Symons J (2014) China and the blunt temptations of geo-engineering: the role of solar radiation management in China's strategic response to climate change. Pacif Rev 27(1):1–26. https://doi.org/10.1080/09512748.2013.807865

Egede-Nissen B (2010) Scientists in the geoengineering discourse: social advocates or neutral umpires? In: Paper presented at the Workshop "ethics of geoengineering: investigating the moral challenges of solar radiation management", University of Montana, 18–20 Oct 2010

Emmerling J, Tavoni M (2018) Exploration of the interaction between mitigation and solar radiation management in cooperative and non-cooperative international governance settings. Global Environ Change 53:244–251

Fairbrother M (2016) Geoengineering, moral hazard, and trust in climate science: evidence from a survey experiment in Britain [journal article]. Clim Change 1–13. https://doi.org/10.1007/s10584-016-1818-7

Fernow, H. (2012). Complex Risks and the Limits of Cost-Benefit Analysis. *Forum Marsilius-Kolleg*(2: Beyond calculation - Climate Engineering risks from a social sciences perspective), 12–18, doi:https://doi.org/10.11588/fmk.2012.0.9408

Fialka J (2020) U.S. Geoengineering research gets a lift with $4 million from Congress (23 January 2020) Science

Flegal J (2018) The evidentiary politics of the geoengineering imagninary. UC Berkeley Electronic Theses and Dissertations

Fleming JR (2010) Fixing the sky. The checkered history of weather and climate control. Columbia University Press, New York, NY

Fleming JR (2012) Will geoengineering bring security and peace? What does history tell us? Sicherheit Frieden 30(4):200–204

Fleming JR (2013) A history of weather and climate control. Geoengineering Our Climate Working Paper and Opinion Article Series

Fletcher JO (1968) Changing climate. In: Corporation R (ed) Santa Monica, CA, RAND Corporation

Fletcher JO (1969) Managing climatic resources vol. P-4000–1. Santa Monica, CA, RAND Corporation

Foley RW, Guston DH, Sarewitz D (2015) Toward the anticipatory governance of geoengineering. Geoengineering Our Climate Working Paper and Opinion Article Series

Galaz V (2012) Geo-engineering, governance, and social-ecological systems: critical issues and joint research needs. Ecol Soc 17(1):24–33

Gannon KE, Hulme M (2018) Geoengineering at the 'edge of the world': exploring perceptions of ocean fertilisation through the Haida Salmon Restoration Corporation. Geo: Geogr Environ https://doi.org/10.1002/geo2.54

Gardiner SM (2011a) A perfect moral storm: the ethical challenge of climate change. Oxford University Press, Oxford

Gardiner SM (2011b) Some early ethics of geoengineering the climate: a commentary on the values of the royal society report. Environ Values 20(2):163–188

Gardiner SM (2013a) The desperation argument for geoengineering. PS: Polit Sci Polit 46(1):28–33

Gardiner SM (2013b) Geoengineering and Moral Schizophrenia. What is the question? In: Burns WCG, Strauss AL (eds) Climate change geoengineering. philosophical perspectives, legal issues, and governance frameworks. New York, Cambridge University Press, pp 11–38

Gardiner SM, Fragnière A (2018) The tollgate principles for the governance of geoengineering: moving beyond the oxford principles to an ethically more robust approach. Ethics, Pol Environ 21(2):143–174. https://doi.org/10.1080/21550085.2018.1509472

Geden O (2019) Targeting carbon dioxide removal in the European Union. Climate Pol 19(4)

Geden O, Scott V, Palmer J (2018) Integrating carbon dioxide removal into EU climate policy: prospects for a paradigm shift. WIREs Clim Change. https://doi.org/10.1002/wcc.521

Goeschl T, Heyen D, Moreno-Cruz J (2013) The intergenerational transfer of solar radiation management capabilities and atmospheric carbon stocks. Environ Resour Econ 56(1):85–104

Goes M, Tuana N, Keller K (2011) The economics (or lack thereof) of aerosol geoengineering. Clim Change 109(3–4):719–744. https://doi.org/10.1007/s10584-010-9961-z

Granger Morgan M, Nordhaus RR, Gottlieb P (2013) Needed: research guidelines for solar radiation management. Issues in Science and Technology, Spring, 37–44

Greene C (2010) Shifting the debate on geoengineering. Science 328(5979):690–691

Greene C, Monger B, Huntley M (2010) geoengineering: the inescapable truth of getting to 350. Solut Sustain Desir Fut 1(5):57–66

Gupta A, Möller I (2018) De facto governance: how authoritative assessments construct climate engineering as an object of governance. Environ Polit https://doi.org/10.1080/09644016.2018.1452373

Güssow K, Oschlies A, Proelss A, Rehdanz K, Rickels W (2013) Ocean iron fertilization: time to lift the research taboo. In: Burns WCG, Strauss AL (eds) Climate change geoenegineering: philosophical perspectives, legal issues, and governance frameworks. Cambridge University Press, New York, USA, pp 242–262

Haas PM (1992) Introduction: epistemic communities and international policy coordination. Int Organ 46(1):1–35

Hale B (2012) The world that would have been. moral hazard arguments against geoengineering. In: Preston CJ (ed) Engineering the climate: the ethics of solar radiation management. Plymouth, UK, Lexington Books, pp 113–131

Hamilton C (2013a) Earthmasters. the dawn of the age of climate engineering. Yale University Press, New Haven, CT

Hamilton C (2013b) The ethical foundations of climate engineering. In: Burns WCG, Strauss AL (eds) Climate change geoenegineering: philosophical perspectives, legal issues, and governance frameworks. Cambridge University Press, New York, USA, pp 39–58

Harding A, Moreno-Cruz J (2016) Solar geoengineering economics: from incredible to inevitable and half-way back. Earth's Fut 4(12):569–577. https://doi.org/10.1002/2016EF000462

Harding A, Moreno-Cruz JB (2019) The economics of geoengineering. In: Letcher TME (ed) Managing global warming. An interface of technology and human issues. Amsterdam, Academic Press, pp 729–750

Harnisch S, Uther S, Böttcher M (2015) From 'go slow' to 'gung ho'? Climate engineering discourses in the UK, the US and Germany. Global Environ Polit 15(2):57–78. https://doi.org/10.1162/GLEP_a_00298

Heutel G, Moreno-Cruz J, Shayegh S (2018) Solar geoengineering, uncertainty, and the price of carbon. Journal of Environmental Economics and Management 87:24–41

Heyen D (2012) An economic perspective on risk of climate change and SRM—limitations of methodology, new concepts and the precautionary principle. Forum Marsilius-Kolleg, 2: beyond calculation—climate engineering risks from a social sciences perspective, 34–40. https://doi.org/10.11588/fmk.2012.0.9408

Heyen D, Wiertz T, Irvine JP (2015) Regional disparities in SRM impacts: the challenge of diverging preferences. Clim Change 133:557–563. https://doi.org/10.1007/s10584-015-1526-8

Heyward C, Rayner S (2013) Apocalypse Nicked! Climate Geoengineering Governance Working Paper Series, 006

Hiller S, Renn O (2012) Public perception of geoengineering. Sicherheit & Frieden 30(4):215–220

Himmelsbach R (2018) How scientists advising the European Commissions on research priorities view climate engineering proposals. Sci Publ Pol 45(1):124–133. https://doi.org/10.1093/scipol/scx053

Hoppe R (2005) Rethinking the science-policy nexus: from knowledge utilization and science technology studies to types of boundary arrangements. Poeisis Prax 3(3):199–215

Horton JB (2013) Geoengineering and the myth of unilateralism: pressures and prospects for international cooperation. In: Burns WCG, Strauss AL (eds) Climate change geoenegineering: philosophical perspectives, legal issues, and governance frameworks. Cambridge University Press, New York, USA, pp 168–181

Horton, J. B. (2015). The emergency framing of solar geoenengineering: time for a different approach. Anthropocene Rev, Prepubl. https://doi.org/10.1177/2053019615579922

Horton JB, Reynolds J (2016) The international politics of climate engineering: a review and prospectus for international relations. Int Stud Rev 18:438–461
Hourdequin M (2018) Geoengineering justice: the role of recognition. Science, technology, and human values, Online first. https://doi.org/10.1177/0162243918802893
Hourdequin M (2019) Climate change, climate engineering, and the 'global poor': What does justice require? Ethics, Policy Environ Online first. https://doi.org/10.1080/21550085.2018.1562525
Hubert A.-M, Reichwein D (2015) An exploration of a code of conduct for responsible scientific research involving geoengineering. IASS Working Paper Series
Hulme M (2003) Abrupt climate change: can society cope? Philos Trans Royal Soc A 361:2001–2021
Hulme M (2014) Can science fix climate change? A Case against climate engineering. Polity Press, Cambridge, UK
Hulme M (2017) Calculating the incalculable: is SAI the lesser of two evils? Ethics Int Aff 31(4):507–512. https://doi.org/10.1017/S0892679417000491
Huttunen S, Hildén M (2014) Framing the controversial: geoengineering in academic literature. Sci Commun 36(3):3–29
Huttunen S, Skytén E, Hildén M (2015) Emerging policy perspectives on geoengineering: an international comparison. Anthropocene Rev 2(1):14–32. https://doi.org/10.1177/2053019614557958
IISD Reporting Services (2018) Summary of the Katowice climate change conference: 2–15 December 2018. Earth Negot Bull 12(747)
IISD Reporting Services (2019) Summary of the Fourth Session of the United Nations Environment Assembly. Earth Negot Bull 16(153)
IPCC (2014b). Climate change 2014: Synthesis report. contribution of working groups I, II and III to the Fifth assessment report of the intergovernmental panel on climate change [Core Writing Team, R. K. Pachauri and L. A. Meyer (eds.)]. Geneva, IPCC
Izrael YA, Zakharov VM, Petrov NN, Ryaboshapko AG, Ivanov VN, Savchenko AV et al (2009) Field studies of a geo-engineering method of maintaining a modern climate with aerosol particles. Russ Meteorol Hydrol 34(10):635–638. https://doi.org/10.3103/S106837390910001X
Jamieson D (1996) Ethics and international climate change. Clim Change 33(3):323–336
Joint Science Academies (2008) Statement: climate change adaptation and the transition to a low carbon society. Amsterdam, InterAcademy Council
Kahan DM, Jenkins-Smith H, Tarantola T, Silva CL, Draman D (2015) Geoengineering and climate change polarization: testing a two-channel model of science communication. Anna Am Acad Polit Soc Sci 658:193–222. https://doi.org/10.1177/0002716214559002
Keith DW (2000) Geoengineering the climate: history and prospect. Annu Rev Energy Env 25:245–284
Keith DW (2013) A case for climate engineering. boston review, Cambridge/London
Keith DW, Duren R, MacMartin DG (2014) Field experiments on solar geoengineering: report of a workshop exploring a representative research portfolio. Philos Trans Royal Soc A 372:20140175
Kerr RA (2006) Pollute the planet for climate's sake? Science 314(5798):401–403. https://doi.org/10.1126/science.314.5798.401a
Kiehl JT (2006) Geoengineering climate change: treating the symptom over the cause? Climatic Change 77(3–4):227–228. https://doi.org/10.1007/s10584-006-9132-4
Kintisch E (2010) Hack the planet. Science's best hope—or worst nightmare—for averting climate catastrophe. Wiley, New York, NY
Klein N (2014) This changes everything—capitalism versus the climate. London, Allen Lane
Klepper G (2012) What are the costs and benefits of climate engineering? And can we assess them? Sicherheit Frieden 30(4):211–214
Klepper G, Rickels W (2014) Climate engineering: economic considerations and research challenges. Rev Environ Econ Pol 8(2):270–289. https://doi.org/10.1093/reep/reu010
Kreuter J (2015a) Discussion of climate engineering in social science literature: technofix, plan B or ultima ratio? The Forum for Climate Engineering Assessment Blog

Kreuter J (2015b) Technofix, plan B or ultima ratio? A review of the social science literature on climate engineering technologies. INSIS Occasional Papers Series

Kreuter J, Matzner N, Baatz C, Keller DP, Markus T, Wittstock F et al (2020) Unveiling assumptions through interdisciplinary scrutiny: observations from the German priority program on climate engineering (SPP 1689). Clim Change https://doi.org/10.1007/s10584-020-02777-4

Lane L (2013) Climate engineering and the anthropocene era. In: Burns WCG, Strauss AL (eds) Climate change geoenegineering: philosophical perspectives, legal issues, and governance frameworks. Cambridge University Press, New York, USA, pp 115–145

Larson EJ (2016) The red dawn of geoengineering: first step toward an effective governance for stratospheric injections. Duke Law & Technology Review 14(1):157–191

Latham J, Bower K, Choularton T, Coe H, Connolly P, Cooper G et al (2012) Marine cloud brightening. Philos Trans Royal Soc A 370:4217–4262. https://doi.org/10.1098/rsta.2012.0086

Launder B, Thompson JMT (2008) Preface. Philos Trans Royal Soc A 366(1882):3841–3842. https://doi.org/10.1098/rsta.2008.0151

Lawrence MG (2006) the geoengineering dilemma: to speak or not to speak. Climatic Change 77(3–4):245–248. https://doi.org/10.1007/s10584-006-9131-5

Lawrence MG, Crutzen P (2017) Was breaking the taboo on research on climate engineering via albedo modification a moral hazard, or a moral imperative? Earth Future 5(2):136–143. https://doi.org/10.1002/2016EF000463

Lawrence MG, Schäfer S, Muri H, Scott V, Oschlies A, Vaughan NE et al (2018) Evaluation climate geoengineering proposals in the context of the Paris agreement temperature goals. Nature Commun 9(3734). https://doi.org/10.1038/s41467-018-05938-3

Lederer M, Kreuter J (2018) Organising the unthinkable in times of crises: Will climate engineering become the weapon of last resort in the Anthropocene? Organization 25(4). https://doi.org/10.1177/1350508418759186

Lenton TM (2011) Early warning of climate tipping points. Nat Clim Change 1:201–209. https://doi.org/10.1038/nclimate1143

Lenton TM, Held H, Kriegler E, Hall JW, Lucht W, Rahmstorf S et al (2008) Tipping elements in the Earth's climate system. PNAS 105(6):1786–1793

Lin AC (2013) Does geoengineering present a moral hazard? Ecol Law Q 40:673–712

Linnér B-O, Wibeck V (2015) Dual high-stake emerging technologies: a review of the climate engineering research literature. WIREs Climate Change 6:255–268

Lloyd ID, Oppenheimer M (2014) On the design of an international governance framework for geoengineering. Global Environ Polit 14(2):45–63

Low S (2017) The futures of climate engineering. Earth's Future 5:67–71. https://doi.org/10.1002/2016EF000442

Luokkanen M, Huttunen S, Hildén M (2014) Geoengineering, news media and metaphors: framing the controversial. Public Understand Sci 23(8):966–981. https://doi.org/10.1177/0963662513475966

Maas A, Scheffran J (2012) Climate conflicts 2.0? Climate engineering as a challenge for international peace and security. Sicherheit & Frieden 30(4):193–200

MacMartin DG, Ricke KL, Keith DW (2017) Solar geoengineerig as part of an overall strategy for meeting the 1.5 °C Paris target. Philos Trans Royal Soc A, 376(20160454). https://doi.org/10.1098/rsta.2016.0454

Macnaghten P, Szerszynski B (2013) Living the global social experiment: an analysis of public discourse on solar radiation management and its implications for governance. Glob Environ Change 23:465–474. https://doi.org/10.1016/j-gloenvcha.2012.12.008

Macnaghten P, Owen R (2011) Good governance for geoenegineering. Nature 479(7373):293. https://doi.org/10.1013/479293a

Mahajan A, Tingley D, Wagner G (2018) Fast, cheap and imperfect? US public opinion about solar geoengineering. Environ Polit 28(3):523–543. https://doi.org/10.1080/09644016.2018.1479101

Markusson N (2013) Tensions in framings of geoengineering: constitutive diversity and ambivalence. Climate Geoengineering Governance Working Paper Series, 003

Markusson N, Wong P.-H (2015) Geoengineering governance, the linear model of innovation, and the accompanying geoengineering approach. Climate Geoengineering Governance Working Paper Series, 020

Markusson N, Ginn F, Ghaleigh NS, Scott V (2013) In case of emergency press here': framing geoengineering as a response to dangerous climate change. WIREs Clim Change 5(2):281–290. https://doi.org/10.1002/wcc.263

Markusson N, Gjefsen MD, Stephens JC, Tyfield D (2017) The political economy of technical fixes: the (mis)alignment of clean fossil and polical regimes. Energy Res Soc Sci 23:1–10

Marland G (1996) Could we/should we engineer the earth's climate? Climatic Change 33(3):275–278

Matzner N, Herrenbrück R (2016) Simulating a climate engineering crisis: climate politics simulated by students in model United Nations. Simulation & Gaming, Early Edition. https://doi.org/10.1177/1046878116680513

Matzner N, Barben D (2018) Verantwortungsvoll das Klima manipulieren? Unsicherheit und Verantwortung im Diskurs um Climate Engineering. In: Janich N, Rhein L (eds) Unsicherheit als Herauforderung für die Wissenschaft. Peter Lang, Berlin et al., pp 143–179

McClellan J, Keith DW, Apt J (2012) Cost analysis of stratospheric albedo modification delivery systems. Environ Res Lett 7:1–8

McLaren D (2015) Comparing measures of justice in climate engineering assessment. In: Paper presented at the climate engineering symposium 2015, Berlin, July 2015

McLaren D (2016) Mitigation deterrence and the 'moral hazard' of solar radiation management. Earth's Future 4:596–602. https://doi.org/10.1002/2016EF000445

Mercer AM, Keith DW, Sharp JD (2011) Public understanding of solar radiation management. Environ Res Lett 6:1–9. https://doi.org/10.1088/1748-9326/6/4/044006

Merk C, Pönitzsch G, Kniebes C, Rehdanz K, Schmidt U (2015) Exploring public perceptions of stratospheric sulfate injection. Clim Change 130:299–312

Merk C, Pönitzsch G, Rehdanz K (2016) Knowledge about aerosol injection does not reduce individual mitigation efforts. Environ Res Lett 11. https://doi.org/10.1088/1748-9326/11/5/054009

Merk C, Pönitzsch G, Rehdanz K (2018) Do climate engineering experts display moral hazard behavior? Clim Pol 19(2):231–243. https://doi.org/10.1080/14693062.2018.1494534

Michaelson J (2013) Geoengineering and climate management. From marginality to inevitability. In: Burns WCG, Strauss AL (eds) Climate change geoengineering. philosophical perspectives, legal issues, and governance frameworks, pp 81–114. Cambridge University Press, New York, USA

Moore JC, Chen Y, Cui X, Yuan W, Dong W, Gao Y et al (2016) Will China be the first to initiate climate engineering? Earth's Future 4:588–595. https://doi.org/10.1002/2016EF000402

Moreno-Cruz J, Keith DW (2013) Climate policy under uncertainty: a case for geoengineering. Clim Change 121(3):431–444. https://doi.org/10.1007/s10584-012-0487-4

Moreno-Cruz J, Smulders S (2010 Revisiting the economics of climate change: the role of geoengineering. Selected Works from Juan B. Moreno-Cruz

Moreno-Cruz J, Ricke K, Wagner G (2015) The economics of climate engineering. Geoengineering Our Climate Working Paper and Opinion Article Series

Morrow DR (2014) Starting a flood to stop a fire? Some moral constraints on solar radiation management. Ethics, Polit Environ 17(2):123–128. https://doi.org/10.1080/21550085.2014.926056

Morrow DR (2014a) Ethical aspects of the mitigation obstruction argument against climate engineering research. Philos Trans Royal Soc A, 372(2031). https://doi.org/10.1098/rsta.2014.0062

Morrow DR (2017) International governance of climate engineering. a survey of reports on climate engineering, 2009–2015. FCEA Working Paper Series

Morrow DR, Kopp RE, Oppenheimer M (2009) Toward ethical norms and institutions for climate engineering research. Environ Res Lett 4:1–8

National Science Foundation Special Commission on Weather Modification (1965). Weather and Climate Modification. In N. S. Foundation (Ed.). Washington, D.C.: National Science Foundation

NAS (1992). Policy implications of greenhouse warming: mitigation, adaptation, and the science base (C. o. S. Panel on Policy Implications of Greenhouse Warming, Engineering and Public Policy, Trans.). Washington, DC: National Academy of Sciences.

NAS (2015a) Climate intervention: reflecting sunlight to cool earth. The National Academy of Sciences, Washington, D.C

NAS (2015b) Climate intervention: carbon dioxide removal and reliable sequestration. Nat Acad Sci, Washington, D.C

NERC (2010) Experiment Earth? Report on a public dialogue on geoengineering. Natural Environment Research Council, Swindon, UK

Nerlich B, Jaspal R (2012) Metaphors we die by? Geoengineering, metaphors, and the argument from catastrophe. Metaphor Symb 27(2):131–147

Neuber F (2018) Buying time with climate engineering? An analysis of the buying time framing in favor of climate engineering. Hochschulschrift. https://doi.org/10.5445/IR/1000084294

Nicholson, S., & Thompson, J. M. T. (2015). Strange Bedfellows: Climate Engineering Politics in the United States. *Geoengineering Our Climate Working Paper and Opinion Article Series*

Nicholson S, Jinnah S, Gillespie A (2018) Solar radiation management: a proposal for immediate polycentric governance. Clim Policy 18(3):322–334

Nightingale P, Cairns R (2014) The security implications of geoengineering: blame, imposed agreement and the security of critical infrastructure. Climate Geoengineering Governance Working Paper Series, 018

Oldham P, Szerszynski B, Stilgoe J, Brown C, Eacott B, Yuille A (2014) Mapping the landscape of climate engineering. Philos Trans Royal Soc A 372:2031. https://doi.org/10.1098/rsta.2014.0065

Parson EA, Keith DW (2013) End the deadlock on governance of geoengineering research. Science 339:1278–1279. https://doi.org/10.1126/science.1232527

Pasztor J (2017b) Toward governance frameworks for climate engineering. Global cooperation in dangerous times. Global Challenges Quarterly Risk Report

Peters GP, Geden O (2017) Catalysing a political shift from low to negative carbon. Nature Clim Change 7:619–621

Pidgeon N, Corner A, Parkhill K, Spence A, Butler C, Poortinga W (2012) Exploring early public responses to geoengineering. Philos Trans Royal Soc A 370(1974):4176–4196

Pidgeon N, Parkhill K, Corner A, Vaughan NE (2013) Deliberating stratospheric aerosols for climate geoengineering and the SPICE project. Nat Clim Change 3(5):451–457

Porter KE, Hulme M (2013) The emergence of the geoengineering debate in the UK print media: a frame analysis. Geogr J 179(4):342–355. https://doi.org/10.1111/geoj.12003

Preston CJ (2012a) Introduction. In: Preston CJ (ed) Engineering the climate: the ethics of solar radiation management. Lexington Books, Plymouth, UK, pp 1–12

Preston CJ (2012b) The extraordinary ethics of solar radiation management. In: Preston CJ (ed) Engineering the climate: the ethics of solar radiation management. Lexington, Plymouth, UK, pp 1–11

Preston CJ, Carr W (2019) Recognitional justice, climate engineering, and the care approach. Ethics, Policy Environ Online first. https://doi.org/10.1080/21550085.2018.1562527

Rabitz F (2016) Going rogue? Scenarios for unilateral geoengineering. Futures 84:98–107

Raimi KT, Maki A, Dana D, Vandenbergh MP (2019) Framing of geoengineering affects support for climate change mitigation. Environ Commun J Nat Cult 13(3):300–319. https://doi.org/10.1080/17524032.2019.1575258

Rasch PJ, Tilmes S, Turco RR, Robock A, Oman L, Chen C.-C et al (2008) An overview of geoengineering of climate using stratospheric sulphate aerosols. Philos Trans Royal Soc A, 366:4007–4037. https://doi.org/10.1098/rsta.2008.0131

Rayner S (2014) To know or not to know? A note on ignorance as a rhetorical resource in geoengineering debates. Climate Geoengineering Governance Working Paper Series, 010

Rayner S (2016) What might Evans-Pritchard have made of two degrees? Anthropology Today 32(4):1–2

Rayner S, Heyward C, Kruger T, Pidgeon N, Redgwell C, Savulescu J (2013) The Oxford principles. Clim Change 121:499–512. https://doi.org/10.1007/s10584-012-0675-2

Reynolds J (2015) A critical examination of the climate engineering moral hazard and risk compensation concern. Anthrop Rev 2(2):174–191. https://doi.org/10.1177/2053019614554304

Reynolds J (2018) Governing experimental responses. Negative emissions technologies and solar climate engineering. In: Jordan A, Huitema D, Van Asselt H, Forster J (eds) Governing CLIMATE CHANGE. Policentricity in action? Cambridge University Press, Cambridge, pp 285–302

Rickels W, Klepper G, Dovern J, Betz G, Brachatzek N, Cacean S et al (2011a) Gezielte Eingriffe in das Klima? Eine Bestandsaufnahme der Debatte zu Climate Engineering. Bundesministerium für Bildung und Forschung, Berlin

Ricke K, Moreno-Cruz J, Caldeira K (2013) Strategic incentives for climate geoengineering coalitions to exclude broad participation. Environ Res Lett 8:1–8

Robock A (2008) 20 reasons why geoengineering might be a bad idea. Bull Atom Scient 64(2):14–18. https://doi.org/10.2968/064002006

Robock A, Bunzl M, Kravitz B, Stenchikov GL (2010) A test for geoengineering? Science 327:530–531. https://doi.org/10.1126/science.1186237

Roshan E, Khabhazan MM, Held H (2019) Cost-Risk Trade-off of mitigation and solar geoengienering: considering regional disparities under probabilistic climate sensitivity. Environ Resour Econ 72:263–279

Russell LM (2012) Offsetting climate change by engineering air pollution to brighten clouds. Bridge 42(4):10–15

Russell LM, Sorooshian A, Seinfeld JH, Albrecht BA, Nenes A, Ahlm L et al (2013) Eastern Pacific emitted aerosol cloud experiment. Bull Am Meteor Soc 94:709–729

Sandler T (2018) Collective action and geoengineering. Rev Int Organ 13(1):105–125

Schäfer S, Stelzer H, Maas A, Lawrence MG (2014) Earth's future in the anthropocene: technological interventions between piecemeal and utopian social engineering. Earth's Fut 2(4):239–243. https://doi.org/10.1002/2013EF000190

Scheer D, Renn O (2014) Public Perception of geoengineering and its consequences for public debate. Clim Change 125:305–318

Schelling TC (1996) The economic diplomacy of geoengineering. Climatic Change 33(3):303–307

Schneider SH (1996) Geoengineering: could—or should—we dot it? Climatic Change 33(3):291–302

Schneider SH (2008) Geoengineering: could we or should we make it work? Philos Trans Royal Soc A 366:3843–3862

Scholte S, Vasileiadou E, Petersen A (2013) Opening up the societal debate on climate engineering: how newspaper frames are changing. J Integr Environ Sci 10(1):1–16

Scott D (2012) Insurance policy or technological fix? The ethical implications of framing solar radiation management. In: Preston CJ (ed) Engineering the climate: the ethics of solar radiation management. Lexington, Plymouth, UK, pp 151–168

Scott KN (2013) Regulating ocean fertilization under international law: the risks. Carbon Clim Law Rev 2:108–116

Scott D (2013) Philosophy of technology and geoengineering. Geoengineering Our Climate Working Paper and Opinion Article Series

Shepherd J, Cox P, Haigh D, Keith DW, Launder B, Mace G et al (2009) Geoengineering the climate: science, governance and uncertainty. The Royal Society, London

Sikka T (2012a) A critical discourse analysis of geoengineering advocacy. Critic Disc Stud 9(2):163–175

Sikka T (2012b) A critical theory of technology applied to the public discussion of geoengineering. Technol Soc 34:109–117

Sikka T (2019) Activism and neoliberalism: two sides of geoengineering discourse. Capit Nat Social, Online first. https://doi.org/10.1080/10455752.2018.1554690

Sillmann J, Lenton TM, Levermann A, Ott K, Hulme M, Benduhn F et al (2015) Climate emergencies do not justify engineering the climate. Nature Clim te Change 5:290–292
Singer C, Matchett L (2015) Climate action gaming experiment: methods and example results. Challenges 6:202–228
Singer C, Milligan T, Rethinaraj TSG (2014) How China's options will determine global warming. Challenges 5:1–25
Smit WA (2006) Military technologies and politics. In: Goodin RE, Tilly C (eds) The Oxford handbook of contextual political analysis. Oxford University Press, Oxford (UK), pp 722–744
Speicher C (2017) Zürcher Startup Unternehmen mit Weltpremiere: CO_2 wird aus der Luft gefiltert (31 May 2017) Neue Zürcher Zeitung
SPP 1689 (2019a) Climate engineering und unsere Klimaziele—eine überfällige debatte. In: Schwerpunktprogramm 1689 der Deutschen Forschungsgemeinschaft "climate engineering: risks, opportunities?" (SPP 1689) (ed). Kiel, Kiel Earth Institute
Steffen W, Grinevald J, Crutzen P, McNeill J (2011) The anthropocene: conceptual and historical perspectives. Philos Trans Royal Soc A 369:842–867
Steffen W, Broadgate W, Deutsch L, Gaffney O, Ludwig C (2015) The trajectory of the anthropocene: the great acceleration. Anthrop Rev 2(1):81–98
Steffen W, Rockström J, Richardson K, Lenton TM, Folke C, Liverman D et al (2018) Trajectories Of The Earth System In The Anthropocene. PNAS 0(0). https://doi.org/10.1073/pnas.1810141115
Stilgoe J (2015) Geoengineering as collective experimentation. Sci Eng Ethics, Online, 1–19. https://doi.org/10.1007/s11948-015-9646-0
Strong AL, Cullen JJ, Chisholm SW (2009) Ocean fertilization: science, policy, and commerce. Oceanography 22(3):235–261. https://doi.org/10.5670/oceanog.2009.83
Sugiyama M, Kosugi T, Ishii A, Asayama S (2016) Public attitudes to climate engineering research and field experiments: preliminary results of a web survey on students' perception in six Asia-Pacific countries. PARI Working Papers
Svoboda T (2015) Geoengineering, agent-regret, and the lesser of two evils argument. Environ Ethics 37(5):207–220
Svoboda T, Buck HJ, Suarez P (2018) Climate engineering and human rights. Environ Polit 28(3):397–416. https://doi.org/10.1080/09644016.2018.1448575
Svoboda T, Irvine PJ, Callies DE, Sugiyama M (2019) The potential for climate engineering with stratospheric sulfate aerosol injections to reduce climate injustice. J Global Ethics https://doi.org/10.1080/17449626.2018.1552180
Szerszynski B (2010) Reading and writing the weather: Climate technics and the moment of responsibility. Theory Cult Soc 27(9):9–30
Szerszynski B (2014) Geoengineering and religion: a history in four characters. *Geoengineering Our Climate Working Paper and Opinion Article Series*
Szerszynski B, Galarraga M (2013) Geoengineering knowledge: interdisciplinarity and the shaping of climate engineering research. Environ Plann A 45:2817–2824
Talberg A, Christoff P, Thomas S, Karoly D (2018) Geoengineering governance-by-default: an earth system governance perspective. Int Environ Agreem 18:229–253. https://doi.org/10.1007/s10784-017-9374-9
Teller E, Wood L, Hyde R (1997) Global warming and ice ages: projects for a physics-based modulation of global change. Lawrence Livermore National Laboratory
Teller E, Hyde R, Wood L (2002) Active climate stabilization: practical physics-based approaches to prevention of climate change. Lawrence Livermore National Laboratory
Thiele S, Fuchs BM, Ramaiah N, Amann R (2012) Microbial community response during the iron fertilization experiment LOHAFEX. Appl Environ Microbiol 78(24):8803–8812
Tingley D, Wagner G (2017) Solar geoengineering and the chemtrails conspiracy on social media. Palgrave Commun 3:12. https://doi.org/10.1057/s41599-017-0014-3
Tollefson J (2012) Ocean-fertilization project off Canada sparks furore. Nature News 490:458–459

Treasury HM (2020) Plan for jobs. Presented to parliament by the chancellor of the exchequer by command of her majesty. London, HM Treasury

Turkaly C, Nicholson S, Livingston D, Thompson JMT (2017) Establishing a climate engineering research clearinghouse. In: Assessment FFCE (ed) FCEA papers and reports. Carnegie Endowment for International Peace, Washington D.C.

Uther S (2014) Diskurse des climate engineering. argumente, Akteure und Koalitionen in Deutschland und Großbritannien. Wiesbaden, Springer Fachmedien

Victor DG (2008) On the regulation of geoengineering. Oxford Rev Econ Pol 24(2):322–336

Victor DG, Morgan MG, Apt J, Steinbrunner J, Ricke K (2009) The geoengineering option: a last resort against global warming? Foreign Affairs 88(2):64–76

UN (1977) Convention on the prohibition of military and any other hostile use of environmental modification techniques. In U. Nations (Ed.). New York, United Nations

Weili W, Ying C (2014) A Chinese perspective on solar geoengineering. Geoengineering Our Climate Working Paper and Opinion Article Series

Weinberg AM (1966) Can technology replace social engineering? Bull Atomic Scient 22(10):4–8

Wendt A (1998) On constitution and causation in international relations. Rev Int Stud 24(Special Issue):101–117

White LJ (1967) The historical roots of our ecologic crisis. Science 155(3767):1203–1207

Whyte KP (2012) Indigenous peoples, solar radiation management, and consent. In: Preston CJ (ed) Engineering the climate: the ethics of solar radiation management. Lexington, Plymouth, UK, pp 65–76

Whyte, K. P. (2019). Indigeneity in geoengineering discourses: some considerations. Ethics, Policy Environ Online first. https://doi.org/10.1080/21550085.1562529

Wibeck V, Hansson A, Anshelm J (2015) Questioning the technological fix to climate change—Lay sense-making of geoengineering in Sweden. Energy Res Soc Sci 7:23–30. https://doi.org/10.1016/j.erss.2015.03.001

Wiertz T (2012) From prediction to critique: the moral hazard debate. Forum Marsilius-Kolleg, 2: beyond calculation—climate engineering risks from a social sciences perspective 49–56. https://doi.org/10.11588/fmk.2012.0.9408

Wiertz T (2016) Visions for climate control: solar radiation management in climate simulations. Sci Technol Human Values 41(3):1–23. https://doi.org/10.1177/0162243915606524

Wong P-H (2015) Confucian environmental ethics, climate engineering, and the 'playing God' argument. Zygon 50(1):28–41

Yusoff K (2013) The geoengine: geoengineering and the geopolitics of planetary modification. Environ Plann A 45:2799–2808

Chapter 3
Ideas and Objects, Meaning and Causation—Frame Analysis from a Modernist Social Constructivism Perspective

Abstract Kreuter presents an elaborate and comprehensive theoretical construct to assess inadvertent framing of Climate Engineering approaches in academic discussion and its potential impact on decision-making. Based on the perspective of modernist social constructivism, Kreuter discusses the role of meaning as both constitutive and causally relevant. Aside from incorporating the notion of co-constitution of technology and society into political analysis, this chapter explores the role of academic communications in the construction of meaning of Climate Engineering. The theoretical construct provides a tool to analyze both legitimizing and delegitimizing moves of meaning-construction through language along four ideal-type frame themes, namely security, complexity, economy and appropriateness.

This chapter presents a deductively developed theoretical construct of abstract categories of analysis which allow the researcher to empirically and systematically answer the central research question: *How is the meaning of CE approaches constructed in the academic discussion through frames which legitimize or delegitimize the use of these approaches to address anthropogenic climate change?* Thus, this chapter is dedicated to defining, on the basis of theoretical literature, which portion of reality is considered relevant for the present study and how this portion of reality can be conceptualized and made accessible to empirical analysis. Thus, theory in the sense of a set of non-contradictory, general statements is important for this study as it helps to select relevant information from the mass of information provided by social reality, to systematize this information according to general rules, and in order to give insights into the relevance of this information by indicating which implications the state of affairs can have (Spindler and Schieder 2010: 23).

The analysis is based on the theoretical perspective of modernist social constructivism. The theoretical perspective, on a meta level, defines the ontological and epistemological 'playing field' in which the analysis is situated: It defines the rules according to which reality and research are assumed, in the present study, to function. However, as these rules are set out, in the general perspective in very abstract terms, further assumptions specific to the research question are derived from STS, practice theory and framing theory in communication studies and political analysis. These assumptions define how the academic discussion on CE is analyzed in the

J. Kreuter, *Climate Engineering as an Instance of Politicization*,
Springer Climate, https://doi.org/10.1007/978-3-030-60340-3_3

Image 3.1 Structure of theoretical approach

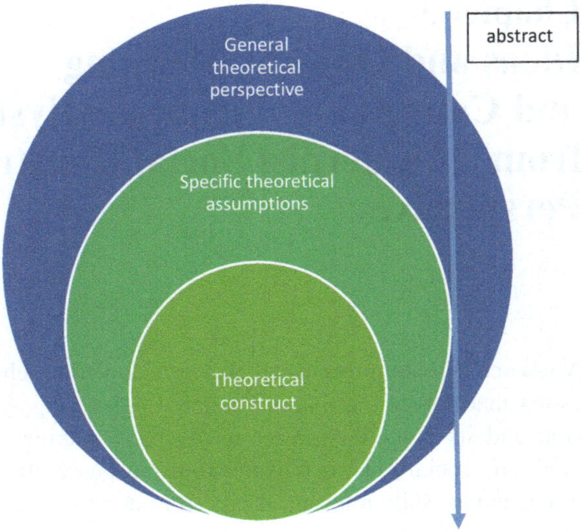

present study and explicates the relevance of this approach. The approach is rendered applicable to empirical analysis through the theoretical construct for analysis. The shape of the construct is derived from frame theory. The content of the construct is comprised of a set of eight ideal-type discursive representations of CE—operationalized as frames—which either legitimize or delegitimize CE as an approach to address the challenge of climate change. The ideal-type frames are developed from IR-theoretical literature as well as empirical discourse and framing analyses on CE and other emerging technologies. Each of them can initiate a specific discursive process, which are outlined below. The theoretical construct will be illustrated through the case of the early academic discussion on nuclear technology in the next chapter before it is operationalized, in the following chapters, for an analysis of the academic discussion of CE. 'Ideal-type' means that each of these frames represents an extreme form of either a legitimizing or delegitimizing frame, respectively. Many other variants of these frames can be conceived of.

This chapter will proceed in decreasing order of abstraction: Firstly, on the most abstract level, I outline the general theoretical perspective of modernist social constructivism, drawing on insights from practice theory. Secondly, more concretely, I present and discuss the theoretical assumptions specific to the research question. Thirdly, and on the basis of the foregoing considerations, I create a theoretical construct providing categories to identify legitimizing and delegitimizing frames of CE approaches in the academic discussion (Image 3.1).

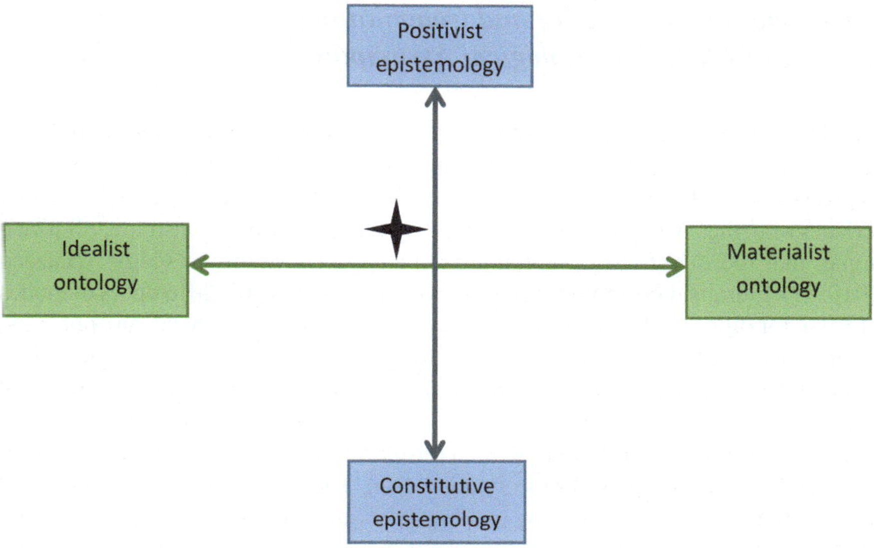

Image 3.2 Epistemological and ontology situation of the present study (Present study indicated through black star)

3.1 Defining This Study's Ontological and Epistemological Position—Modernist Social Constructivism

The theoretical construct is based on the general theoretical perspective of modernist social constructivism. This perspective presents a set of assumptions on how reality is constituted—ontological assumptions—and on how knowledge is constituted—epistemological assumptions. I understand modernist social constructivism, with Adler, as an "attempt […] to build a bridge between the widely separated positivist/materialist and idealist/interpretive philosophies of social science" (Adler 1997: 323). I outline the ontological and epistemological assumptions of modernist social constructivism in order to make transparent the general theoretical perspective espoused here and to disclose the reasons behind this choice. I call it a choice on the basis that the conflict between different perspectives with regards to ontology and epistemology can be considered "a debate not to be won, but a pluralism to live with" (Waever 1996: 156; see also Wendt 1998; Gschwend and Schimmelfennig 2007).[1] Modernist social constructivism as applied in this study is dedicated to a moderately idealist ontology as well as a moderately positivist epistemology (see Image 3.2).

[1]This pluralism is united in the pursuit of truth about international politics, albeit applying different methods and assuming different conditions to be fulfilled to approximate truth (Wendt 1998: 101; Behnke et al. 2012: 19).

3.1.1 Discourse, Practice and Communities of Practice—Ontological Assumptions

Ontology, or the study of being, asks for the nature of being and reality. Ontological assumptions in IR theories make statements about how "entities [...] that populate global politics" (Balzacq 2010: 58) emerge, develop and change, i.e. what the characteristics of beings in this world are and how (political) reality is constituted. Balzacq argues that "constructivism is committed to a pluralist-idealist ontology" (Balzacq 2010: 57). Pluralism in ontology refers to assumptions about the differences of modes of being for different things in the world such as humans, on one hand, and numbers, on the other. Idealism in ontology refers to the assumption that ideas constitute reality. Materialism, in contrast, refers to the notion that reality is constituted by material objects. In IR, a strictly idealist ontology means that, for example, not the number of nuclear warheads is relevant for the constitution of reality, but the existence of ideas such as human rights norms or nationalism. The very diverse strands of social constructivism share the following ontological assumption in common: Reality and the beings in the political world are not exclusively materially and objectively given, but also *in a way and to a certain degree* constructed through ideas (Adler 1997: 322). Accordingly, constructivism can be viewed situated on a 'middle ground' between ontological materialism and ontological idealism: Constructivism "is the view that *the manner in which the material world shapes and is shaped by human action and interaction depends on dynamic normative and epistemic interpretations of the material world*" (Adler 1997: 322).

The degree of relevance of ideas vis-à-vis material objects in constituting reality—i.e. how far to the left in image 3.2 a constructivist study is situated—depends on the specific approach assumed. In the present study, I assume the following type of interaction of ideas and material objects: Ideas give meaning to material objects and guide action with regards to these objects. This means that material objects in this world are relevant for this brand of constructivist study: The things in the world and their physical characteristics are given. However, what they *mean* for people, and, accordingly, what action is taken with regards to them, is socially constructed.

> Although, for example, a glacier is, in physical terms, a mass of ice with certain properties, it depends on its discursive representation whether it counts as an object of scientific inquiry, an emblem of pure nature or the witness of dangerous global warming [...]. (Methmann et al. 2013a: 5)

I go further than that and assume that even the term 'a mass of ice' is a discursive representation, just one that seems to be more straight-forward and neutral than the rest. However, it still does something to what a glacier is *for us*. Neither words nor pictures nor film are capable of capturing the entirety of the physical entity of a glacier. Whenever we try to capture it through language or image, we already represent it in a certain way and highlight certain aspects. This representation is, on one hand, dependent on what a glacier means, but also, on the other, influences again this very meaning. This does not, however, mean that there is not a physical entity out there that we call 'glacier', i.e., that social reality was completely independent of material objects. It also does not mean that the glacier does not have specific characteristics independent of human action: I agree that glaciers will melt at a specific rate dependent on environmental conditions, e.g. atmospheric temperature and precipitation, irrespective of whether humans understand, know about or believe in this phenomenon. But the socially constructed meaning of the material object guides action with regards to it. For example, if I consider a glacier as 'an emblem of pure nature', then I might find the claim rather compelling that action needs to be taken in order to preserve the glacier. This will, in turn, influence how I interact with the material objects of this world—e.g., whether I go walking around on the glacier or not, knowing that this might negatively affect it. This is why this study is situated a little to the left of the center of axis 'ontology': Because, while the physical materiality of objects set the frame of social reality, how actors interact with these givens is guided through constructed meanings, and this interaction, in turn, can change material constitution of reality, if only in very small ways.

Discourses play a significant role in this process of construction of meaning: "The world is not objectively accessible, but instead is imparted *through language* through discursive processes and symbolic interpretations" (Uther 2014: 60).[2] This means that much of the meaning-construction of social reality, which the brand of social constructivism espoused here relies upon, is done through what many authors call discourse. The frames that are studied in this analysis are considered as elements in discourse. Discourse, here, is understood, following Maarten Hajer, as "an ensemble of ideas, concepts, and categories through which meaning is given to social and physical phenomena, and which is produced and reproduced through an identifiable set of practices" (Hajer 2006: 67). This captures the simultaneous constructedness and construct*ive*ness of the social meaning of material objects: Through language and practice humans both capture the meaning material objects have, and change or reproduce this meaning. Similarly, Dryzek defines discourse as "a shared way of apprehending the world. Embedded in language, it enables those who subscribe to it to interpret bits of information and put them together into coherent stories or accounts" (Dryzek 2005: 9). This means that meaning is constructed in discourses in "supra-individual and rather unconscious" ways (Methmann et al. 2013a: 6).

At the same time, human actors, in this study, will not be considered as entities subordinate to a super-human structural power of discourse. Actors, according to Uther's discussion, are both constructed and constructing (see Uther 2014: 55). Following her approach, actors are considered to be the central active agents in any meaning-making situation, and thus not helpless pawns in the flow of discourse: "The actor is not passively subdued to the discourse, as, for example, the subject of Foucault, but it actively participates in shaping the discourse through its own interpretive achievements" (Uther 2014: 66).[3] This two-fold characterization of actors as

[2]Original quote: Die Welt ist nicht objektiv zugänglich, sondern wird durch diskursive Prozesse und symbolische Deutungen *sprachlich* vermittelt. (Uther 2014: 60), translated by JK.

[3]Original quote: Der Akteur ist nicht nur passiv dem Diskurs unterworfen, wie etwa das Subjekt bei Foucault, sondern gestaltet den Diskurs durch eigene Interpretationsleistungen aktiv mit. (Uther 2014: 66), translated by JK.

both constructed and constructing is captured by practice theory. The terms 'practice' and 'discourse' are considered to not be antithetical to each other: Practices are "meaningful patterns of action, which, in being performed more or less competently, simultaneously embody, act out, and possibly reify background knowledge and discourse in and around the material world" (Adler and Pouliot 2011: 4). Thus, (non-linguistic) practices are understood to interact with (linguistic) discourses and in this interaction construct or deconstruct meanings. Other authors reconcile linguistic and non-linguistic elements of meaning-construction by subsuming both under the term 'discourse'(see Anshelm and Hultman 2015: 11; Methmann et al. 2013a: 6) As Dryzek (2006: 3) argues, "discourses are a matter of practice as well as words, for actions in the social realm are always accompanied by language that establishes the meaning of action" (see also Adler 2008: 198). I assume, in line with these authors, that meaningful action—or practice—and meaningful talk—or discourse—construct or deconstruct the ideational meaning of an object. Further, I assume that practice is meaningful action which acts out certain meanings about the world, thus making it causally relevant.

Practice theory provides tools to put actors center stage as practice as meaningful action depends on the existence of an actor. Adler and Pouilot describe practice theory as an "invitation […] to conceive of the social as bundles of ideas and matter that are linguistically, materially, and intersubjectively mediated in the form of practices" (Adler and Pouliot 2011: 14). Practices as "meaningful patterns of action" (Adler and Pouliot 2011: 4) can more specifically be defined as a type of action, which, in turn, is a type of behavior. As Adler and Pouliot (2011: 5) describe: "[A]ction is behavior imbued with meaning. Running in the streets aimlessly is mere behavior, running after a thief is an action endowed with meaning." For an action to be a practice, it needs to fulfill additional criteria: "Practices[…] are patterned actions that are embedded in particular organized contexts and, as such, are articulated into specific types of action and are socially developed through learning and training" (Adler and Pouliot 2011: 5). Accordingly, a practice is behavior endowed with meaning, "embedded in an organizational context, repeated over time and space, constituted by knowledge […], and articulated as part of a complex set of other social performances […]" (Adler and Pouliot 2011: 6). The knowledge in question is in part tacit, and it not only constitutes practices, but is also bound up in them. Adler and Pouliot call this 'background knowledge': "expectations, dispositions, skills, techniques, and rituals that are the basis for the constitution of practices and their boundaries" (Adler and Pouliot 2011: 17). This background knowledge "does not create uniformity of a group of community, but organize[s] their differences around pervasive understandings of reality" (Adler and Bernstein 2005: 296).

Ontologically, practice theory is understood by its proponents to provide a way out of the dichotomy of agency versus structure, or to conceptualize the two-fold characterization of actors as both constructed and constructing (see above). The agency-structure-dichotomy describes an analytical problem similar to the one of hen and egg: Should agents inside of the structure be understood to have final causal power over phenomena in the social world, or is it the structure, which was in turn constructed by agents, that finally influences what happens? For example, should developments in global climate policy be accorded to the UNFCCC (structure) or to efforts of the government of small island states (actor)? According to the structuration theory by Giddens (1984), actors and structures determine and influence each other. Practice theory can give a more detailed insight into the workings of this reciprocal constitution:

> [P]ractices are both *individual (agential)* and *structural* […]. When 'disaggregated', practices are ultimately performed by individual human beings and thus they clearly are what human agency is about. […] Recursively, in and through practice, agents lock in structural meaning in time and space. (Adler and Pouliot 2011: 16)

Adler argues that these agents, these

> 'carriers' of social structures across functional and geographical boundaries are not necessarily states or societal networks, but 'communities of practice' […] – like-minded groups of practitioners who are informally as well as contextually bound by a shared interest in learning and applying a common practice. (Adler 2008: 196)

In the analysis of the discursive construction of CE, the tool of the Communities of Practice (CoP) approach is applied. The CoP approach allows for the analysis of actors which transcend national borders, such as transnationally active scholars, researchers and scientists as well as political and economic actors, all of whom can conceivably influence the social construction of CE—both of its ideational meaning and its causal materiality. The focus in this study lies on academic actors as a community of practice. It needs to be kept in mind, however, that communities of practice can be characterized as both actor and structure:

> Communities of practice are intersubjective social structures that constitute the normative and epistemic ground for action, but they also are agents, made up of real people, who – working via network channels, across national or organizational lines, and in the halls of government – affect political, economic, and social events. (Adler 2008: 199)

The Epistemic Communities (EC) approach (see Haas 1992; Adler and Haas 1992), an approach "considered to be the most influential" (Beck 2015: 286) in the debate on the role of academic and other experts in global governance, can, according to Adler, be understood as the description of the role of a very specific type of community of practice (Adler 2008: 199). This approach provides the background for the explication of one of the central ontological assumptions of this study, namely that the academic discussion is particularly relevant in the construction and enactment of the legitimacy or illegitimacy of CE as measure to address climate change.

To sum up, the moderately idealist social-constructivist ontology defining the understanding of reality in this study means that specific communities of practice can—in interaction with other factors—constitute the meaning of an object and, by enacting this meaning through practice—again, in interaction with other factors—cause material change in the configuration of objects in the world. For example, a community of practice constituting the "normative and epistemic ground for action" (Adler 2008: 199) on cyber security such as the German Chaos Computer Club regularly constitutes the meanings of objects such as virtual networks or computer viruses by defining, for the former, their vulnerability to attack, and, for the latter, their threat potential. This is done, for example, in interviews or reports. The community also enacts these meanings through practice, e.g. by interacting with federal offices and agencies in order to change the material configurations in our reality according to the meanings applied to objects, e.g., changing the material basis of digital infrastructures in order to make them more resilient.

In this study, the academic community researching CE constitutes the community of practice that is most interesting as it is assumed to play a significant role in the early construction of the legitimacy or illegitimacy of CE as a measure to address climate change.[4] This community constitutes the legitimacy or illegitimacy of these approaches through framing them, for example in academic publications, which present a form of practice as these publications carry the framings of CE into the public, thus enacting these framings. However, it is important to note that this framing is, in most publications, not the main intention of the respective authors: Other than epistemic communities in their original conception, the group of academics researching and discussing CE approaches is not united in a common policy endeavor. The main intention of this diverse group of actors is the publication and discussion of research results. I claim, however, that even this disinterested activity can, when it concerns a set of physical objects that remain to be materially developed, influence the legitimacy or illegitimacy of these objects—in this case, CE approaches—even if this was not the intention of the actor.

3.1.2 Meaning as Both Causal and Constitutive—Epistemological Assumptions

Up to this point, the ontological perspective of this study, namely the interaction of ideas and material objects in the constitution of social reality through practice—here named moderately idealist social-constructivist ontology—was outlined. In the following paragraphs, the epistemological perspective will be described. Epistemology is defined as the study of knowledge. Epistemological assumptions of IR theories make statements about what can be known about the entities in global politics and how it can be known. Epistemologically, "constructivists are united in an opposition to empiricism" (Balzacq 2010: 58). This means that according to constructivism, experience cannot be the final source of truth, because empirical findings, which are the results of experience, do not exist outside of or prior to theory; instead, theoretical considerations and assumptions are always co-constitutive of the facts that are found.

This assumption creates a common ground for different strands of constructivism. The modernist (or conventional) and the post-modernist (or critical) variant (see Balzacq 2010: 58; Wendt 1998: 101) disagree on the question of whether and how the "truth about international politics" can be accessed (Wendt 1998: 101) and, in particular, whether positivist methodology can be of use or not. This disagreement is closely connected to the so-called Third Debate of IR theory (Lapid 1989) which

[4]This assumption will be explicated further below.

divides theoretical approaches into positivist, on the one hand, and post-positivist, on the other hand. One important issue in this debate (Wendt 1998: 101) is whether social science is aimed mainly at *explaining* or at *understanding* social phenomena.[5] The aim of *understanding* a social phenomenon is "to recover the individual and shared meanings that motivated actors to do what they did" (Wendt 1998: 102). Theoretical approaches following this goal are called 'constitutive theories' (Wendt 1998: 105) The aim of an *explanation* of a social phenomenon, in contrast, is "to find causal mechanisms and social laws" (Wendt 1998: 102). Studies following this logic usually apply methods which borrow strongly from the natural sciences, based on the assumption that an objective or "outsider's" view on the phenomenon is possible (Hollis and Smith 1990: 1). Wendt calls these approaches 'causal theories' (Wendt 1998: 105). While the former try to give an account of the current *status* of a phenomenon, the latter aims to provide a description of its causal *dynamics*. Wendt argues that two different types of questions are relevant to be asked in research by these two perspectives: Causal theoretical studies aim to answer 'Why' and 'How' questions such as "'why did Gorbachev move to end the Cold War?'" (Wendt 1998: 104) by developing statements of the shape of "'X causes Y'" (Wendt 1998: 105). Constitutive theoretical studies aim to answer 'How-possible' or 'What' questions such as "'How was it possible for Stalin, a single individual, to exercise so much power over the Soviet people?'" or "'What kind of political system is the European Union'" (Wendt 1998: 105), aiming to achieve insight into the "conditions of possibility" of social phenomena (Wendt 1998: 106), the elements and relations between those elements that a phenomenon is made up of, instead of through which factors it was caused.

Post-modernist constructivism aims to understand social phenomena by analyzing how meaning is constructed or 'constituted' (Wendt 1998: 104); it is "decidedly interpretivist" (Balzacq 2010: 58). This means that this perspective does not attempt to provide explanations of causal mechanisms bringing about social phenomena, but instead to provide understandings of how their meanings have been constituted: "Their goal is to show how the properties of a system are constituted" (Wendt 1998: 105). In contrast, modernist constructivists are more positivist in the sense that they "attempt to explain both the causal and constitutive effects of unobservables in world politics (e.g. structures or processes)" (Balzacq 2010: 58). Positivist methodology is considered critically, but not rejected completely. This means that both the analysis of meanings and the analysis of causality are important for modernist constructivists: In IR, they are trying to give accounts of causes and effects of unobservables as well as understand the way that they are constructed, that meaning is ascribed through language and action.

Again, practice theory provides one way of reconciling two different strands of constructivism: Epistemologically, practice theory is an attempt to reconcile causal and constitutive approaches to IR theory and is thus in line with the modernist constructivist approach taken here:

---

[5]For comprehensive and profound discussions of this debate, see Lapid (1989), Wendt (1999), Hollis and Smith (1990).

> [...] [P]ractices are both *material* and *meaningful*. On the one hand, practices are material insofar as they are doings enacted in and on the world. [...] At the same time, practices also are shot through with meaning [...] Through social interaction, people attribute meaning to their activities and build on these to interact further. (Adler and Pouliot 2011: 15)

This study is dedicated to a modernist constructivist approach: It aims to empirically analyze *construction and deconstruction* of legitimacy of CE approaches as measures to address climate change through framing in the academic discussion against the backdrop of the assumption that the constructed legitimacy of CE has a *causal* impact on decision-making concerning these approaches. Thus, it is assumed that frames in the academic discussion have both "causal and constitutive effects" (Balzacq 2010: 58) on world politics—*causing*, together with a number of other factors, specific decisions to be taken by *constituting* the meaning of CE approaches. As the analysis of the constitution—or construction—of legitimacy or illegitimacy of CE approaches is conducted here on the basis of the assumption that this constructed meaning of these approaches is, in turn, causally relevant for decision-making, this study is dedicated to a slightly more positivist than interpretivist epistemology. The focus of this study does not lie on assessing the causal impact of the constructed meaning of CE—which is assumed to exist as, otherwise, the present study would be irrelevant—but instead on tracing the construction or construction of this meaning through framing.

3.2 The Roles of Actors, Technology, Legitimacy and the Academic Discussion—Specifying This Study's Perspective

While the section above presented the general theoretical perspective of this study, further assumptions need to be made transparent which went into the design of the theoretical construct and the analysis. These assumptions will be laid out in this section. They are made in order to support the central ontological claim of this thesis, namely that (I) *the legitimacy of CE approaches, for which the early academic discussion sets the stage, can influence future decision-making concerning these approaches.* This claim has not been tested empirically—instead, it provides the basis on which the empirical study of the framing of CE is based. The central epistemological claim of this thesis posits that (II) *an analysis of frames in the academic discussion on CE can give insights into the construction of the meaning of CE with regards to its influence on future decision-making.* This claim will, likewise, not be tested empirically, but provides the justification for the designing of the shape of the theoretical construct on the basis of frame theory. This will be set out in the following sections.

The following assumptions have been made in this study in order to justify the ontological claim (I) above:

1. Phenomena such as anthropogenic climate change become political—are politicized—if and when they are moved from the realm of necessity to the realm

of contingency. Only after this politicization has taken place, can measures to address the problem can reasonably be presented.
2. The global political sphere is populated by a more diverse set of actors than exclusively state or intergovernmental organizations.
3. Technologies can be conceptualized as one type of governance measures to address a political problem, which means that their research, development and use can be understood as a political decision.
4. The legitimacy of technological governance measures influences decision-making concerning their research and use.
5. The academic discussion is particularly relevant in construction of the meaning of CE approaches.

3.2.1 *Politicization Is a Precondition to Meaningfully Offering Measures to Address a Phenomenon*

The first assumption is that political problems do not appear out of nowhere, but instead are formed when a phenomenon is 'politicized' in the discourse theoretical rendition of the term.[6] This assumption has to be made as it would otherwise make no sense to claim that measures to address climate change are presented in any arena in a meaningful way, and, conversely, also not to study them. In this sense, the process of politicization captures the opening-up of issues to debate: "issues are politicized when they become the subject of deliberation, decision making and human agency where previously they were not" (Hay 2007: 81). Hay defines discursive politicization (Hay 2014: 309) as "associated with the extension of the capacity for human influence and deliberation which comes with disavowing the prior assignment of an issue—or issue domain—to the realm of fate or necessity" (Hay 2007: 81). An example is the removal of the identity of a state's leader from God-given necessity (as in hereditary monarchies, for example) to the sphere of deliberation and democratic choice.[7] Climate change has already been moved to this stage of politicization: The very idea of mitigation measures to combat climate change indicates that the adverse effects of climatic changes are not considered a necessary development, but one that can be counteracted and influenced. As Anshelm and Hultman put it: "Climate change is political" (Anshelm and Hultman 2015: 3).

Wood and Flinders (2014: 154) argue that "a process of *discursive policisation* [sic] would […] involve the promotion of a topic as a public issue where *competing interpretations exist* as choices" (Wood and Flinders 2014: 161). Discursive politicization, then, describes a process in which an issue is presented as one that is

[6] See Chap. 1.

[7] Technologies can be a cause for this change, as Wood and Flinders stress, through "the spillover effects of recent scientific and technological developments that offer new opportunities to control issues that were previously thought to be beyond human control" (Wood and Flinders 2014: 154). Here, however, technology will not be analyzed as the cause of politicization, but rather as the effect of depoliticization.

subject to "antagonism, conflict, difference and choice" (Wood and Flinders 2014: 162). There is not one clear solution to the problem, but rather a number of choices that need to be considered through deliberation. In that way, discursive politicization is the process which reveals "the possibility that society can be constituted differently" (Wood and Flinders 2014: 162). Conversely, discursive processes also have the "power to transform some contingent understandings into taken-for-granted truths, the seemingly natural order of things, so pervasive and powerful as to be inescapable" (Dryzek 2006: 20), i.e., to move issues back into the realm of necessity.

As outlined above, discursive politicization takes place with regard to climate change in the shape of calls for opening up the debate and "established concepts, identities, and issue framings" (Brown 2015: 16). The ideal-type frames that are analyzed in this thesis legitimize or delegitimize a technology to be developed and applied as a tool to address the political problem of climate change. They can, in accordance with Wood and Flinders (2014), nurture processes of depoliticization, i.e. the moving back of the issue from deliberation into a narrower field in which options are limited. This, however, is not because the issue is considered part of the 'natural order of things'. Instead, after the ideal-type processes that will be presented below have taken place, a specific set of measures to address the challenge is considered proper procedure. It is important to point out that both legitimizing or delegitimizing frames might depoliticize: The former would present one or more CE approaches as a necessary measure to address the challenge of climate change, thus sidelining other approaches. The latter could, theoretically, lead to a moratorium, thus rendering these approaches 'unthinkable' and removing them from the portfolio of potential measures.

3.2.2 *The Political Sphere Is Populated by a Diverse Set of Actors*

The second assumption, which is common to many theoretical perspectives in IR, is that not only states or state actors are relevant in international politics, but also a diverse set of other actors such as civil society, transnational corporations or academic experts. Without this assumption, a study of the academic discussion of CE would not be sensible in order to find out anything about potential political decision-making on CE. The term 'global governance' "refers to the entirety of regulations put forward with reference to solving specific denationalized and deregionalized problems or providing transnational common goods" (Zürn 2012: 730). Thus, it aims to account for the empirical observation that, in global politics—the process of negotiating and mediating between various, often conflicting interests—and global policy-making—the content of governance tools to address global challenges—not only classic nation-states and international institutions play a role, but also various other actors, such as transnational corporations and civil society. Thus, the term is often applied to denote the widening of the perspective from traditional IR studies, which focus on inter-state relations, to a more comprehensive approach of assessing various factors influencing the way in which 'denationalized problems' are solved. As a political science research perspective, it has been applied in numerous studies on the global challenge of anthropogenic climate change and its governance (see, for example, Biermann 2010).

3.2.3 Measures to Address Political Problems Can Be Technologies

I assume in this study that development and use of technology can be presented as a governance measure or as a course of action or method to be pursued in order to address a specific political problem. Governance refers to "the systems of authoritative norms, rules, institutions, and practices by means of which any collectivity, from the local to the global, manages its common affairs" (Ruggie 2014: 5). Following this definition, CE governance has two dimensions in abstract terms: The first dimension is the governance *of* CE approaches.[8] The second dimension of CE governance is the governance of climate change *through* the application of CE approaches. This dimension is inherent in the very definition of CE as "the *deliberate large-scale manipulation of the planetary environment to counteract anthropogenic climate change*" (Shepherd et al. 2009: 1). CE is defined by the intent to address the social problem[9] of climate change, and can thus be understood to be part of 'technological politics', a term from STS denoting the idea that technology can "supplant other ways of building, choosing, acting, and enforcing which are more commonly considered political" (Winner 1977: 237; see also Hilpert 2016: xxi). In the case of CE, these technological approaches can be perceived as (one among various) potential governance measures aimed at "solving [a] specific denationalized and deregionalized problem[…]" (Zürn 2012: 730), namely the problem of anthropogenic climate change. Here, CE approaches are understood as technological (as opposed to, e.g., legalistic) governance measures as well as objects of governance.

Technology in Politics

This conceptualization of CE approaches as potential technological governance measures is in line with the understanding of technology as "politics by other means" (Bijker 2006: 686). The term technological politics „is meant to emphasize […] that the rule of technological circumstances in the modern era does in fact supplant other ways of building, maintaining, choosing, acting, and enforcing which are more commonly considered political"(Winner 1977: 237). There are "instances in which the invention, design, or arrangement of a specific technical device or system becomes

[8] See Chap. 2.

[9] 'Social problem' will be defined here as a collective action problem which "exists whenever the common or collective interest of a group or society is not best served by the narrow pursuit by individuals of their own (perceived) self-interest" (Hay 2007: 2).

a way of settling an issue in a particular community "(Winner 1980: 123). This relevance of technology for governance is being recognized today:

> The increasing importance of new technologies for […] problem solving (e.g. energy supply, environmental problems or better pharmaceuticals) […] meets a key societal expectation of modern governments – i.e. to foster development through government policies. (Hilpert 2016: xxi)

The meaning of the term *technology* is ambiguous (see Winner 1977: 8; Bijker 2006: 682). As philosopher Martin Heidegger (1954) points out, the term 'technique' ("Technik") can be defined instrumentally or anthropologically: as a means for a purpose or as a human action.[10] Similarly, 'technology' can be defined as means to achieve an end, but also as human practice. Winner points out that 'technology' comprises a class of objects considered technological—"tools, instruments, machines, appliances, weapons, gadgets" (Winner 1977: 11)—as well as a body of technological activities—"skills, methods, procedures, routines" (Winner 1977: 12). For example, the CE approach of SAI can be defined as a means to achieve the purpose of countering anthropogenic global warming, or as an activity by certain human actors. More recently, Bijker differentiates between the meaning of technology as "physical objects or *artefacts*",[11] as "human *activities*" and as "*knowledge*" (Bijker 2006: 682), thus between material objects, human practices with regards to these objects and knowledge that goes into creating, using and otherwise handling these objects. Differently, Arthur differentiates three definitions of technology (Aronson 2010: 259): Accordingly, technology can be understood as "*a means to fulfill a human purpose*", "*an assemblage of practices and components*" or "the entire *collection of devices and engineering practices available to a culture*" (Arthur 2009: 28). Thus, he builds up a hierarchy, starting out with the single object, then moving to the object(s) entangled in social practices, and then to the entirety of these entanglements in society. These types of layered definitions have found application in political science studies (see, for example, Methmann and Rothe 2012: 332).

As pointed out above, the first definition seems straightforward in the case of CE: These approaches are, by definition, aimed at fulfilling a human purpose, i.e. influencing the climate in order to combat anthropogenic climate change. This definition already points in the direction of CE as a governance measure. However, the policy of regulating car traffic in big cities, aimed at reducing greenhouse gas emissions, also meets the criterion of 'fulfilling a human purpose'. Perceived only in the terms of the first definition, every governance measure would be a technology and every technology a governance measure in the broadest sense. Thus, it is important not to lose sight of the materiality of technology as pointed out by Bijker when he calls technologies 'artefacts'. Arthur's second definition captures both artifacts and activities, all of which are entangled in some way and make up a somewhat coherent whole that can be meaningfully differentiated from other coherent wholes. In the case of CE, one would be able to define a technology of CE which is made up of a bundle of

[10] Technik ist ein Mittel für Zwecke. […] Technik ist ein Tun des Menschen. (Heidegger 1954: 13), translation by JK.

[11] While I apply the American spelling of 'artifact', Bijker uses the British English spelling 'artefact'.

technological artifacts—e.g. blimps to move aerosol into the stratosphere—as well as knowledge and activities—as the blimp does not by itself move into the stratosphere to spray aerosols. This is all the more fitting as the umbrella term 'CE' encompasses a number of different approaches which make use of different artifacts and different practices and knowledge. Arthur calls this entirety of components and activities a 'domain':

> They are not invented; they emerge, crystallizing around a set of phenomena or a novel enabling technology, and building organically from these. They develop not on a time scale measured in years, but on one measured in decades – the digital domain emerged in the 1940s and is still building out. And they are developed not by a single practitioner or a small group of these, but by a wide number of interested parties. (Arthur 2009: 145)

Whether a CE domain will develop in the future is an open question—and if it does, around which specific approaches it would develop. If it does, the current early discussion will have had a constitutive function, to which the third definition points: CE would not exist in isolation, like a stain of ink on an otherwise white sheet of paper. It would be embedded into a vaster system of technologies and technological activities such as fossil fuel combustion technology, which has had a large stake in the creation of the problem of climate change to begin with, or satellite technology, which makes it possible to calculate a global average temperatures and thus monitor the global climate in the first place. Thus, Arthur's third definition accommodates the notion that 'technology', today, points to a "vast, diverse, ubiquitous totality that stands at the center of modern culture" (Winner 1977: 9). Following this three-pronged differentiation, CE as defined above is, in this study, considered a possibly emerging technological domain. This is the way in which the term of 'individual technologies' will be used: as defining different technological domains—none of which, as mentioned above, exists today on a scale to significantly reduce anthropogenic climate impacts. The broader term 'technology' is understood in the sense of Winner's ubiquitous totality.

Technology in Society: the Notion of Co-production

The main weakness of Arthur's account with respect to the purpose of this thesis is that he does not consider the interactions between technology and society in any detail. Arthur understands technologies to be self-creating in the way that he traces the evolution of individual technologies as the process in which new technologies are "birthed […] from previous ones" (Arthur 2009, 15). Thus, a new technology is not considered to come out of the blue, but to develop out of existing earlier technologies. Unlike Arthur, Bijker does not take for granted this "self-creating" character of technology (Arthur 2009: 2). He criticizes this stance of 'technological determinism'.

> Technological determinism […] comprises two elements: it maintains that (1) technology develops autonomously, following an internal logic which is independent of external influences; and that (2) technology shapes society by having economic and social impacts. (Bijker 2006: 683)

According to Bijker, technological determinism is the basis for one of two contrasting descriptions of technology: the 'standard image' of science and the constructivist view. According to the standard image, which is based on the assumption of technological determinism, "scientific knowledge is objective, value-free and discovered by specialists. Technology, similarly, is an autonomous force in society and technology's working is an intrinsic property of technical machines and processes" (Bijker 2006: 683). The constructivist view, on the other hand, contends that the meaning of technology is not autonomous, but constructed through interaction: "The interaction within and among relevant social groups can give different meanings to the same artefact" (Bijker 2006: 684).[12]

The notion of co-production captures the assumption that the materiality and objectivity of technology is influenced by social configurations and decisions, and at the same time, society is shaped by the technologies people use and apply. Neither technology nor society is considered to be self-creating, but rather as co-produced by each other. Thus, technologies are not influenced only by the laws of physics which draw the boundaries of what is technologically possible, but also by social processes, which shape which technologies are researched, and how they are designed and developed (Jasanoff 2004a: 2; Bijker 2006: 684): The role of a technology in the world and which use it is put to is not considered the necessary outcome of physical laws, but they do set the boundaries of the possible. Inside of these boundaries, social factors are relevant (Markusson and Wong 2015: 12). This is by no means meant in a deterministic way:

> [The] aim is not to provide deterministic causal explanations of the ways in which science and technology influence society, or vice versa; [...]. Rather, it is to make available resources for thinking systematically about the processes of sense-making through which human beings come to grips with worlds in which science and technology have become permanent fixtures. (Jasanoff 2004b: 38).

This notion of co-production allows for the consideration of CE approaches as *objects* of governance, on one hand: This is the direction of society influencing technology described by Jasanoff. This perspective includes, for example, the regulation of nuclear weapons technology through a sophisticated and complex system of norms, rules and institutions containing, for example, the Non-Proliferation Treaty and the International Atomic Energy Agency (Thakur 2015). Conversely, however, this notion also allows for the analysis of individual technologies as governance *measures* when technology is developed and used to influence social order. If this is the case, technologies do not blindly influence social order but are picked and chosen to do just that. For example, the problem of low voter turnouts at democratic elections has been, in some countries, addressed through the introduction of online voting systems (Hall 2012). In this case, online technology was used to govern a social problem. The constructivist perspective of co-production allows one to meaningfully

---

[12]In accordance with the carefully positivist epistemology espoused in this study, Bijker's criticism of the 'standard image' of scientific knowledge is not entirely followed in this study: Hence, I take the position that scientific knowledge is based, to some degree, on objective laws. However, I assume that social factors also play a role in the generation of scientific knowledge.

analyze the social construction of technologies as governance measures: They are not assumed to be self-determined—or only determined by the laws of physics—but rather viewed as chosen in social processes of decision-making, just like any other governance measure or regulation. It is conceivable that CE approaches are, in political decision-making, chosen as a measure to help regulate the global challenge of climate change.

Analyzing Co-production: The Multilevel Perspec*tive*

The interdependent development and influence of technology and society is the focus of the multilevel perspective (Geels 2007; Verbong and Geels 2007; Geels 2004, 2012, 2014; Schot and Kanger 2018): At the core of the model lies the socio-technical system, which can be "defined as configurations of actors, technologies and institutions" that "underlay contemporary societies" (Schot and Kanger 2018: 1045) and "encompass production, diffusion and use of technology" (Geels 2004: 900) to fulfill specific functions. These functions are, for example, the "provision of energy, mobility, food, housing, communication, water, healthcare, education, finance etc" (Schot and Kanger 2018: 1045).

The multilevel model is applied, for example, by Smit (2006: 734), who uses it to analyze the co-evolutionary development of military technologies after WWII. This model allows for Jasanoff's 'thinking systematically' about the roles of technologies and politics and their interdependency in the constitution of social order and for the situation of individual technological approaches, such as CE, in their wider social context. Smit argues that the co-evolution can be analyzed on different levels:

> [Socio-technical] landscapes, at the macro level in this approach encompass material and immaterial elements: material infrastructures, political cultures and coalitions, social values, world-views, and paradigms. They are beyond the direct influence of actors and evolve over time, partly through accumulating developments at the lower levels. In this approach, the meso-level is the level of *socio-technical regimes* which describe the dominant practices, rules, and assumptions in technological innovation processes. The micro level (or *niche* level), describing individual actors (including organizations) and local practices in technology developments, often is a source for initiating change at the higher levels. (Smit 2006: 734)

This means that, in the context of political analysis, technology and society are part of a socio-technical landscape, which "is a landscape in the literal sense, something around us that we can travel through; and in a metaphorical sense, something that we are a part of, that sustains us" (Rip and Kemp 1998: 334). Thus, the technological approaches of CE are situated at the micro level of the socio-technical landscape, being discussed right now by individual actors which might initiate change by moving CE into the portfolio of governance measures applied to address the challenge of climate change. Through this, they would influence the socio-technical regime. This regime, according to Rip and Kemp, can be defined as

> the rule-set or grammar embedded in a complex of engineering practices, production process technologies, product characteristics, skills and procedures, ways of handling relevant artifacts and persons, ways of defining problems – all of them embedded in institutions and infrastructures. (Rip and Kemp 1998: 338)

Thus, the regime is the software counterpart to the hardware of the technical system, which constitutes a composition of materials, skills, artifacts etc. like, for example, the railroad system comprised of rails, trains, personnel etc. The regime informs one about how the system works and gives rules as to how it should work.

Applying this notion of the socio-technical landscape on the basis of the idiom of co-production allows us to map out the context in which a set of technological approaches such as CE is situated: They would not come into the world out of the blue, and they would not be limited to physical effects on this world. Instead, they are constructed by various actors in a complex system made up of technological artifacts and technological systems, but also rules and norms as well as a social regime ordering the landscape, and they might be chosen as governance measures to address specific challenges. Also, their effects on social configuration become apparent, as these approaches would cause ripples across the entire socio-technical landscape, potentially changing it altogether. In this perspective, technology in general, as well as individual technologies or technological artifacts, make up an integral part of governance as "the entirety of regulations put forward with reference to solving specific denationalized and deregionalized problems or providing transnational common goods" (Zürn 2012: 730). This is in line also with Winner, who pointed out that technologies can be "inherently political" (Winner 1980: 123) for example "in instances in which the invention, design, or arrangement of a specific technical device or system becomes a way of settling an issue in a particular community" (Winner 1980: 123).

However, it is still an open question how to assess what the *role* of technology or individual technologies like CE is in specific empirical governance phenomena. For example, how would we assess what the role of surveillance cameras is for the governance of terrorism threats? A question like this cannot easily be addressed in global governance study, which traditionally focusses on the role of human actors—nation-states and state-related institutions, but also non-state actors such as transnational corporations or civil society—as well as the social structures in which they act and which they construct. This is clear, for example, from Armeni's definition of the term, which understands governance as "a system of norms, processes and structures established, and implemented, by a constellation of states and non-state actors [...] formally and informally operating within the international order" (Armeni 2015: 878). Thus, global governance is concerned with structures and actors. Technology and technologies seem not to fit either set. Thus, Pouliot, for example, argues that "[e]nmeshed in social relations, material objects often acquire an epistemic life of their own that may affect, in turn, the very people that constructed them" (Pouliot 2010: 298). Winner goes even further than that, arguing: "Rather than insist that we immediately reduce everything to the interplay of social forces, [this point of view] suggests that we pay attention to the characteristics of technical objects and the meaning of those characteristics" (Winner 1980: 123). How, though, can these 'characteristics' of technology be conceptualized?

Emergence in Focus: The Collingridge Dilemma

A first conceptualization of the 'activity' of technology in the political context that aims to account for this is provided by the so-called "material turn" (Mutlu 2013).

This research perspective puts emphasis on the "social and political lives" (Mutlu 2013: 385) of objects, looking at "the co-productive relationship between the origins and everyday functioning of objects". The assumption behind this perspective is that material objects and social order have an interdependent relationship: Objects influence how social life is shaped, and conversely, social actors and structures influence how objects are constituted. For example, nuclear warheads can have an influence how diplomats act as "material objects can acquire agency through enmeshment in social relations" (Pouliot 2010: 295) as well as being planned, thought up and created by human actors. At the core of these notions stands actor-network theory (ANT). In this perspective, "material objects are not placed in a dualism in relation to other actors and systems. Instead, objects function as nodes that mediate the interaction between actors and systems" (Mutlu 2013: 387).[13] This perspective allows for the research on political phenomena by focusing on the object itself: How does, for example, an automobile mediate between actors and the system in the context of mobility? The use of a car by an actor affects the system, as it reinforces the social need for roads and traffic regulation. The contestation of the car as a mode of transportation can affect the system as well, for example if new norms and ideas such as car-free days or areas are introduced (Willsher 2017). Thus, this perspective allows us to assess specific interlinkages between actors and systems by pinpointing the analysis around one specific object. Such an object-centered perspective has been applied, for example, in the political analysis of biometric identification technology (Jacobsen 2015), non-lethal weapons (Anais 2013), nuclear warheads (Pouliot 2010) and other objects.

For this study, the insights from ANT are important in particular as they add to the relevance of studying the early stages of technology development—in the case of CE, the very situation when these technological approaches are mainly ideas that have not yet taken much of a material form—in short, the situation in which we currently find ourselves. The way they are socially constructed right now is relevant, in turn, for the way in which they might shape our world in the future. This notion is captured, famously, by technology researcher David Collingridge:

> attempting to control a technology is difficult, and not rarely impossible, because during its early stages, when it can be controlled, not enough can be known about its harmful social consequences to warrant controlling its development; but by the time these consequences are apparent, control has become costly and slow. (Collingridge 1980: 19)

This so-called Collingridge dilemma (Markusson and Wong 2015: 5) is particularly salient in the case of CE. The fear of creating a monster has been communicated by various academic and other actors, for example through the slippery slope argument, which "describes situations in which conducting research on climate engineering technologies might increase the likeliness of deployment" (Zürn and Schäfer 2013: 272). Each step of development increases the likelihood of the next being taken:

[13] Various other connecting points between ANT and political study of global governance have been drawn in the literature (Bueger 2013; Nexon and Pouliot 2013; Barry 2013; Cudworth and Hobden 2013).

"The momentum grows until full-scale development eventually becomes unstoppable" (Hulme 2014: 69). At the other end of the discussion stands the notion that a moratorium might keep us from understanding—and being able to use, if necessary—technologies potentially important in addressing climate change while there is still enough time to consider them from all perspectives and assess all their implications (Keith 2013: 13/88; Long et al. 2015: 30).

Thus, in Mutlu's words, I am interested not in the stages of 'continuity' or 'transformation' in the relationship between objects (or actants), actors and systems, but in the stage of 'emergence'. Continuity describes a stage at which "the actant's role is stable" and the system is somewhat sustained. To return to the example of automobiles, one could make the argument that the fossil fuel powered car as an actant has been, for the last decades, in this stage. Whether the socio-technical system of mobility, which, in Europe and the USA, relies heavily on individual car-ownership, will move into a stage of transformation depends, for example, on the development of hybrid and electric cars, the establishment of a useful infrastructure for this type of vehicle, the development of the social discourse on individual versus collective mobility and many other factors. Transformation, "or the reinvention of the system" is, according to Mutlu "inevitable. In other words, unstable systems inevitably breakdown and re-generate, albeit through different alignments and assemblages" (Mutlu 2013: 391). For the case of CE, the first stage, that of 'emergence', is of particular interest: "Emergence is the initial moment an actant is introduced […]" (Mutlu 2013: 391). It is questionable whether the actants of CE (e.g. blimps to spray aerosols) have been introduced already, i.e. if emergence already takes place when the actant is being talked about, or if the actant needs to be out in the world as a material object to warrant that description—blimps exist, of course, but so far, not as CE artifacts. In the latter case, CE would be somewhere before emergence, without the necessity that this technology will ever materially emerge, but the possibility exists. So far, CE is mainly socially constructed through talk and model calculation. This is where the constructivist view on technology by Bijker becomes important: In pointing out that it is important to understand how the technology as a governance measure is constructed.

This construction, so far, is done mainly by academics in debates and academic publication. One criticism levered against ANT, however, is that, through the very focus on material objects, does not account well enough for human agency, interest and conflict: "[…] when actor-network theory confronts the nature of power, as it often does, it side-steps the very questions about people, institutions, ideas and preferences that are of great political concern" (Jasanoff 2004b: 23). It is, however, a particularly interesting point in STS scholarship that science is, just like politics, characterized by processes of contestation. Jasanoff claims: "[…] Latour's networks exercise power by displaying curiously little of the moral and political conflicts that normally accompany the creation and maintenance of systems of governance" (Jasanoff 2004b: 23). Latour himself asserts that "[i]f a criticism can be levelled at ANT it is […] the complete indifference for providing a model of human competence" (Latour 1996: 373).

To sum up, the focus of this study lies not on potential political implications of various CE approaches, but instead on the preconditions for their potentially coming into the material world in the near future. Potential implications mentioned in the literature[14] are, for example, the possibility that one actor might use specific CE approaches unilaterally to their own benefit, disregarding potentially detrimental impacts on other actors (Horton 2013; Lloyd and Oppenheimer 2014), whether mitigation efforts will suffer from the existence of a 'Plan B' (Reynolds 2015), whether the model of liberal democracy might suffer from use of specific CE approaches (Szerszynski et al. 2013; Horton et al. 2018) or that the entire global political system might change and a "*coherent sovereign*" might emerge (Wainwright and Mann 2012). Conversely, as CE approaches can be understood as governance measures as outlined above, it is interesting to assess the preconditions of the deployment of CE: How do CE approaches become thinkable options to address climate change?

3.2.4 The Constructed Legitimacy of Technological Governance Measures Can Influence Decision-Making

Legitimization is understood to be the "process by which legitimacy is acquired" (Trout 1975: 251). The notion of legitimacy that is at issue here is decidedly descriptive: The study aims to show which frames can be traced that legitimize a technology as a governance measure, and which frames exist which delegitimize said technology. The analysis of legitimacy in its descriptive conception was founded by sociologist Max Weber, who defines the term as the prestige of being considered binding ("das Prestige der Verbindlichkeit") (Weber 1980 [1922]: 16). This moved the focus towards the "social functions of legitimacy rather than in the normative justification of governance" (Steffek 2003: 253). In the case of the present thesis, the question is whether CE as a measure to address anthropogenic climate change is considered something that should be done, or must not be done. The relevance of discourse is highlighted by Steffek as the arena in which reasons are given and rationally debated for or against the norm in question (Steffek 2003: 263). Accordingly, legitimization is "the socio-political process of legitimacy formation through actions by various organizations and individuals. Central features are the formation of expectations and visions as well as regulative alignment […]" (Bergek et al. 2008: 581). The authors stress the importance of expert legitimization, especially in the early stages of technological innovation (Bergek et al. 2008: 585). This is the case for CE:

> The science and proposals of geoengineering can be considered 'upstream'. That is to say that significant research and development on them has not yet taken place; many of their possible impacts have not yet been explored; and as yet there are few salient media or public discourses. This makes geoengineering proposals very sensitive to appraisal as knowledge

---

[14]The possibility and probability of these implications actually coming about is highly contested in the literature.

> of both their technical and social science is immature. (Bellamy et al. 2012a: 3; see also Carr et al. 2014)

Legitimization is, according to Bergek et al., in a first step provided by experts, e.g. through technology assessments, as they present "arguments about expected performance" (Bergek et al. 2008: 588). With regard to international cooperation in the environmental sector, Karen Litfin stresses the potential of science in itself—the "universal legitimator of the modern era"—as a source of legitimization: "The authority of science rests on its claim to objective, disinterested and verifiable knowledge, where political authority rests on the consent of the governed and is exercised on the basis of power" (Litfin 2000: 122).

Which role, then, does discourse play in the construction of legitimacy in this descriptive understanding? Steffek addresses this question from a normative perspective when he describes his "account of modern legitimation through rational discourse" (Steffek 2003: 263). It is normative insomuch as the discourse is considered rational, and the result of it is legitimacy. The aim of this thesis, however, is a descriptive account of legitimization and delegitimization processes in the early discussion of technological governance measures. A descriptive perspective is presented by Cornelissen et al. who, from the point of view of organization theory, argue that "[metaphorical and analogical] framing [...] [can be] a way of fostering understanding and creating legitimacy for a change" (Cornelissen et al. 2011: 1702). This means that by presenting the issue in the context of a comparable, but different case (analogy) or by referring to images, contexts and cases which "resonate with broader value orientations in society" (Cornelissen et al. 2011: 1702), a specific measure can be legitimized. More to the point, discourse theorist Norman Fairclough argues that discursive elements such as frames, analogies and metaphors have "the capacity to 'naturalise' ideologies, i.e., to win acceptance for them as non-ideological 'common sense'" (Fairclough, cit. in Jaworski and Coupland 2006: 29). Similarly, Bellamy et al. describe that frames create closure "around particular sets of hidden values or assumptions [... which] produces variably limited ranges of decision options which seem preferable given those framing effects that are privileged" (Bellamy et al. 2012a: 25–26). With reference to their own analysis of CE appraisals, they argue that

> the outputs of geoengineering appraisals can be considered examples of different types of 'decision justification', whereby appraisals can exert inadvertent, tacit or deliberate influence of power on decision making through their various framings and prescriptive policy advice. (Bellamy et al. 2012a: 26)

Discourse, then, is relevant to legitimacy as the latter is considered dependent on the perceptions of the subject: A policy is legitimate if the subject *perceives it as* legitimate. Hurd, in line with Weber, defines it as follows: "*Legitimacy* [...] refers to the normative belief by an actor that a rule or institution ought to be obeyed. It is a subjective quality [...] defined by the actor's *perception* of the institution" (Hurd 1999: 381).

The relevance of discourse for a changed perception of legitimacy of a governance measure can be based on two theoretical perspectives: First, prospect theory as applied in psychology, communications studies and political science analyses, and second, normative change research from IR analysis. Prospect theory aims at explaining how individuals decide in situations of risk, i.e. which option they consider the most desirable. Discourse can influence the perception of legitimacy of options because the way in which options are presented influences their desirability, as psychologists Amos Tversky and Daniel Kahneman have shown. They analyze the effect of framing on decision problems[15] and find that

> Individuals who face a decision problem and have a definite preference (i) might have a different preference in a different framing of the same problem, (ii) are normally unaware of alternative frames and of their potential effects on the relative attractiveness of options […]" (Tversky and Kahneman 1981: 457).[16]

As discourse is here considered to be comprised of a number of elements, frames among them, it is assumed that discourse is relevant for the way subjects perceive the legitimacy of a policy. Thus, this theoretical strand focuses on what happens inside an individual's head when it changes its preferences.

Normative change research,[17] on the other hand, aims at explaining what happens on the international level when political change occurs. In a seminal paper[18] from 1998, Martha Finnemore and Kathryn Sikkink describe the 'life cycle' of norm influence. It is divided in three stages named norm emergence, norm cascade and internalization. In the first stage, "norm entrepreneurs attempt to convince a critical mass of states (norm leaders) to embrace new norms" (Finnemore and Sikkink 1998: 895). Thus, norm entrepreneurs play a central role: "Norm entrepreneurs are critical for norm emergence because they call attention to issues or even 'create' issues by using language that names, interprets, and dramatizes them" (Finnemore and Sikkink 1998: 897). When this has happened successfully, norm leaders try to convince other states to follow the norm: the norm cascades. "At the far end of the norm cascade, norm internalization occurs" (Finnemore and Sikkink 1998: 895). This means that the validity and legitimacy of the norm is taken for granted. However, not every norm succeeds to be internalized. For the role of discourse for the legitimacy of CE

[15]A decision problem is defined, according to the authors, "by the acts or options among which one must choose, the possible outcomes or consequences of these acts, and the contingencies or conditional probabilities that relate outcomes to acts" (Tversky and Kahneman 1981: 453).

[16]This analysis stands in the context of prospect theory, developed by Tversky and Kahneman (1979, 1992) to explain the phenomenon that empirical studies have shown that people do not always pursue their expected utility in situations of risk, as expected utility theory would predict. Rather, the choice depends on whether the outcome of the decision is considered as a relative gain or loss instead of just an absolute change in wealth. "More than 30 years later, prospect theory is still widely viewed as the best available description of how people evaluate risk in experimental settings" (Barberis 2013: 173). Criticism of the theory targets, inter alia, the assumption that decision-making under risk can "be defined accurately by rigid quantitative models" (Nwogugu2006: 451) such as prospect theory.

[17]Overviews over the very rich literature on normative change can be found in Engelkamp and Glaab (2015: 201).

[18]For an evaluation of the text and its relevance for the field, see Engelkamp and Glaab (2015, in particular p. 202 and 203).

approaches, the stage of norm emergence is most interesting, as the case is only just beginning to develop. The stages of norm cascade and norm internalization, should they ever be reached, lie in the future. Thus, norm change research presents an alternative toolkit—to the theoretical approach applied here, combining prospect theory with practice theory—for assessing whether a specific conceptualization of CE approaches is dominant in the academic discussion.

There are various reasons why norm change research approaches are not useful for this study: First, academic actors active in CE research cannot be conceptualized as norm entrepreneurs of a norm such as the one presented here: Very few, if any, researchers in this community openly advocate such as a norm. Further, CE approaches are discussed in very small communities to date, and no domestic support for a norm such as the one formulated above can, so far, be ascertained. Due to these reasons, it is difficult to claim that even the early stage of norm emergence has been reached when it comes to CE approaches. Thus, a norm change analysis might make sense in the future, but does not do so at this point in time. To sum up, through this assumption I claim that the discursive representation of CE approaches through frames influences their legitimacy and, through this, might impact decisions that will be made in the future concerning these approaches.

3.2.5 The Academic Discussion Sets the Stage for the Construction of the Meaning of Climate Engineering

I claim here that the academic discussion (meaning the discussion among and between researchers from natural as well as social sciences, engineering sciences, humanities or other academic fields) is particularly important in constructing the legitimacy, as well as the broader meaning, of CE approaches, and through this, potentially influencing decision-making concerning these approaches. I do so as CE approaches do, so far, mostly not exist materially, i.e., they are not yet "physically-tangible technological object[s]" (Asayama et al. 2017: 82). They exist, to a large degree, only in the minds of people and in the calculations for climate models. This means that a large array of decisions concerning their research, development and use remain to be made. These decisions will be influenced, as claimed above, by the legitimacy of these approaches. This legitimacy, as the meaning of these approaches more broadly, will be shaped to some degree by the academic discussion of CE, as these approaches are, to date, little discussed by decision-makers (Horton et al. 2016).

Succinctly, Huttunen and Hildén state: "Due to the novelty, the uncertainties, and the required scale of any real-world application of geoengineering, researchers play a crucial role in creating the basis for the decisions that determine the future of geoengineering technologies" (Huttunen and Hildén 2014: 5). Similarly, for the case of NETs, Beck and Mahony claim that the IPCC has "'world-making' power, potentially shaping the world to their own image and creating new political realities"

(Beck and Mahony 2018: 1). Gupta and Möller go even further, arguing that "sources of de facto governance, such as authoritative assessments, have normalized and institutionalized CE research, thereby contributing to ordering an emerging, contested field of inquiry characterized by multiple uncertainties and unknowns" (Gupta and Möller 2018: 497). Low argues that "[r]esearchers from across the natural, applied, and social sciences—at least for now—play constitutive roles in framing issues and setting agendas at the boundary spaces between science, society and policy" (Low 2017: 69).

An Idea From Academia

The idea of CE was arguably born in research circles and was introduced to public and political circles only as a second step:

> […] geoengineering emerged in the 1980s as a scientific/technological silver bullet for climate change which was quickly marginalized within the scientific community because of issues, among others, of cost, feasibility, uncertainty, and risk, as well as philosophical objections to the attempt to play 'God.' Despite the fact that little has changed to ameliorate these concerns, a substantial portion of the scientific community has recently embraced the idea of geoengineering and introduced it into both public and policy discourse. (Bowden 2010: 70)

This means that the very first construction of the idea of CE was done in academia. Even today, academic actors are consistently the most vocal group throughout the discourse on CE (Buck 2013: 173; Matzner and Barben 2018). But the influence of academia is not simply based on sheer volume. The EC approach in global governance study recognized in the 1990s the influence of academic and other experts for policy-making in situations of uncertainty.

Arguably, science and scientists, or academic experts, were introduced into political study traditionally as a provider of scientific truths, a repository of the resource of knowledge for decision-makers to tap at need (see Beck 2015: 287). In this vein, Litfin argues that "[t]he authority of science rests on its claim to objective, disinterested and verifiable knowledge, where political authority rests on the consent of the governed and is exercised on the basis of power" (Litfin 2000). This notion is not rarely re-iterated by academic actors themselves, in an attempt to differentiate their own domain of action from 'messy' political conflicts of interest and value judgements (Hoppe 2005: 199). This kind of 'boundary work' (Gieryn 1983) is, according to Jasanoff, done both by academic and by political actors to "enlarge [their] own control over social decisions" (Jasanoff 1987: 199). Hoppe argues that STS, for example through the notion of 'boundary work' has helped "puncture" this "myth": According to him,

> STS brought science from Olympian heights of abstraction, noble curiosity-driven speculation, innovative but stringent experiment, and Humboldtian institutional autonomy, down to earth as a social movement […] driven by local, practical, and sometimes openly political interests, portraying it as entrepreneurial, fiercely competitive, wildly speculative and methodologically pluralist […]. (Hoppe 2005: 205)

I claim that this assessment overstates the role of individual interest of researchers in their personal gain. Similarly, I claim that not all boundary work is done in order to increase one's own control over social decisions. However, this assessment does point out the fact that, instead of being a coherent whole producing a 'final truth', science is made up of various actors who contest each other – as well as each other's data selection, methods, assumptions and conclusions (Jasanoff 2004a: 3; see also Jasanoff 2015: 37; Miller 2001: 481).

This observation can be made in the discussion on CE in particular, which is characterized by an outstanding degree of polarity in some respects: "Most peer-reviewed climate engineering literature [...] emphasizes either the potential dangers of climate engineering or the climate change consequences of refraining from considering the research" (Linnér and Wibeck 2015: 255). This polarity, according to the authors, is "not prevalent in the literature on earlier emerging technologies" (Linnér and Wibeck 2015: 255). This observation is what makes it interesting to study the academic discussion on CE: The ways in which various frames of CE are presented, contested and possibly dominant can, when this discussion is picked up in political circles, influence the legitimacy of these or some of these approaches. I do not go as far as Jasanoff or Miller, discussing the construction of truth-claims in itself, but rather I discuss the construction of the legitimacy or illegitimacy of CE technology on the basis on these truth claims, arguing that scientists do not only communicate knowledge, but also frame it in specific ways, which differ in their effect on the political decision-making taking place on the basis of these communications.

An Epistemic Community of Climate Engineering?

The EC approach from global governance research—"located at the intersection of actor-oriented social constructivist and neoliberal IR theories" (Beck 2015: 286)—offers a tool to systematize the influence of academic experts in global governance. Further, it is useful to assess in which way the approach can be supplemented through STS insights. The EC approach aims at explaining how policies—i.e. the content of specific governance measures—develop under conditions of uncertainty, i.e. when it is difficult for politicians to foresee the consequences of various policy options:

> These forms of uncertainty give rise to demands for particular sorts of information. The information [...] consists of depictions of social and physical processes, their interrelation with other processes, and the likely consequences of actions that require application of considerable scientific or technical expertise. (Haas 1992: 3–4)

Already in the original conceptualization of the EC approach, it is noted that this information does not consist, however, solely in the delivery of raw data: Rather, "it is the product of human interpretation of social and physical phenomena" (Haas 1992: 4), i.e. "epistemic communities are able to frame social reality" (Cross 2013: 149). This means that epistemic communities do not replace decision-making by political actors, but rather frame the issue at hand *before* decision-making takes place:

> By identifying the nature of the issue-area and framing the context in which new data and ideas are interpreted, epistemic communities bound the range of collective discourse on

> policy, as well as guide decision makers in the choice of appropriate norms and appropriate institutions within which to solve or manage problems. (Adler and Haas 1992: 375)

This framing through academic actors takes place, in particular, in the context of environmental phenomena such as climate change, as these are fraught with considerable uncertainty:

> [...] decision makers are seldom certain of the complex interplay of components of the ecosystem and are therefore unable to anticipate the long-term consequences of measures design to address one of the main environmental issues under current consideration. (Haas 1992: 13)

The effect of framing is, according to framing theory, expected to be particularly large when the frames are communicated by 'credible sources' such as academic experts: "[...] frames delivered by credible sources are more likely to shift opinions" (Chong and Druckman 2007a: 112). Various empirical studies have identified a particularly strong influence of expert framing in cases of decisions concerning technology (see, for example, Donk et al. 2012: 7; Lee et al. 2005: 254). The framing of and through science is the focus of a number of publications in frame analysis (see, for example, Nisbet 2010a; Nisbet and Mooney 2007; Gamson and Modigliani 1989; Nisbet et al. 2003). It must be noted that academia follows very strict rules of objectivity and reliability in public communication, e.g. in academic publications. Thus, strategic framing—the attempt to define the problem in a specific way to achieve a certain strategic goal (see Lakoff 2014)—can be expected to be rare. Framing in a broader sense, however, cannot be completely prevented, as otherwise, the object of interest cannot be clearly identified and delineated. This is all the more applicable in academic discussion on CE, where participants commonly make arguments to justify their work on such a contested topic (see, for example, Crutzen 2006: 212). The relevance of epistemic community influence on CE decision-making has been noted before:

> Due to the novelty, the uncertainties, and the required scale of any real-world application of geoengineering, researchers play a crucial role in creating the basis for the decisions that determine the future of geoengineering technologies. (Huttunen and Hildén 2014: 5; see also Horton 2015: 2)[19]

The considerations above point to a particular relevance of epistemic communities in decisions concerning the issue of climate change. Epistemic communities, as the providers of this kind of knowledge under conditions of uncertainty (Haas 1992: 4), are defined through four characteristics that a group of actors has to fulfill in order to be called an epistemic community:

> (1) a shared set of normative and principled beliefs, which provide a value-based rationale for the social action of community members; (2) shared causal beliefs, which are derived from their analysis of practices leading to or contributing to a central set of problems in their domain and which then serve as the basis for elucidating the multiple linkages between possible

[19] As discussed before, I do not assume that the academic discussion is the only factor determining CE decision-making. However, it is a very important one, and thus it can be expected that, if a dominant frame can be found in academic discussion, it will be this very frame which is taken into the governance discussion.

> policy actions and desired outcomes; (3) shared notions of validity – that is, intersubjective, internally defined criteria for weighing and validating knowledge in the domain of their enterprise; and (4) a common policy enterprise – that is, a set of common practices associated with a set of problems to which their professional competence is directed, presumably out of the conviction that human welfare will be enhanced as a consequence. (Haas 1992: 3)

The researchers involved in the academic discussion of CE can be described as fulfilling criteria (2) and (3): Concerning (2), they all subscribe to the causal belief that climate change is caused through certain chemicals, as well as to the causal belief that certain technological approaches subsumed under the term 'CE' might be able to alleviate some of the effects of climate change. Criterion (3) is fulfilled as the discussion takes place in academic journals, which are subject to the criterion of peer-review for weighing contributions to the discussion. Aside from this, norms of reliability and replicability are relevant in all areas of scientific research. What, however, is meant by the shared set of normative and principled beliefs? I understand it as the agreement with a certain normative view. In this understanding, criterion (1) as well is fulfilled, as the various voices in the CE discussion are united in their rationale that climate change is a challenge that should be addressed (see Kreuter 2015b). This criterion takes into account that science is, as Miller put it, not devoid of norms, ideologies and beliefs. Criterion (4), however, poses a problem to this study: The participants in the academic discussion on CE are not, as discussed before, joined in a common policy enterprise. While there are some voices calling for the quick research and even development of CE approaches in order to be ready to act when it is deemed necessary (see Caldeira and Keith 2010), there are many who are very critical of these approaches (see Robock 2008). Further, most participants in the discussion on CE arguably do not even *intend* to influence decision-making: The boundary work that they engage in aims at differentiating between scientific research, which aims only at finding facts, from political decision-making, at which point the legitimacy or illegitimacy of CE approaches to address climate change is supposed to be decided upon.

I argue, however, that the insights from STS should be considered here and the EC approach modified accordingly. In this, I follow Cross who argues that "we must pay more attention to *internal* dynamics within an epistemic community" (Cross 2013: 138): First, that there is, indeed, much contestation in science, which is important for the scientific endeavor, and that making a homogeneous, shared policy project a condition for the study of epistemic communities restricts this study to a very small set of actors. Second, that Bijker's constructivist view of technology indicates that researchers influence the construction of the legitimacy of CE technology even if they do not intend to do so, by framing it inadvertently. Thus, strategic intent is not a necessary condition for the influence of scientific actors: "The way problems are framed by experts shapes our view of social needs and the kinds of political responses that count as scientifically 'sound' and politically necessary" (Beck 2015: 288). I argue that this is the case in particular under conditions of uncertainty on the part of political decision-makers. By taking social constructivism seriously in this way, the EC approach is opened up for the study of more loosely set groups of researchers. In this way, the EC approach makes sense at it considers researchers as a special actor in the political sphere, but needs to be modified to allow for a more loosely set of actors than originally considered.

This modification is useful as the EC approach can still help understand the role of academic experts in defining governance: "Epistemic communities are at the forefront of recognized trends towards transnational governance" (Cross 2013: 138). Specifically, with regards to this study, the approach offers a useful typology of processes of influence by academic actors which allows us to locate, in a hypothetical process of introducing CE as a governance measure into the socio-technical landscape, the point at which we are now and what role academic actors play at this point. Similarly to the policy cycle (Jann and Wegrich 2007), which maps out the different stages in the life cycle of various governance measures, Adler and Haas map out four steps of "policy evolution": "policy innovation, diffusion, selection, and persistence" (Adler and Haas 1992: 373). The influence of epistemic communities is considered particularly salient at the stage of policy innovation, as "epistemic communities play an evolutionary role as a source of policy innovations" (Adler and Haas 1992: 374). As such, the discussion on CE in academic circles right now can be the source of a new policy towards climate change, that of intentionally modifying climate on a large scale, instead of the dominant approach to date, i.e. the reduction of greenhouse gas emissions. By discussing CE approaches, the epistemic community can "alter perceptions and frame the context for collective responses to international problems" (Adler and Haas 1992: 376), in this case legitimize or delegitimize CE as a response to climate change. Research in the area of the influence of researchers on CE understanding is comparatively limited (Bellamy 2013) but there are some insights (Belter and Seidel 2013; Oldham et al. 2014; Bellamy et al. 2012a; Huttunen and Hildén 2014: 21; Linnér and Wibeck 2015). Researchers have usually followed an inductive approach, exploring what is there in the discourse on CE. The study presented here differs from these earlier studies on several counts: Firstly, and most importantly, the research of this project is guided by political theory on legitimizing talk, and thus by a deductively developed set of categories which allow the differentiation of legitimizing and delegitimizing frames. Secondly, for this study, both text and meta data were assessed.

The assumptions presented above, taken together, provide justification of the central ontological claim of this thesis, namely that *the legitimacy of CE approaches, for which the early academic discussion sets the stage, can influence future decision-making concerning these approaches.* In order to systematically assess the academic discussion on CE with regards to this research question, a theoretical construct was designed which allows the researcher to identify specific discursive processes which can influence the meaning of CE and subsequently decisions made concerning these approaches. This construct will be elaborated in the following section. The content of the theoretical construct—eight ideal-type frames, four legitimizing and four delegitimizing CE—is deductively developed from strands of social constructivist IR theory and informed by the findings of state-of-the-art empirical studies, both on CE and on other emerging technologies.

3.3 Analyzing Frames of Security, Complexity, Economy and Appropriateness

In order to be able to answer the research question, two theoretical steps remain to be taken: firstly, we must define in more detail the concept of 'frame' chosen to operationalize discursive processes in order to render them identifiable in the empirical analysis. In other words, to be able to decide when a specific discursive process is in evidence and when it is not. Secondly, we must deductively flesh out the different types of discursive processes that could, on the basis of IR theory, influence decision-making on CE by legitimizing or delegitimizing these approaches. In the following section, frame theory is used to operationalize discursive processes. First, frame theory is described in general. Then, the relevant mechanism of framing effects on decision-making is outlined. In the next section, the types of frames which can have a legitimizing or delegitimizing effect are presented. In the final section of this chapter, the theoretical construct developed on the basis of these considerations is laid out.

3.3.1 Identifying Discursive Processes Through Frame Analysis—The Structure of the Theoretical Construct

Frame theory was used for this study to operationalize discursive processes which can legitimize or delegitimize CE approaches. This means that, in order to be able to find out whether one of the legitimizing and delegitimizing processes is present in the empirical material, ideal-type frames as indicators for each of these processes were defined. Then, the empirical material was analyzed in order to find out whether and which ideal-type frames are present. The focus lies on frames—as opposed to, for example, metaphors, storylines, arguments or other elements of discourse (see Uther 2014: 68). Following Markusson (2013: 6), frames are understood to be part of discourse and present micro-level objects of discourse, which itself is understood as a macro-level phenomenon. 'Framing' is, according to Rein and Schön (1993: 146) "a way of selecting, organizing, interpreting, and making sense of a complex reality to provide guideposts for knowing, analyzing, persuading, and acting." This means that framing is one way to give "meaning [to] social and physical phenomena" (Hajer 2006: 67). Individual texts present frames, and these frames, together with other discursive elements and practices, make up the discourse.

Frame theory was chosen over other tools for several reasons: Firstly, frame theory, as conceptualized by Entman, is dedicated to constructivist ontology, thus conforming with the general theoretical perspective espoused in this study (see Edelman 1993: 232; Entman 1993: 54). Secondly, frame theory decidedly aims at providing accounts of how social action—such as decision-making—can be influenced by elements of discourse. Through framing, actors make a portion of physical or social reality accessible to themselves and to others and structure *perception* as well as *action*: "A frame is a perspective from which an amorphous, ill-defined, problematic situation can be made sense of and acted on" (Rein and Schön 1993: 146). In this process of framing, a certain perspective is assumed, which highlights certain aspects of reality and sidelines others:

> Framing essentially involves *selection* and *salience*. To frame is to *select some aspects of a perceived reality and make them more salient in a communicating text, in such a way as to promote a particular problem definition, causal interpretation, moral evaluation and/or treatment recommendation* for the item described. (Entman 1993: 52)

This does not only indicate that aspects of reality are highlighted or sidelined, but also how this is done, i.e. how a frame is structured: It defines a problem, isolates origins and causes, provides the basis for moral judgments and/or offers ways to approach and solve the problem. This means that, according to framing theory, depending on how a specific problem—like climate change—is framed, the preferred solution to the problem can change.

> An interpretive perspective […] investigates the 'problematization' […] of climate change: how global warming is rendered as an object and problem to be acted upon, how different meanings of climate change (e.g. as a global environmental threat, as an economic problem or as a security risk) emerge and become dominant, how the different meanings affect the toolkit of feasible political solutions, and what the effects of such practices are. (Methmann et al. 2013b: 4)

For example, if climate change is presented as a problem of inefficient politics, the solution of CE can seem reasonable: If political negotiation cannot help, maybe technology can. If, on the other hand, climate change is presented as a problem of human production and consumption patterns, CE does not seem like a reasonable solution. In this case, the option of socio-economic change would make more sense.[20] This means that framing can influence the perceived legitimacy of policy options as they "link causal accounts of policy problems to particular proposals for action and facilitate the normative leap from 'is' to 'ought'" (Rein and Schön 1993: 148). This mechanism indicates how frames can influence the legitimacy of governance measures: By presenting not only a specific problem, but also outlining a solution to this problem. I will go into further detail below when discussing the framing effect.

Framing as a Way of Creating Meaning

The concept of 'framing' originated in psychology. Sociologist Erwing Goffman moved the topic into social sciences in the 1970s (Goffman 1974), where it was picked up by communication analysis and later informed political study (Benford and Snow 2000: 611). I use an approach to frame analysis based on political communications scholar Robert Entman as well as social policy analyst Martin Rein and philosopher Donald Schön, who are inspired by sociologist Erwing Goffman. Additionally, the deliberations on the application of framing analysis for political science by Dennis Chong and James N. Druckman influence my approach. These authors

[20] For the influence of framing on decision-making, see Kahneman and Tversky (1984); Tversky and Kahneman (1981, 1974).

were chosen as their approaches are compatible with each other, providing consistent information on different levels of abstraction for the study of frames. Entman as well as Rein and Schön address how frames work and where they can be found, making arguments for their relevance in political study. Chong and Druckman elaborate on those considerations by describing the process of framing in detail, outlining the effect of frames on decision-making and creating an approach for the analysis of this effect which can be operationalized for various empirical cases. Also, all these authors share in common that they allow for the political study, not only of framing by political elites such as elected representatives, but also through other elite actors, such as academic experts. First of all, however, the sociological basis of frame analysis by Goffman needs to be addressed.

Goffman sets the stage for the analysis of frames in social sciences in his seminal book *Frame Analysis*, in which he analyzes the "organization of experience" (Goffman 1974: 13) and finds that "[w]hen the individual in our Western society recognizes a particular event, he tends, whatever else he does, to imply in his response (and in effect employ) one or more frameworks or schemata of interpretation of a kind that can be called primary" (Goffman 1974: 21). Primary frameworks, according to Goffman, fulfill the task of "rendering what would otherwise be a meaningless aspect of the scene into something that is meaningful" (Goffman 1974: 21). For example, while a person running is not socially meaningful, if framed as a thief distancing themselves from the scene of the crime, the action is rendered meaningful. Meaning, accordingly, does not exist without these primary frames. These, in Goffman's understanding, can take various shapes, but always fulfill the same functions:

> Primary frameworks vary in degree of organization. Some are neatly presentable as a system of entities, postulates, and rules; others – indeed, most others – appear to have no apparent articulated shape, providing only a lore of understanding, an approach, a perspective. Whatever the degree of organization, however, each primary framework allows its user to locate, perceive, identify, and label a seemingly infinite number of concrete occurrences defined in its terms. (Goffman 1974: 21)

Thus, frames allow us to identify the meaning of events and actions. Goffman explicitly does not intend to make any claims about "the organization of society" (Goffman 1974: 13), i.e., the influence of frames on social contexts. However, he argues that the frameworks that social groups use to render events meaningful influence their shared world view (Goffman 1974: 27). This means, for a social constructivist, that frames influence how the social world is constituted.

Goffman seems to assume that there is one 'officially' true meaning of an action or event, which can then be distorted by various actors through diverse means, e.g. what he calls 'keying' and 'fabrication' (Goffman 1974: Chaps. 3 and 4). However, he does concede that the 'official' framework to understand and explain a phenomenon can change under circumstances of tension: "Certainly individuals exhibit considerable resistance to changing their framework of frameworks. A public stir—or at least a ripple—is caused by any event that apparently cannot be managed within the traditional cosmology" (Goffman 1974: 28–29). In such a case, the official frame can shift if a new frame can help better to understand the situation by fitting more easily with the empirical phenomena confronting the individual. For example, if I hear a knock on the door, I might understand it as a person requesting entry. However, when I open the door, finding nobody on the doorstep, I reframe the event, understanding that the knock came, instead, from a loose tree branch moving in the wind and brushing against the door (example from Goffman 1974).

Frames in Social Analysis

What is missing in Goffman's account—and deliberately so, as he does not intent to provide a social science methodology, but rather an account of how individual experience is organized—is an understanding of how frames are *defined* in a social context, *where* they occur in communication,[21] how they *evolve, change and shift* in this context, and how frames can *influence political action*, such as decision-making. Thus, the accounts of Entman as well as Rein and Schön are useful to guide my analysis. Their approaches are decidedly socially constructivist in nature: "The […] process of […] framing socially constructs the situation, defines what is problematic about it, and suggests what courses of action are appropriate to it" (Rein and Schön 1993: 153; see also 150). A frame, in this understanding, is *defined* as "[…] a perspective from which an amorphous, ill-defined, problematic situation can be made sense of and acted upon" (Rein and Schön 1993: 146). Similarly, Hoppe finds that "[t]o infuse a polyvalent world with meaning, sense and purpose, and to make action and judgement possible at all, people need […] frames as a sort of mental grappling hook" (Hoppe 1999: 207). This means that some framing always takes place, as otherwise reality is not accessible to understanding. The activity of framing, then, is "[…] a way of selecting, organizing, interpreting, and making sense of a complex reality to provide guideposts for knowing, analyzing, persuading, and acting" (Rein and Schön 1993: 146). This takes place mainly through communication in various ways.

Entman identifies four *locations* in which such frames occur: "[…] the communicator, the text, the receiver, and the culture" (Entman 1993: 52). These locations differ because communicators and receivers, as conscious actors, can influence and shape the frames that they use, while the text can be understood as the 'carrier' or medium of frames: "*Communicators* make conscious or unconscious framing judgements in deciding what to say, guided by frames (often called schemata) that organize their belief systems" (Entman 1993: 52). While communicators use frames in order to render intelligible the information that they want to convey, the decision over which frame to use can be conscious and strategic, but it can also be subconscious, meaning that communicators themselves are influenced by their world view. Communicators, thus, are not the final and last source of frames. Communicators of interest in the study outlined here are natural and social scientists, as well as scholars from humanities analyzing CE. Receivers are influenced by the communicator's frame, but not solely by this: "The frames that guide the *receiver's* thinking and conclusion may

[21]Goffman studies 'frames in thought', i.e. "an individual's (cognitive) understanding of a given situation" (Druckman 2001: 227–228). What is of interest here, however, is the academic discussion on CE, and, accordingly, 'frames in communication': "The frame that the speaker chooses may reveal what the speaker sees as relevant to the topic at hand" (Druckman 2001: 227).

or may not reflect the frames in the text and the framing intention of the communicator" (Entman 1993: 52–53). Thus, the receiver's conception of the issue can also be impacted by other factors, e.g. frames picked up in other communications by other communicators. Decision-makers could, for example, be influenced not only by academic publications, but also media reports, public opinion analyses or NGO communications. As discussed earlier, here it is assumed that the framing done by experts is a highly important factor in shaping the perception—and subsequent decision-making—in the context of CE.

While communicators and receivers are important actors in the process of framing, the text is the medium conveying the frames: "The *text* contains frames, which are manifested by the presence or absence of certain keywords, stock phrases, stereotyped images, sources of information, or sentences that provide thematically reinforcing clusters of facts or judgements" (Entman 1993: 52). The texts, for the study at hand, are academic journal publications from natural and social sciences as well as humanities. Content analysis tools can be used to study frames in texts, as these can make visible which elements are present and which are absent in a specific text. The entirety of frames is, by Entman, called the 'culture': "The *culture* is the stock of commonly invoked frames; in fact, culture might be defined as the empirically demonstrable set of common frames exhibited in the discourse and thinking of most people in a social grouping" (Entman 1993: 53).

As neither communicators nor receivers nor culture can be understood as the final source of frames, it is important to understand how frames evolve, how they change and shift. The basis of frames is, according to Rein and Schön, the so-called 'appreciative system':

> The social philosopher Geoffrey Vickers coined the term *appreciative system* to refer to the systems of values, preferences, norms, and ideas on the basis of which we frame the grand actuality of the world and thereby make it coherent to ourselves. (Rein and Schön 1993: 146–147)

The appreciative system, then, is the totality of mental notions which are used to make sense of the material world. Rein and Schön argue that policy-relevant frames are constructed on one or more of three levels in the social world: "personal life, scientific or scholarly inquiry, and policy-making. Moreover, unless one wants to assume a world of policy uninhabited by people and lacking in knowledge, these three levels of framing must be related to each other" (Rein and Schön 1993: 146). The role of scientists and scholars for policy-relevant framing becomes apparent here. The question of emergence of frames is further assessed in Entman (2004), Chong and Druckman (2007a: 117), Gamson (2005).

The Influence of Frames on Decision-Making: The Framing Effect

The most interesting aspect of frames for this study is their influence on decision-making concerning CE. The functions of frames, which are already visible in the definition of 'frame', indicate this influence by outlining what frames do in social communication:

> Frames "*define problems* – determine what a causal agent is doing with what costs and benefits, usually measured in terms of common cultural values; [...] *diagnose causes* – identify the forces creating the problem; [...] *make moral judgements* – evaluate causal agents and their effects; [...] and *suggest remedies* – offer and justify treatments for the problems and predict their likely effects." (Entman 1993: 52)

This means that frames are understood not only as a structuring mechanism for individuals to understand reality; but also as influential in the perceptions of actors concerning the causes of phenomena as well as how they should be dealt with best—the governance approach that should be chosen to address the respective problem.[22] In this way, frames connect the descriptive and the normative level (Rein and Schön 1993: 148), which enables them to influence the legitimacy of individual courses of action as well as influence the decision on a specific course.

The social science notion of the influence of framing on beliefs and actions—known as 'framing effect' (Chong and Druckman 2007a: 104; Druckman 2011: 282)—is based on the work of psychologists Amos Tversky and Daniel Kahneman,[23] who found, in empirical studies, that "[i]ndividuals who face a decision problem and have a definite preference [...] might have a different preference in a different framing of the same problem [...]" (Tversky and Kahneman 1981: 457; see also Kahneman and Tversky 1984). Most interesting to this study is the influence of frames not on individual decisions, but on political choices involving "individuals and groups, with conflicting goals, reaching collective agreements about how to allocate scarce resources" (Druckman 2011: 288). The relevance of frames in the context of such political choices has been observed in a number of studies (Benford and Snow 2000; Chong and Druckman 2007a, b; Druckman 2011; Jachtenfuchs 1996a; Levy 2003; Nelson et al. 1997; Rein and Schön 1993; Schlipphak 2011; Sniderman and Theriault 2004; Gamson and Modigliani1989). For example, Rein and Schön stress the relevance of frames in policy decisions:

> Policy frames and their underlying appreciative systems are revealed through the stories participants are disposed to tell about policy situations. The problem-setting stories, frequently based on generative metaphors, link causal accounts of policy problems to particular proposals for action and facilitate the normative leap from 'is' to 'ought'. (Rein and Schön 1993: 148)

Similarly, Entman finds that framing in this understanding "[...] has important implications for political communication. Frames call attention to some aspects of reality while obscuring other elements, which might lead audiences to have different reactions" (Entman 1993: 55).

---

[22] Accordingly, the study of framing effects in political science is particularly relevant for decision-making theory, as the empirical evidence of these effects challenge the assumption of rational choice (see, for example, Druckman 2004: 673), e.g. in democratic elections.

[23] As Entman elaborates: Kahneman and Tversky (1984) offer perhaps the most widely cited recent example of the power of framing and the way it operates by selecting and highlighting some features of reality while omitting others (Entman 1993: 53). Similarly, Druckman argues that "[t]he obvious starting point to build a theory of framing effects is Tversky and Kahneman's [...] prospect theory." (Druckman 2004: 674)

The process behind framing effects differs for two different types of frames: valence or *equivalency frames*, and emphasis or *issue frames* (Chong and Druckman 2007a: 114). An equivalency "effect occurs when different, but *logically equivalent*, phrases cause individuals to alter their preferences [...]. This typically involves casting the same information in either a positive or negative light" (Druckman 2004: 671). For example, a program to counter the spread of a lethal disease can be framed in terms of the lives that will be saved or in terms of lives that will be lost if the program is adopted. Both frames can present logically equivalent phrases, which, however, can cause individuals to alter their attitude with regard to the program (Tversky and Kahneman 1981: 453). In contrast, "issue frames focus on qualitatively different yet potentially relevant considerations"(Druckman 2004: 672). This means that these frames highlight different aspects of the issue. An example is the framing of climate change as a security problem on the one hand and as an economic challenge on the other: Both frames focus on different aspects of the phenomenon of climate change. As Sniderman and Theriault point out, "the concept of framing, for the study of political choices, typically refers to characterizations of a course of action in terms of an alternative 'central organizing idea' [...]" (Sniderman and Theriault 2004: 136).

This is important as the way the frame works and the process of the framing effect differ between issue and equivalency frames, and accordingly, issue frames can be expected to have an effect in other circumstances than equivalency frames. Druckman finds that counter-framing (i.e. the competition of several alternative frames) as well as heterogeneous discussion (i.e. discussion of several participants briefed with differing frames) reduces the effect of equivalency framing. Even homogenous discussion can—among experts—reduce and eliminate these effects (Druckman 2004: 682). Issue framing effects weaken, according to Druckman and Nelson (2003: 737), similarly under conditions of frame competition and in heterogeneous discussion. However, issue framing effects are not reduced in homogeneous discussions (Druckman and Nelson 2003: 738). Sniderman and Theriault (2004: 144), in a study on public opinion, similarly find that framing effects are "dramatic" when the issue is framed in one specific, clearly dominant way. This means that frames in academic publications can be expected to be particularly influential if they are found to be dominant in the text and no or little counter-framing takes place. The empirical study of this thesis assesses whether there is a dominant issue frame in the academic discussion of CE.

Chong and Druckman, who build on Entman's work and make it more accessible for political science analysis, identify a number of variables influencing the success of framing (Chong and Druckman 2007a: 111). They include characteristics of the communicator, of the receiver and of the text itself. Concerning the communicator, Chong and Druckman argue that frames from credible sources—like scientists—are more influential than others: "[...] frames delivered by credible sources are more likely to shift opinions [...]" (Chong and Druckman 2007a: 112). Concerning the receiver, the predisposition, knowledge and individual motivation to evaluate competing frames are relevant variables. "Perhaps the clearest limit on framing effects is provided by individual predispositions such as values [...]. [...] In general, strong predispositions reduce framing effects by increasing one's resistance to disconfirming information" (Chong and Druckman 2007a: 111). However, the knowledge of the receiver before receiving the frame is important also, albeit being less clear in terms of the direction of influence (see also Druckman and Nelson 2003: 732):

> Studies of another individual-level moderator – knowledge – have produced conflicting results. […] After controlling for prior attitudes, knowledge enhances framing effects because it increases the likelihood that the considerations emphasized will be available or comprehensible to an individual. (Chong and Druckman 2007a: 112)

Additionally, the motivation of the receiver to weigh conflicting frames against each other is of importance. If the receiver is not motivated to do so, he or she will just stick to his/her preconceived notions (Chong and Druckman 2007a: 110). In the text, the absolute and relative frequencies of frames impact their effect: If a frame is repeated constantly, it is expected to be more influential on the attitude of the receiver as its accessibility and availability increases (see Chong and Druckman 2007a: 111). This is the absolute frequency. If a frame is presented in a competitive environment, for example if it is not the only, dominant frame present, its influence on the recipient decreases (Chong and Druckman 2007a: 111). The competitive environment refers to the way in which there is competition between alterative frames or if there is one single dominant frame.

While Chong and Druckman analyze framing "within the broader democratic process that links politicians and other opinion leaders to the public" (Chong and Druckman 2007a: 104), my focus is on framing in the process that conveys meaning from academic publication to decision-makers. However, I assume that Chong and Druckman's elaborations are valid also for my study: In both cases, communicators are "elite actors" (Chong and Druckman 2007a: 117) with particular credibility. In both cases, the framing influences action, and this influence depends on various variables. For the analysis at hand, factors influencing frame acceptance located with the text and the communicator are empirically assessed, as I intend to find out whether academic discussion legitimizes or delegitimizes CE through framing. Thus, I assess frequencies and competition of frames and frame elements in the text as well as the name and disciplinary focus of the communicator. The influence on the receiver is assumed to exist, as indicated above. Also, I accept Chong and Druckman's assumption that the factors relevant with the receiver (namely, predisposition, knowledge and motivation) influence frame acceptance (Chong and Druckman 2007a), not testing them further, as the acceptance of frames by the audience is not part of the empirical study presented here. I assume, however, that there do exist predispositions in the receiver—as there are preconceived values on technology or climate—but that there are only few dispositions specifically concerning CE, as these approaches are little known outside of academic circles.

Thus, what is interesting for this study are *frames in communication* and not *frames in thought*. The former refers to "the words, images, phrases, and presentation styles that a speaker […] uses when relaying information about an issue or event to an audience" (Chong and Druckman 2007b: 100). This is what researchers do when they publish articles: They relay information about an issue to an audience. Frames in thought refer to "an individual's cognitive understanding of a given situation" (Chong and Druckman 2007b: 101) and are influenced by these frames in communication in processes of framing effects (Druckman 2001: 228). If one specific frame can be found to be highly frequent and dominant in the academic discussion, it can be expected that this frame has an effect on the receiver—in this case, the decision-maker. The relevance of frame dominance and competition develops from the idea that different frames, in a socially constructivist view, create different social realities:

> Framing is problematic because it leads to different views of the world and creates multiple social realities. Interest groups and policy constituencies, scholars working in different disciplines, and individuals in different contexts of everyday life have different frames that lead them to see different things, make different interpretations of the way things are, and support different courses of action concerning what is to be done, by whom, and how to do it. (Rein and Schön 1993: 147)

Frames can differ and compete, both between and within professions and disciplines. For example, in the academic discussion on CE, frames describing climate change as a security problem compete with frames describing the phenomenon as an economic challenge. This is what Rein and Schön call 'policy controversy'. They differentiate it from another type of policy discord with regards to framing which they call 'policy disagreement':

> […] policy *disagreements* arise within a common frame and can be settled in principle by appeal to established rules. But policy *controversies* cannot be settled by recourse to facts alone, or indeed by recourse to evidence of any kind. Because they derive from conflicting frames, the same body of evidence can be used to support quite different policy positions. (Rein and Schön 1993: 148)

This means, conversely, that 'controversies', in this understanding, are not issue of politics alone, but also of academic debate: As recourse to empirical facts alone cannot resolve the controversy, researchers, working with the same empirical data, can defend various positions without contradicting themselves, depending on the frame which they subscribe to.

In such a case of controversy, based on competing frames, it is particularly interesting which frame turns out to be dominant, as this frame defines the situation for all participants:

> Frame competition is the struggle between differing problem definitions, the latter being the basis for the emergence of interests and preferences. In a political system, the struggle among competing frames is the struggle for power, the power to define a situation authoritatively for all participants in the system and thus pre-structure the way interests can be articulated, claims made and policy decisions be taken. (Jachtenfuchs 1996a: 29; see also Morth 2000: 174)

It is important to note that framing is assumed to exist in any kind of human communication and not necessarily be used in a strategic way: I accept that objectivity is the ideal of any academic publication. Thus, strategic framing can be assumed to be much less prevalent here than, say, in political communication. However, even in this highly standardized and objectified type of communication, framing does take place, as decisions are made by the researchers concerning the issue that they want to analyze, the indicators that they want to observe, the metrics that they intend to apply etc. All of these fulfill the function of defining the problem and its causes and thus, frame the issue. In short, I claim that, in communication, any actor frames (see Entman 2010: 223). Thus, I do not follow the understanding of 'frame' common to studies of political communication, in which it is considered an intentional strategy adopted by actors to achieve their goals (see, for example, Lakoff 2014), but rather understand framing as a common practice in any kind of communication to select and convey in communication a specific portion of the real world to make statements about. Frames, then, are "general organizing devices" (Nisbet 2010a: 45) in communication.

Even beyond that, in the academic debate on CE, recommendations are made concerning the right way to approach the issue, as researchers justify their interest in a topic as controversial as this, by arguing that it has become necessary to analyze these tools in order to combat climate change (see, for example, Crutzen 2006: 221). Entman makes a similar observation concerning framing in the media: "Journalists may follow the rules of 'objective' reporting and yet convey a dominant framing of the news text that prevents most audience members from making a balanced assessment of a situation" (Entman 1993: 57). While academic standards are certainly more restrictive than those imposed by the standards of objective reporting, framing is not made impossible, because academic publications cannot live without specific decisions which frame the object of analysis and the methods of analyzing it. As Huttunen and Hildén point out: "it is clear that neither the problems nor their solutions are purely technical in the sense that research could provide unambiguous advice on optimal solutions" (Huttunen and Hildén 2014: 4). Similarly, Nisbet describes a change of understanding science communication, in which "one can detect a growing recognition that communication is not simply a translation of facts, it is a negotiation of meaning" (Nisbet 2010a: 41). He goes so far as to claim that "[s]cience debates are little different from political controversies" (Nisbet 2010a: 42). I do not agree with this claim. Instead, I assume that objective facts are the focus of science debates, but these cannot be conveyed without using frames to some degree, which can change meanings.

Dominance of Frames

As frequency and dominance of frames are to be assessed in this study, it remains to be decided how to identify dominant frames. Entman provides instructions:

> From a framing perspective, dominant meaning consists of the problem, causal, evaluative, and treatment interpretations with the highest probability of being noticed, processed, and accepted by the most people. To identify a meaning as dominant or preferred is to suggest a particular framing of the situation that is most heavily supported by the text and is congruent with the most common audience schemata. (Entman 1993: 56)

This implies that there are two dimensions of the dominance of a frame: First, the internal consistency of the text in presenting and supporting this particular frame, as opposed to various, equally important frames; and second, the external acceptance of the frame by the audience, which depends on the congruence of the frame with the world view prevalent in the audience. The audience, in the case at hand, consists of two parts: firstly, the community of social and natural scientists, which accept specific frames and incorporate them into their own writing; secondly, decision-makers relevant for CE governance, which accept frames and incorporate them into policy decisions. Specifically, a frame might be considered dominant if it is supported consistently and without competition in academic publications, and if it can be found to have made its way into political decisions concerning CE governance and/or climate policy. Accordingly, I aim to find out whether there are frames in the academic discussion which become dominant over time, or if the discussion becomes more competitive with time. As the project presented here studies the academic discussion on CE, assuming its influence on decision-makers as a given, the internal consistency has been analyzed, but not the external acceptance of a frame.

Entman puts the influence of dominant frames into perspective by arguing that "[c]ertainly people can recall their own facts, forge linkages not made explicitly in the text, or retrieve from memory a causal explanation or cure that is completely absent from the text" (Entman 1993: 56). This is definitely true. However, I agree with Entman that this does not prove that the idea that frames affect decisions is wrong, as these linkages and earlier attitudes are, themselves, products of earlier framing:

> [...] [I]f the old attitude resistant to the new framing treatment was itself (as is quite likely) influenced by framing effects at earlier times, the resistance provides scant logical basis for empirical or normative generalizations about citizen susceptibility to framing. (Entman 2010: 332)

Moreover, for new topics, such as CE, the influence of dominant frames rises with the complexity of the issue at hand: "[...] on most matters of social or political interest, people are not generally so well-informed and cognitively active, and [...] framing therefore heavily influences their responses to communications [...]" (Entman 1993: 56). As Sniderman and Theriault (2004: 142) point out, for issues that are uncommonly obscure, the observation that there is a framing effect is "too easy to demonstrate". As climate change and CE are highly complex, science-driven matters which make their understanding difficult, the influence of frames in academic debate can be expected to be particularly strong in this case (Corner and Pidgeon 2015: 425).

To sum up, the present study assesses the academic discussion on CE to find out whether and which dominant frames exist. Thus, the discussion between academics as 'credible' sources of frames was chosen. This is feasible under the assumption that, while strategic framing can be expected to be rare in academic discussion, frames as necessary means to communicate about material things exist in all kinds of communication, including academic communication with its strict rules of standardization and objectivity. Concerning the communicator, meta data was collected on the author of the texts, the disciplinary focus of the author and the year of publication. The focus of analysis lies on frame frequency and frame dominance or competition. Before I turn to the empirical analysis, however, one last theoretical question needs to be addressed:

Which discursive processes, to be identified in the empirical analysis through ideal-type frames, can legitimize or delegitimize CE as an approach to address climate change?

3.3.2 Mapping Out the Theoretical Construct—Eight Ideal-Type Frames of Climate Engineering

The eight ideal-type discursive processes making up the theoretical construct were derived from various strands of social constructivist IR theory and informed by empirical frame analyses on CE and other emerging technologies. The theoretical strands of IR theory have been selected on the basis of (1) their providing a social constructivist account of discursive processes which might influence political decision-making with a potential to depoliticize the discussion (see above) and (2) their conformity with the findings of inductive studies on the discourse and framing of CE. I follow Nisbet, who, on the basis of studies on the framing of nuclear energy, nanotechnology and climate change, argues that "[s]cholars have a tendency to 'reinvent the wheel' in identifying and labelling the frames that exist in any debate" (Nisbet 2010b: 46). He proposes "a set of frames that appear to reoccur across science-related policy debates" (Nisbet 2010b: 51) and that were originally defined and isolated by Gamson and Modigliani in an empirical study of framing of nuclear energy (Gamson and Modigliani 1989). In the present paper, Nisbet's set serves as a basis for the construction of a frame set to be assessed in empirical study.[24] This set conforms with Max Weber's differentiation of causes of determination of social action (see below) and combines empirical findings of earlier frame analyses of CE—in both the academic and news media discussion—with political science approaches to the social construction of governance through discourse such as securitization (Buzan et al. 1998: 26; Balzacq 2010: 63) and ecological modernization (Hajer 1997: 25–26). Thus, I connect theoretical concepts with the findings of empirical study to establish a set of frames which can affect decision-making and which are then, in empirical analysis, isolated in the academic discussion of CE to see if one of them is dominant (Image 3.3) (Table 3.1).

Four Frame Themes Based on Max Weber's Causes of Social Action

The theoretical construct applied in this study is based on eight ideal-type discursive processes which can influence this decision with regards to CE approaches.

[24] An alternative frame set construction to the one introduced by Gamson and Modigliani (1989) is presented by Steve Rayner, who differentiates three "organizational cultures" in environmental decision-making based on cultural theory (Rayner 1991: 88; see also Rayner 1995). These 'cultures' can be considered equivalent to 'frames' (Jachtenfuchs 1996b: 48), as they also, like frames, represent different problem definitions and problem-solving rationales. Rayner's differentiation was not adopted here as Gamson and Modigliani's frame set offers itself better to the connection to political science theory on the influence of discourse on governance choices, and because their frames were derived from empirical analysis on a similarly technological case of decision-making.

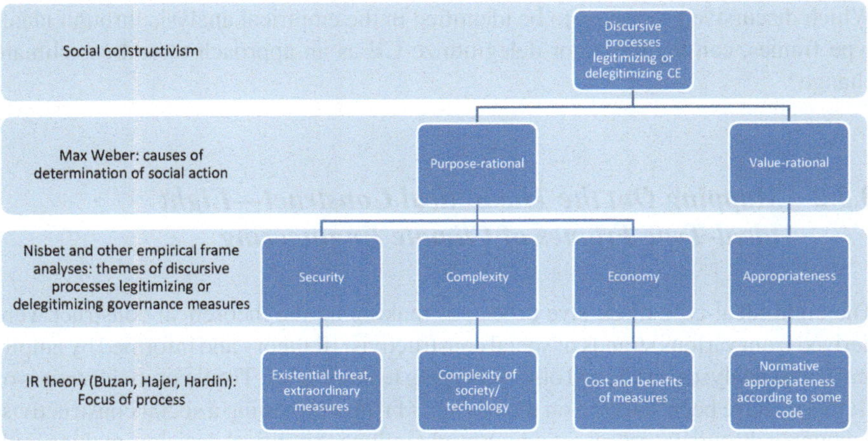

Image 3.3 Theoretical basis of decisions in designing the theoretical construct

Table 3.1 Overview over four frame themes and their basis in IR and frame theory

Frame theme	Rationale for decision-making	Theoretical basis	Corresponding frame(s) in Nisbet (2010a: 46), based on Gamson and Modigliani (1989: 24)
Security	Logic of necessity: decision based on what is necessary to prevent an existential threat	Securitization theory (Balzacq 2010)	• Pandora's box/Frankenstein's monster/runaway science
Complexity	Logic of simplicity/complexity: decision based on what is considered the simplest measure to solve a problem	Tragedy of the commons (Hardin 1968)	• Social progress • Scientific/technical uncertainty
Economy	Logic of cost/benefit: decision based on what is considered the cheapest measure	Ecological modernization (Hajer 1997)	• Economic development/competitiveness
Appropriateness	Logic of appropriateness: decision based on what is considered to be the 'right' approach according to some code	Value-rational determination of action	• Morality/ethics

The choice of these eight discursive moves was informed by Max Weber's differentiation of causes of determination of social action. Weber defines social action as action "which, according to its meaning as held by the actor or actors, refers to the behavior of *others* and is guided by it in its course".[25] (Weber 1980 [1922]: 1) Thus, the decisions of researching, developing or using CE approaches can be considered a social action as it refers to the behavior of others—e.g. of reducing GHG emissions or changing energy production systems. Max Weber differentiates between four types of causes of determination for social action: purpose-rational ("zweckrational"), value-rational ("wertrational"), affectional ("affektuell") and traditional ("traditional") (Weber 1980 [1922]: 13).[26] Purpose-rational action is determined by the expectation of a specific usefulness of this action for a purpose. Thus, if the decision to use CE approaches is made on purpose-rational considerations, the decision-maker expects that these approaches will serve a specific purpose—such as reducing costs or preventing danger. Value-rational action is determined by the belief in the normative value of the action independent of its success. For decisions concerning CE, this would mean that these approaches were considered the 'right' thing to do according to some normative code—e.g. ethical or religious.

The theoretical construct applied in this study presents three purpose-rational ideal-type discursive processes and one value-rational ideal-type process, each in their two extreme forms: legitimizing CE approaches, on one hand, and delegitimizing these approaches, on the other. They are constituted as 'ideal types' in an understanding indebted to Max Weber. He argues: "Sociology develops […] concepts of *types* and seeks general rules of events, in contrast to History, which aims at the causal analysis and attribution of *individual, culturally* relevant actions, structures, personalities."[27] (Weber 1980 [1922]: 9). For example, Balzacq, casting securitization as an ideal type, argues: "[…] an ideal type is a useful instrument in producing analytical statements about our experience in the world" (Balzacq 2015: 104). He, like the present study, follows Weber and maintains that "the construction of an ideal type aims to put in relief the unique traits that certain phenomena have in common, which are brought together, in an idealized fashion, under the same concept" (Balzacq 2015: 104).

---

[25]Translated by JK; Original quote: Soziales Handeln' […] soll ein solches Handeln heißen, welches seinem von dem oder den Handelnden gemeinten Sinn nach auf das Verhalten anderer bezogen ist und darin in seinem Ablauf orientiert ist. (Weber 1980 [1922]: 1)

[26]Translated by JK; Original quote: Wie jedes Handeln kann auch das soziale Handeln bestimmt sein 1. zweckrational: durch Erwartungen des Verhaltens von Gegenständen der Außenwelt und von anderen Menschen und unter Benutzung dieser Erwartungen als ‚Bedingungen'oder als ‚Mittel' für rational, als Erfolg, erstrebte und abgewogene eigne Zwecke,—2. wertrational: durch bewußten Glauben an den—ethischen, ästhetischen, religiösen oder wie immer sonst zu deutenden—unbedingten Eigenwert eines bestimmten Sichverhaltens rein als solchen und unabhängig vom Erfolg,—3. affektuell, insbesondere emotional: durch aktuelle Affekte und Gefühlslagen,—4. traditional: durch eingelebte Gewohnheit (Weber 1980 [1922]: 13)

[27]Translated by JK; Original quote: Die Soziologie bildet […] *Typen*-Begriffe und sucht generelle Regeln des Geschehens. Im Gegensatz zur Geschichte, welcher die kausale Analyse und Zurechnung *individueller, kultur*wichtiger, Handlungen, Gebilde, Persönlichkeiten erstrebt (Weber 1980 [1922]: 9)

Other than Weber, who aims to find out the causal motives of actions (Weber 1980 [1922]), the present study focuses on outlining processes which can influence the legitimacy of CE approaches. Thus, the ideal-types developed here present models of processes that would take place in such as way only if CE was represented, in discussion, with regards to one specific purpose, e.g. security, or with regards exclusively to its value to some normative code. This, of course, does not take place in reality. Ideal types are "remote from reality"[28] (Weber 1980 [1922]: 10), as Weber himself argues: Ideal types appear in this "absolutely ideal *pure* form just as little in reality as a physical reaction which is calculated under the presumption of absolutely empty space.[29] These ideal-types are relevant, though, in order to enable to identify elements of the discussion which conform with that type, and might, accordingly, influence decision-making. 'Ideal' is, thus, not understood in a normative sense, thus not as the best process to take place. Instead, it is to be understood as conforming perfectly to one of the types of process outlined below. The most important feature of an ideal type, according to Balzacq, is "internal coherence, which is to say that the features that an ideal type encompasses are logically connected. In short, contradictions are fatal to ideal types" (Balzacq 2015: 105).

The four types or themes of processes can be understood, in frame theory language, as equivalent to different issue frames, as they present content not in logically equivalent terms—as equivalency frames—but rather focus on different aspects of the content: "issue frames focus on qualitatively different yet potentially relevant considerations (Druckman 2004: 672). Each of the themes can spark either a legitimization or a delegitimization process regarding CE. In the four legitimizing processes, CE approaches are presented in the way of a measure to address a specific politicized problem (e.g., climate change). In the four delegitimizing processes, CE approaches themselves are considered the central problem. Each set of thematic processes (one legitimizing and one delegitimizing) is characterized by a specific understanding of the central problem. The way in which the central problem is characterized influences, in turn, which governance measures are considered useful to address the problem (see Bellamy et al. 2012a: 14; Markusson 2013: 12; Szerszynski and Galarraga 2013: 2818; Huttunen and Hildén 2014: 5; Asayama 2014: 89). All four discursive processes are operationalized as frames. Each frame theme presents ideal-type frames of two variants: one legitimizing CE (and thus supporting the all-out research and development of these approaches), and one delegitimizing CE (and thus supporting, for example, a moratorium on these approaches) (Table 3.2).

The two variants of the *security* frame share in common that both of them motivate action on the basis of the purpose of security, of protecting a central referent object against threats. In the case of the present study, this action is decision-making with regards to CE approaches. The central problem is, in a security frame, characterized as an existential threat, which justifies the extraordinary measure that can

[28]Tranlsated by JK; Original quote: wirklichkeitsfremd (Weber 1980 [1922]: 10).

[29]Translated by JK; Original quote: […] in dieser absolut idealen *reinen* Form vielleicht ebensowenig je in der Realität auftreten wie eine physikalische Reaktion, die unter Voraussetzung eines absolut leeren Raums errechnet ist (Weber 1980 [1922]: 10).

Table 3.2 Overview over discursive processes addressed in this study

Type of determination of social action following Weber	Purpose-rational						Value-rational	
Thematic set	Security		Complexity		Economy		Appropriateness	
Types of processes	Legitimizing CE	Delegitimizing CE	Legitimizing CE	Delegitimizing CE	Legitimizing CE	Delegitimizing CE	Legitimizing CE	Delegitimizing CE

help protect against this threat. In Entman's differentiation of frame elements, this extraordinary measure is presented as the value of the frame element 'proposed remedy'. The two variants differ in what is identified as the central problem or the existential threat: climate change—or some other phenomenon against which CE might help protect—on the one hand, or CE itself, on the other. In the former case, CE as a measure is justified through representing climate change—or some other problem—as an existential threat, endowed with urgency to act. Other measures could, likewise, be legitimized through this representation: For example, climate change induced migration could be used to justify increased border protection (Oels 2016: 191). In this case, however, the framing would not conform to the ideal-type of the security frame legitimizing CE. In the latter case, CE is delegitimized by presenting these approaches themselves as an existential threat to a valued referent object: 191for example, to society or to the survival of humankind. In this variant, the frame presents the not-doing—or not-researching or not-developing—of CE in the element 'proposed remedy'. As CE approaches do not exist yet to a scale relevant to the global climate, it is not the dismantling of the technology which is in focus, but rather the preventing its installation.

The two variants of the *complexity* frame motivate action—or, again, in the case of this study, decision-making concerning CE—on the basis of the purpose of simplicity, of providing simple solutions to complex problems. Accordingly, the central issue is characterized as a complex challenge which can be condensed into a problem that has a simple answer. The measure proposed to address the problem is legitimized on the basis of its simplicity. In the case of the legitimization of CE, these approaches are presented as a simple fix in opposition to complex 'social engineering' approaches to addressing the problem. It functions only under the assumption that a simple solution can exist for something like climate change. Conversely, in the delegitimizing variant, this frame presents CE approaches themselves as a complex problem: one with many unknown and, to some degree, unknowable side-effects and implications on a number of levels. Accordingly, even a measure that might, at first, seem to be a 'crisp' and 'clean' solution, is unveiled as having complex physical as well as social implications, which means that no answer is ever simple. The remedy proposed in this variant is, again, to not apply these complex approaches as they would produce more problems that they might solve.

The two variants of the *economy* frame—the last of the three frames motivating action on the basis of some purpose—focus on the purpose of reducing quantifiable costs and increasing benefits. Accordingly, the central problem poses costs to some entity—either a specific entity such as a company, or a vague one such as society. The measure to address the problem is, conversely, presented as one incurring relatively lower costs or even benefits with regards to the problem. In the case of legitimizing CE, climate change—or some other problem—is presented to cause economic costs and risks. CE approaches are legitimized on the basis of their comparative cheapness. Other measures might also be legitimized through this reasoning, such as institutional change and market mechanisms (Hajer 1997: 40). In the delegitimizing variant of this frame, CE itself is presented to incur high costs, and accordingly, is proposed not to be applied in any way (Table 3.3).

Table 3.3 Differentiation by problem definition of frames legitimizing CE approaches

Frame	Ideal type: Legitimizing CE	Ideal type: Delegitimizing CE
Security	Problem: Climate change threatens a valued referent object (e.g. society)	Problem: CE threatens a valued referent object
Technology	Problem: Climate change poses a complex socio-technical challenge	Problem: CE poses a complex socio-technical challenge
Economy	Problem: Climate change produces economic costs	Problem: CE produces economic costs
Appropriateness	Problem: It is inappropriate to allow for unchecked climate change	Problem: It is inappropriate to conduct CE

Finally, the two variants of the *appropriateness* frame do not, as mentioned above, refer to a specific purpose in motivating action concerning decision-making about CE. Instead, in these frames, CE is legitimized or delegitimized based on its perceived conformity to—or violation of—a specific set of values. Thus, the central problem is characterized as violating this set of values, and the measure to address it is the 'right thing to do' in order to conform to this set. For the purpose of the frame analysis, it is of no importance which set of values is referred to, or on what basis its validity is argued to rest—i.e., whether it is based on a transcendental notion of 'deriving from God' or on some other source. Accordingly, the sets of values invoked can be widely different, deriving for example from religious ethics or from environmental justice considerations. The variant legitimizing CE, then, presents the central problem as one violating the set of values, and CE as a measure that might remedy this violation. Conversely, in the delegitimizing variant, CE itself is considered the violating phenomenon, and in order to prevent this violation, preventing CE is considered the appropriate measure. All of the eight frame types represented in the frame set depoliticize the debate on CE by closing down the array of options. The legitimizing CE-variants close down the discussion in favor of CE approaches—by taking all other options off the table save CE. The delegitimizing variant does the same against CE—by taking CE approaches off the table. Both variants, then, reduce the number of options in the discussion. The end-point of this process would see a decision choosing one specific option, through which the discussion is closed finally.

The four thematic sets cover five of Nisbet's eight frames. The other three, 'public accountability', 'middle way' and 'conflict/strategy' (Nisbet 2010a: 46) are not represented in this construct as (1) they do not easily lend themselves to legitimizing or delegitimizing CE discursively, and (2) they are rarely or not represented in the frames identified inductively in past studies. However, in order to ensure that frames or frame elements which are not expected in the scope of the theoretical construct applied in this study can be identified in empirical analysis, the methodological approach allows for additions to be made to the codebook whenever text is found which cannot be coded with the existing construct, but requires a new code—and thus, a new category—to be added.

Theme 1: Security

The discursive processes of the thematic set of security frames are based on securitization theory. Securitization is a constructivist approach that describes one "distinctive pattern of depoliticisation [sic]" (Wood and Flinders 2014: 164), namely the process through which actors can transform an issue into a security issue by presenting it as such through language, which justifies the use of exceptional measures to address it (Buzan et al. 1998: 23; Balzacq 2010: 56; Van Wijk and Fischhendler 2015: 6). In Balzacq's words, securitization "predominantly examines how security problems emerge, evolve and dissolve" (Balzacq 2010: 56). Security is viewed as a social construction based on the use of language. Classic extraordinary measures are, for example, the dedication of large sums of money, military intervention or restrictive policy measures (Oels and von Lucke 2015: 47–48). I focus on a certain kind of securitization process which can justify the use of exceptional technological measures (see Bigo 2013: 129; 2014). In this way, securitization theory can provide the discursive conditions of legitimation of technologies as exceptional measures. The delegitimizing variant of this process focusses, conversely, on the presentation of the technological measure itself as an existential threat.

Securitization theory constitutes a rich array of approaches and concepts that have been introduced to IR in the last two decades.[30] I focus on two main approaches to studying securitization:[31] the original approach by the Copenhagen school and the sociological model by the Paris school, represented here mainly by Balzacq and Bigo. According to the Copenhagen school, "the process of securitization is what in language theory is called a speech act" (Buzan et al. 1998: 26). The notion of speech act is famously discussed by Austin, who distinguishes constative from performative utterances (Austin 1962: 94). Simply put, constative utterances, on the one hand, are utterances that contain a statement that describes a state of affairs and that can be found to be true or false. Performative utterances, on the other hand, "do not 'describe' or 'report' or constate anything at all, are not 'true or false'; […] the uttering of the sentence is, or is a part of, the doing of an action" (Austin 1962: 5) This means that performatives are utterances that change the state of affairs and the world (see Buzan et al. 1998: 26). A classic example of a speech act is the utterance of wedding vows, which change the world in the way that, before the utterance, the person speaking is not married, and afterwards he/she is. According to the Copenhagen school, by constructing something as an existential threat to a valued referent object through a speech act, a securitizing actor, e.g. "a state representative[,] moves a particular development into a specific area, and thereby claims a special right to use whatever means are necessary to block it" (Waever 1995: 55; see also Buzan et al. 1998: 26).

[30] For a very helpful overview over securitization theories, see, for example, Balzacq et al. (2016).

[31] For more fine-grained differentiations of different approaches to securitization see, for example, Balzacq et al. (2015); Balzacq (2015).

Balzacq criticizes the Copenhagen school approach on various grounds (see Balzacq 2010: 60), foremost among which is the focus on the speech act approach (Balzacq 2005: 171). According to his argument, the speech act understanding is not adequate to what is actually meant by securitization: "It is useful to think of security pronouncements not as speech acts that are successful to the extent that rules are followed by the agents, but as discursive techniques" (Balzacq 2010: 64) which are often part of a routine and happen unknowingly. Accordingly, to Balzacq, the main question of securitization studies is "*under what conditions* the social content and meaning of security produces threats" (Balzacq 2010: 64). Balzacq offers an alternative understanding of securitization as "a pragmatic act, i.e. a sustained argumentative practice aimed at convincing a target audience to accept, *based on what it knows about the world*, the claim that a specific development is threatening enough to deserve an adequate policy to curb it" (Balzacq 2010: 60). Based on this, he develops a sociological model of securitization analysis which understands the process to be a pragmatic act:

> In this light, securitization is a process whereby
>
> - patterns of *heuristic artefacts* (metaphors, image repertoires, analogies, stereotypes, emotions, etc.)
> - are contextually mobilized by a recognized agent, who
> - works persuasively to prompt a *target* audience *to build a coherent network of implications* (feelings, sensations, thoughts and intuitions), that concurs with the enunciator's reasons for choices and actions, by
> - investing the referent subject with such an *aura of unprecedented threatening complexion* that
> - a *customized political act must be undertaken immediately* to block its development. (Balzacq 2010: 63)

The aim is to bring the "context in which securitization occurs" (Balzacq 2010: 63) to the fore.

It is this focus on the context that makes this understanding of securitization interesting both as a discursive process in its own right as well as a model or 'conceptual skeleton' for the construction of the other ideal-type processes (Balzacq 2011). In all of these processes, a modernist constructivist approach is adopted in so far as it is not the "intrinsic attribute of the enunciation" (Balzacq 2010: 62) alone which is considered crucial for the development of the act. Instead, it is also highly relevant *who* utters the enunciation—in the present case, academic experts. Also, it is not necessarily the intention of the securitizing actor to invest the referent object with "unprecedented threatening complexion". Instead, the 'who', the 'how' and the 'what' of securitization together with the sociocultural context provide this threatening complexion, which, in turn, may lead to a specific political act or decision. In the present study, the 'who' is provided by academic actors as a group of actors with special relevance in the discussion of CE. The 'how' is defined by the specific context of communication that is provided by peer-reviewed academic publication and is assessed through frames. The 'what' is provided by the structure of the four thematic sets of ideal-type discursive processes, e.g. securitization. Balzacq et al. thus describe securitization theory as one that

> seeks to explain the politics through which (1) the security character of public problems is established, (2) the social commitments resulting from the collective acceptance that a phenomenon is a threat are fixed and (3) the possibility of particular policy is created. (Balzacq et al. 2016: 494)

In the case of the present study, accordingly, I am interested in (1) whether and how a security character of climate change is established in the CE discussion, (2) whether any social commitments are fixed resulting from the perception of climate change as a threat regarding CE, and (3) whether the possibility of policy introducing CE as a measure to address climate change is created. However, not only the legitimization, but also the delegitimization of CE through securitization is conceivable. Thus, conversely, I am also interested in (1) whether and how the security character of CE is established, (2) whether any social commitments are fixed resulting from the perception of CE as a threat, and (3) whether the possibility of policy blocking CE is created.

A number of studies—on the securitization of climate change, on one hand, and on the discourse of CE, on the other—indicate the promise of applying this ideal type process in an analysis of the academic discussion of CE. The political science analysis of climate change and climate policy from the perspective of security studies is popular (see, for example, Brzoska and Oels 2011; Corry 2012; Dalby 2014; Detraz and Betsill 2009; Grauvogel and Diez 2014; Lucke et al. 2014, 2016; Oels and von Lucke 2015; McDonald 2012, 2013; Methmann and Rothe 2012, 2013; Oels 2012; Parsons 2010; Rothe 2011; Scott 2008; Trombetta 2008; Mayer 2012) (for an overview, see Brzoska 2009; Lucke et al. 2016). For example, Cohen finds that "over the last few years new imagery […] frames the challenge [of climate change] as tantamount to a protracted state of armed hostility" (Cohen 2011: 199). Also, STS scholars analyze the connection between security systems and the biosphere (see, for instance, Masco 2010: 29). However, questions have also been raised concerning the suitability of the approach to analyze the issue of climate change (Trombetta 2008; Corry 2012; Oels and von Lucke 2015). One line of argument observes that, while a number of speech acts have declared climate change to be an existential threat, there were no exceptional measures (Oels and von Lucke 2015: 43). One reason for this might be that climate change, while threatening, is not commonly considered an imminent danger, but rather a risk (see Beck 1998; Kessler and Daase 2008; Corry 2012). It might also be possible, however, that the extraordinary measures have not been taken *yet* (see Methmann and Rothe 2013: 107). CE could be conceptualized as a potential future extraordinary measure to address climate change (see Methmann and Rothe 2012: 331).

The present study does not aim to ascertain whether the discussion of CE is a case of securitization with extraordinary measures still pending or rather a case of riskification. Instead, it aims to assess whether any security claims are being made in the academic discussion at all—be they focusing on more specific threats or vaguer risks—and whether these claims present a dominant framing of CE. This is in line with Balzacq's understanding of securitization: There, the decisive question is not whether or not extraordinary measures have been taken, but rather whether or not an issue has been presented as a threat and in which context this presentation has taken place. Inductive studies give first indications that such an endeavor might be fruitful: They show that such concepts as 'crisis' or 'emergency' are present in the CE discussion inside and beyond the academic arena.[32] For example, the emergency argument for CE presents an urgency to deal with the issue of climate change through reference to the potential of global catastrophe which might be averted through the application of CE measures (Hulme 2008: 11; Gardiner 2013b: 28; Asayama 2014: 90; Horton 2015: 1). Concepts of catastrophe or necessity are found in media analyses on CE (Scholte et al. 2013; Nerlich and Jaspal 2012; Buck 2013; Luokkanen et al. 2014), and notions of emergency are found in analyses of academic publications (Bellamy et al. 2012b: 605; Linnér and Wibeck 2015: 260; Huttunen and Hildén 2014: 15).

The security framing of CE has been discussed in the recent literature extensively, pointing to some interesting characteristics of this representation (Markusson 2013; Markusson et al. 2013; Anshelm and Hansson 2014a; Bellamy et al. 2012a, b). *Tipping elements* in the earth system and the related notion of abrupt climate change (Bellamy and Hulme 2011), for instance, are considered elements of this representation: "If such 'points of no return' leading to deemed crippling climate states can be identified, and should they be expected to be imminently exceeded, geoengineering might be considered the only option and therefore a necessity" (Markusson et al. 2013: 284). CE, in this representation, is not "accompanied with promises of a better world. The spokespeople of geoengineering do not offer future prosperity; instead, their legitimacy is based on negative expectations" (Anshelm and Hansson 2014b: 109; see also Asayama 2014; Asayama et al. 2017). In general, a representation of necessity of CE is encountered in instances of the security framing (see also Heyward and Rayner 2013).

It is important to remember here that a frame is considered "a valence-neutral organizing device […] meaning that it may lead to pro-issue, anti-issue, and neutral positions" (Nisbet 2010a: 45). The security frame can provide justification for both support and opposition to CE:,

> [j]ournalists who employ a Security framing conceive of either anthropogenic climate change, or geoengineering in response to this climatic change, as a threat to local, national or international security: the break-down of society, and the end of society as we know it standing as the underlying concern. (Porter and Hulme 2013: 349)

Academic publications, according to Linnér and Wibeck (2015), tend to highlight either the dangers of doing nothing or the dangers of considering climate change (see also Kreuter 2018). To capture this polarity, the present study pitches two extreme frame types for each theme against each other, assessing for each of the two extremes whether and with which frequency it is present in the discussion.

---

[32] Various operationalizations are applied in these studies, some focusing on frames, others on metaphors or arguments. All of those are considered, here, part of discourse. I will subsume all of these perceptions under the term 'security framing' for the sake of simplicity.

Table 3.4 The two ideal-type security frames structured following Entmann (1993, 52)

Ideal-type frame	Legitimizing CE	Delegitimizing CE
Rationale	Logic of necessity: decision based on what is necessary to prevent an existential threat	
Problem definition	Climate change poses existential threat/serious risk	CE poses existential threat/serious risk
Analysis of cause	Political failure to address climate change challenge	Academic/economic/other proposals to discuss/develop/apply CE approach(es)
Evaluation of CE	Pro	Counter
Proposed remedy	CE	Not CE

To summarize, securitization theory provides tools to analyze a process through which technological measures can be legitimized discursively. The conditions for legitimation through technological securitization, then, are the discursive presentation of the issue at hand as an existential threat which is characterized by urgency and the presentation of the respective technology as a way to counter that threat.[33] In order to conduct the empirical analysis, I operationalize securitizing moves as frames: The *security* frame defines a problem in terms of the protection against *threats* or risks. It introduces the rationale of *necessity* as the central criterion in decision-making: A valued object needs to be protected, and thus, extraordinary measures—which would otherwise be considered unthinkable—are justified. Thus, an extraordinary measure is justified if it serves to protect the valued object (e.g., society). In Nisbet (2010a), this frame is represented through the frames of 'Pandora's box/Frankenstein's monster/runaway science', which define the issue in terms of 'catastrophe' or 'fatalism' and a "call for precaution in face of possible impacts or catastrophe" (Nisbet 2010a: 46) (Table 3.4).

Theme 2: Complexity

As mentioned above, the structure of the process of securitization was used as a template for the other discursive processes which can be expected to have an influence on the legitimacy of CE. On this basis, a process called technocratization has been modelled and operationalized. This is a process whereby actors discursively transform an issue into one that can be reduced to a technological challenge. In

[33]Rothe analyzes the connection between securitization and technology the other way around: He bases his study on the theoretical strand of 'New Materialism'. This is a theoretical strand that expresses "opposition to the dominance of linguistic constructivism" ("Widerstand gegen die Dominanz des linguistischen Konstruktivismus", (Rothe 2015: 97, translated by the author) in security studies. Accordingly, a "renewed attention" is given "to actual material changes and processes" (Coole 2013: 452). Rothe argues that technologies as non-linguistic, material things and practices, not just discourses, can and do securitize climate change (Rothe 2015: 97). In this study, then, technologies are studied as the consequence of securitization, and in Rothe's paper, they are studied as the point of departure, as is common in IR studies from the theoretical perspective of new materialism (Connolly 2013; Coole 2013). Lundborg and Vaughan-Williams (2015) offer an enlightening approach to reconcile material-centered and language-centered studies in IR.

that way, technocratization, like securitization, can be understood as a process of depoliticization as it reduces the possible options to deal with the issue at hand and may diminish the political 'wiggle room', as it presents a technological approach as the easiest and thus, the 'right' solution. The presentation of technology as an uncontroversial, uncomplicated and objective way to address social problems is described through the idea of a 'technological fix' presented in the 1960s by Cold War physicist Alvin Weinberg:

> In view of the simplicity of technological engineering, and the complexity of social engineering, to what extent can social problems be circumvented by reducing them to technological problems? Can we identify Quick Technological Fixes for profound and almost infinitely complicated social problems, 'fixes' that are within the grasp of modern technology, and which would either eliminate the original social problem without requiring a change in the individual's social attitudes, or would so alter the problem as to make its resolution more feasible? (Weinberg 1966: 5)

'Social engineering', in his understanding, describes the activities which are aimed at solving social problems, i.e. "politics and political philosophy" (D. Scott 2013: 2). He assumes that behind social problems, there are underlying issues of technology and engineering which can be solved through "crisp and beautiful technological solution[s]" (Weinberg 1966: 4).[34]

The main differences to securitization lie in the way the issue is presented and in the nature of the approach to address it. In securitization, the issue at hand is presented as a threat, a security issue. In the process of technocratization, the issue is presented as one that is, at heart, concerned with complexity and simplicity. In the case of climate change, the presentation of the issue as being one of political deadlock can be observed, for example, by pointing to the political failure to achieve a negotiation solution through the UNFCCC process (Kintisch 2010: 40). Such a line of argumentation is pursued, for example, by researchers speaking out for the study and development of CE measures (Crutzen 2006: 211; Launder and Thompson 2008: 3841; Keith 2013: 37; Hulme 2014: viii; Bowden 2010: 75; McLaren 2013: 1). Also, in the process of technocratization, the technologies proposed to solve the problem are not part of a security strategy to approach a social issue, but rather a technological strategy to approach a social issue. Similar and related processes are descripted using the concepts of 'scientisation' (Wood and Flinders 2014: 163) and 'technocratization' (Methmann et al. 2013c: 250).

An early critic of the empirical instances of discursive technocratization is Garrett Hardin in his seminal work on the tragedy of the commons. He identifies a class of "no technical solution problems" (Hardin 1968: 1243) like, for example, overpopulation, and argues that these cannot be solved through a technical solution, namely "a change only in the techniques of the natural sciences, demanding little in the way of change of human values or ideas of morality" (Hardin 1968: 1243). A similar stance is presented by Morozov (2014) who criticizes 'solutionism':

---

[34] At the same time, Weinberg is not blind to the shortcomings of this approach as they often generate new problems. However, he concludes that "the social engineer's problems are never really solved" (Weinberg 1966: 8) and that technological fixes are the best that can be done.

> Recasting all complex social situations either as neatly defined problems with definite, computable solutions or as transparent and self-evident processes that can be easily optimized – if only the right algorithms are in place! – this quest [to ameliorate] is likely to unexpected consequences that could eventually cause more damage than the problems they seek to address. (Morozov 2014: 5)

Morozov outlines the depoliticizing potential of solutionism: By reducing a problem to its 'soluble' characteristics, deliberations on this problem are no longer necessary, only the 'right' action to solve it. In the process of technocratization, then, a move is made to represent a specific challenge as a technological one with a clear, crisp solution, instead of a 'messy' social or political problem wherein every possible means of addressing it entails new challenges of distribution of material and immaterial goods. However, I do not argue here that there was a group of researchers or other 'solutionists' bent on intentionally changing our way of life in that way. Instead, technocratization serves as an ideal-type discursive process which I traced, operationalized as two frames, in the empirical material in order to assess whether frames or frame elements of this kind are present in the academic discussion.

The idea of a 'technological fix' is discussed regularly in the discourse about this problem (see, for example, Corner and Pidgeon 2010: 30; Borgmann 2012: 189; Scott 2012: 158; Nerlich and Jaspal 2012: 135). The notion of the 'technological fix' is embedded in a specific view on technology, namely that of technological progressivism. According to this view, "the consistent application of science and technology is humanity's greatest hope for improving human life" (Scott 2013: 2).[35] The characteristics for legitimization of a technological measure to address a problem through technocratization, then, are the discursive presentation of the issue at hand as technological in nature rather than social or political,[36] and the presentation of and support for a technological solution through a technologically progressivist actor or group of actors.[37]

Traces of a carefully technological progressivist group of actors are visible in the case of CE. In 2010, *Science* journalist Eli Kintisch identified what he calls the "geo-clique" (Kintisch 2010: 8), a group of researchers including atmospheric scientist Ken Caldeira and environmental scientist David Keith which has been concerned about the issue of CE as a possible response to climate change threats as early as 2007. The group of scientists studying CE has been differentiated further by Keith and Caldeira themselves: They propose a "red team/blue team approach" to researching CE technologies, "wherein one team is tasked with showing how an approach can be made to work, and another team is asked with showing why the approach cannot produce a system that can actually diminish environmental risk at an acceptable cost"

---

[35] While this, at first glance, may seem like an approach that enjoyed the height of prominence in the early twentieth century and has been discarded in the meantime, Scott argues that it is still relevant today (Scott 2013: 2; see also Stirling 2014: 85).

[36] Arthur defines a technological problem as "a set of requirements to be met" (Arthur 2009: 111).

[37] As the group of actors is expected to be spread over a number of societal institutions, the concept of technological frame introduced by Bijker (2006: 696) is of more analytical value than the concept of technopolitical regime introduced by Hecht (cit. in Bijker 2006: 698).

(2010: 61) Kintisch understands the difference between the two teams to lie in their "temperament and role":

> Blue team members, such as Lowell Wood, have personalities that incline them to invent ways to alter the atmosphere. Keith leans blue. Russian climate scientist Yuri Izrael and his team also are solid Blue-Teamers. Red Team members, such as a plucky climate modeler called Ray Pierrehumbert, were generally skeptical of geoengineering and strove to find flaws in the blue team's work. Caldeira was bluish-purple. (Kintisch 2010: 8f.)

This means that he considers the dividing line to be whether the researcher in question is in general opposed to CE (red team) or support the idea (blue team). This, in turn, depends on whether the scientist in question believes that humankind should influence the Earth's system actively or not. Kintisch describes the two opposing perspectives:

> The battle lines on geoengineering have begun to take shape. On one side are the modern-day romanticists, who consider geoengineering an a priori violation of humans' role as planetary citizens to let nature be natural and take a humble place within it. […] And then there are the rationalists, who believe that to minimize suffering, it just may be more technological hubris that our species needs. (Kintisch 2010: 233)

Philosopher Clive Hamilton has picked up this differentiation, further specifying it by differentiating between researchers who are strongly supportive, cautiously supportive or deeply opposed to CE (Hamilton 2013a: 136). Similarly, climatologist Mike Hulme points to the central role of Blue Team scientists in the approach taken concerning CE in recent years, particularly their influence in media reporting as well as "in scientific research networks, at parliamentary hearings and on governmental committees" (Hulme 2014: 14f.). Whether this group of researchers can be shown to follow a philosophy of technological progressivism philosophy in the sense of technocratization remains to be seen.

Various inductive studies on the discussion on CE have found discursive elements which are connected to the process of technocratization focusing on complexity (Markusson 2013; Huttunen and Hildén 2014: 15; Porter and Hulme 2013: 345; Scholte et al. 2013: 10). In an analysis of the public debate on CE, Anshelm and Hansson (2014b) argue that

> [p]olitics is seen as obstructing efficient climate change management, while technology and science are depicted as unproven but uncomplicated, although this picture is supplemented by the reservation that geoengineering can only complement reduced CO_2 emissions. (Anshelm and Hansson 2014b: 110)

The notion of simplicity versus complexity becomes apparent in this argument: The technological measure of CE might be represented as a simple fix, while political or social 'engineering' to address the problem can be considered complicated and 'messy'. In the discursive move of technocratization, there is an

> implicit assumption that geoengineering will not generate the same type of political conflicts of interest and deadlocks as the calls for renewables and CO_2 emission decreases have. Geoengineering is without argumentation or problematization assumed to be pure technology devoid of all political consideration. (Anshelm and Hansson 2014b: 112)

Table 3.5 Two ideal-type complexity frames structured following Entmann (1993, 52)

Ideal-type frame	Legitimizing CE	Delegitimizing CE
Rationale	Logic of simplicity: decision based on what is considered the simplest solution	
Problem definition	Climate change presents a complex, 'messy' problem with many implications	CE poses a complex problem with many implications.
Analysis of cause	Over-complexity of social and political processes of decision-making	Over-eagerness of reducing social problems to 'slim' technological challenges
Evaluation of CE	Pro	Counter
Proposed remedy	CE	Not CE

This perception is connected to a specific mechanistic perception of earth which Luokkanen et al. (2014) identify in a media analysis:

> This set of metaphors builds on the concept of a machine or machine-like properties: earth is like a machine and interventions on earth are like interventions in a machine's mechanism. Mechanistic metaphors set the issue of climate control in terms of mechanistic systems such as thermostats or computers or a simple tool such as a sunshade. (Luokkanen et al. 2014: 973)

Again, the findings from empirical studies show that a thematic discursive process "may lead to pro-issue, anti-issue, and neutral positions" (Nisbet 2010a: 45). Accordingly, elements of representations of CE as complicated and messy in its own right have been found (Porter and Hulme 2013: 347; Scholte et al. 2013: 6; Huttunen and Hildén 2014: 4).

What remains to be seen is whether only fragments of complexity frames can be traced in the discussion, or whether complete or, at least, partly complete, complexity frames can be unearthed, and whether those are dominant. To do so, I operationalize technocratizing moves as frames, summarizing the considerations above: The *complexity* frame defines a problem in terms of its characterization as an entanglement of social problems and technological challenges. It introduces the rationale of complexity as the central criterion in decision-making; Thus, a measure is justified if it can be considered the simplest solution to the problem. CE is evaluated according to the criterion of *simplicity*. If it is, on this basis, presented as a simple fix to the complex problem of climate change to replace the "social engineering" (Weinberg 1966: 5) measure of (apparently unsuccessful) climate governance, this legitimizes the measure of CE. If, however, in this frame, CE is considered a complex issue in itself, this delegitimizes these approaches. Corresponding frames from Nisbet's frame set are 'social progress' and 'scientific/technical uncertainty' (Nisbet 2010a: 46). The 'social progress' frame defines an issue as "solution to problems" (Nisbet 2010a: 46), while the 'scientific/technical uncertainty' frames captures the issue of uncertainty, i.e. the challenge of unforeseeable complexity of technological measures (Table 3.5).

Theme 3: Economy

A third ideal-type process that can lead to the legitimization of CE technologies as a way to deal with climate change is what I call processes of economization. In this type of process, climate change is presented as an issue of economic cost, risk and benefit.

The process of economization is one in which actors discursively reduce an issue into an economic problem, i.e. a part of the entirety of quantifiable, tradeable goods and services that make up the economy. Some authors argue that this representation of environmental problems in general is rather old (see Lohmann 2013: 73). "In the beginning of this process, 'sustainable development' was conceptualized, seemingly reconciling economic growth and environmental protection, and has become the dominant perspective of the UN system" (Mert 2013: 29). It is based, in the present study, on the notion of 'ecological modernization', the dominant frame in the social construction of environmental policy-making according to Maarten Hajer (1997: 4): "[...] ecological modernization, first and foremost, introduces concepts that make issues of environmental degradation calculable" (Hajer 1997: 25–26). This means that, in this frame, nature and natural objects and systems such as the climate system are considered quantifiable "public good[s] or resource[s]" (Hajer 1997: 28). This allows for the calculation of costs and benefits arising from these along with those arising from various measures intended to maintain these 'goods' (see Stephan 2013: 60). According to ecological modernization, "it is both necessary to understand environmental problems as economic (external) costs, and use the toolkit of economic instruments to internalize these costs" (Methmann et al. 2013c: 250)

Economization is a process of depoliticization as it reduces the possible options to deal with the issue at hand and thus removes the issue from the realm of political deliberation. It differs from both securitization and technocratization in the way the issue is presented and in the justification for the approach to address it. In securitization, the issue is presented as a threat, a security issue. If technocratization takes place, it is presented as a technological issue. The measures that are taken to approach it, in securitization, are justified through the invocation of an existential threat and the necessity to counter that threat; in technocratization, through the invocation of the objectivity and simplicity of technological solution; and in economization, through cost-benefit-calculation. Economization can justify various types of measures to address the challenge at hand, e.g. the opening of markets or institutional learning (see, for example, Hajer 1997; Anshelm and Hultman 2015). Accounts of such discursive processes are found, in particular, in literature taking a critical theory perspective. Interesting for the present study are processes of economization which can justify technological measures, such as the 'Industrial Fatalism' discourse[38] as criticized by Anshelm and Hultman (2015) following Beck (1995). According to the

[38]While the authors are explicitly interested in ‚discourses', not ‚frames', their focus on interest lies on various frame elements in the discussion: "we found that the following [...] areas were of particular interest for our study: a description of the fundamental problem; the historical construction

authors, this process legitimizes extraordinary technological measures on the basis of framing climate change as an economic problem:

> Within the discourse of Industrial Fatalism, the primary answer to the question, 'How do we handle the severe unintended consequences that the industrial system generates?' is arrived at through intensified large-scale technological development and more efficient innovations. (Anshelm and Hultman 2015: 24)

According to this representation, climate change is a problem that can only be solved "according to the logic of linear modernization" (Anshelm and Hultman 2015: 19). Not all economization processes (i.e. discursive processes in which an issue is presented and understood as an economic problem) justify extraordinary technological measures. This is why 'Industrial Fatalism', as an extreme, is particularly interesting here. If such a process could be traced in the case of CE, a justification of CE is most probable on the grounds of protecting and furthering economic prosperity and growth. An example of a discursive process of 'Industrial Fatalism' is, according to Anshelm and Hultman, the support nuclear power technology enjoyed in Sweden in the face of various environmental challenges:

> When a number of environmental problems (including biocides, DDT and air pollution) that had the potential to undermine the modern industrial production system surfaced in the 1960s, the dominant political parties and environmental groups in Sweden all agreed that it was possible to resolve these problems with the help of modern industrial solutions, such as nuclear power [...]. (Anshelm and Hultman 2015: 19)

The response to these new risks, in a process of Industrial Fatalism, is "more of the same: increased, new and better technology" (Anshelm and Hultman 2015: 174).

Other authors have also approached the issue of economization of climate change also. Methmann et al. state that economization is "rooted in the neoliberal revolution that took place since the 1970s and 1980s" (Methmann et al. 2013c: 250). Similarly, the 'techno-market imaginary' isolated by Levy and Spicer (2013: 664) in NGO, business and state publications on climate change, relies on the "invocation of innovation, entrepreneurship, venture capital and carbon markets". Beyond these very critical stances concerning the economic perspective, the understanding of climate change as an economic issue is rather common, particularly since the publication of economist Nicholas Stern's famous *The Economics of Climate Change*, commonly known as the *Stern Review*, in which he estimates the economic costs of climate change and the benefits of early action (Stern 2006).

In the academic discussion of CE, the economy frame is represented, to some degree, by the "economic view of the 'Risk-benefit' frame [...]. The suitability of geoengineering is thus assessed by pondering the associated costs, benefits, and risks related to climate change, geoengineering, and other mitigation and to some extent also adaptation options" (Huttunen and Hildén 2014: 11). In the media discussion,

---

of the relationship between industrial, economic, political and atmospheric conditions; proposals for action; and responsibility" (Anshelm and Hultman 2015: 11). These 'areas' correspond, to some degree, with the frame elements of problem definition, causal evaluation, suggested remedy and moral judgement (Entman 1993: 52).

however, these frames have been identified specifically (Porter and Hulme 2013: 348; Buck 2013: 170/171; Scholte et al. 2013: 7). For example, Porter and Hulme find that "[a]rticles employing an Economics framing are often concerned with issues such as how much technologies cost, who is going to pay for them, their cost efficiency and the way in which they may support national and international economies and industries" (Porter and Hulme 2013: 348). In a critical discourse analysis of CE advocacy, Sikka argues that a move has been made of "discursively positioning opposition to geoengineering as opposition to the economy" (Sikka 2012a: 170)

What remains to be seen is whether economization moves or fragments of such moves can be traced in the academic discussion on CE. To operationalize economization moves in terms of frames and to sum up, the *economy* frames define the problem in terms of (quantifiable) costs and benefits. They introduce the rationale of *cost-effectiveness* into decision-making. Thus, the measure that is cheapest or has the least cost—of any kind—is justified in these frames. Both variants of the economy frame are captured by Nisbet's 'economic development/competitiveness' frame. Through it, issues are defined by "economic investment, market benefits or risks; local, national, or global competitiveness" (Nisbet 2010a: 46) In Gamson and Modigliani's (1989) account of the nuclear discourse in the USA, the 'progress' frame captures the economy frame variant legitimizing technology, while the 'not cost effective' frame captures its opposite (Table 3.6).

Theme 4: Appropriateness

Unlike the other three themes, the fourth, appropriateness, does not introduce the legitimization or delegitimization of a specific measure based on purpose-rational considerations. In the security frames, legitimization or delegitimization is based on whether or not a phenomenon or measure threatens or supports the purpose of security. In the complexity frames, the purpose of simplicity is in focus. In the economy frames, the main purpose is low cost (and large benefits). In contrast, when frames of appropriateness are active and successful, social action is determined, in Weber's words, not by the pursuit of some purpose, but rather by the belief in some set of normative values, irrespective of the success of the action to serve a purpose (Weber 1980 [1922]: 30). A corresponding discursive process represents the central

Table 3.6 Two ideal-type economy frames structured following Entmann (1993, 52)

Ideal-type frame	Legitimizing CE	Delegitimizing CE
Rationale	Logic of cost/benefit: decision based on what is considered the cheapest/most profitable solution	
Problem definition	Climate change presents an expensive problem	CE poses an expensive approach
Analysis of cause	Failure to recognize the economic implications of climate change	Failure to recognize the hidden costs of CE
Evaluation of CE	Pro	Counter
Proposed remedy	CE	Not CE

problem as one of wrongness with regards to a specific set of values, thus justifying measures to amend that wrong. For example, evaluated according to the set of values represented by human rights, it is wrong to treat a person in such a way as to cause that person bodily harm. Accordingly, measures to address the problem are justified if they conform to these values, and thus can be considered ‘the right thing to do’. If, for example, a CE approach is considered to alleviate a wrong that has been done by emitters of GHG, for example the loss of home and livelihood of people who are affected by sea-level rise, it might be represented as the ‘right thing to do’ in the circumstances according to notions of environmental justice.

Discursive elements of this kind have been identified in various studies on the discussion on CE. For example, in an analysis of UK media framing, Porter and Hulme find both a ‘Morality’ and a ‘Justice’ framing. The ‘morality’ framing, in their understanding, “examines the relative ‘rightness’ and ‘wrongness’ of deploying, or not deploying, geoengineering technology, to advance or resist the idea of intervening in the climate system” (Porter and Hulme 2013: 349). The set of values that is often referred to is Christian or other religious ethics: “In fact the theme of religion prevails as the Morality framing often features religious connotations, references and language. Guilt, blame, judgement and punishment are, for example, biblical themes that play out particularly strongly within this framing” (Porter and Hulme 2013: 349). Accordingly, discussion focuses on whether climate change and CE represent sinful behavior according to this specific set of values (Porter and Hulme 2013: 349). Based on such a discussion of climate change, a certain obligation to conduct CE to right these wrongs might be, and in some cases is, inferred. Conversely, however, CE itself can be presented as wrongful behavior, with regards to religious values for example due to hubris and overestimation of human power over nature. From this, an obligation not to consider CE might be derived.

A ‘Justice’ framing is described by the authors to focus on “what is to be gained or lost from geoengineering, who the beneficiaries and losers of this transaction will be, and the appropriateness of this distribution according to principles such as equality, need and fairness” (Porter and Hulme 2013: 350). In this book, these two framings are considered to belong to the theme of appropriateness, as they both provide a set of values to guide decision-making, and only differ in which specific set of values is referred to. Both framings can cut both ways:

> The Justice framing operates at many scales, concerned with local injustices through to international injustices. Within the latter of these scales, the subject of international development features heavily where, again, geoengineering is conceived of as both a means through which to advance development efforts, and as a threat to its sustainability. (Porter and Hulme 2013: 350)

In a similar analysis of UK newspaper articles, Scholte et al. 2013 find that the discursive element ‘norms and values’ is present in 60% of all cases (Scholte et al. 2013: 9). The ‘Natural Balance’ frame identified by Huttunen et al. in the academic discourse does not differentiate two directions, one legitimizing and one delegitimizing CE, but focusses on presenting CE as “an extension of the behavior that caused climate change in the first place, and hence is an addition to the problem, not a solution”

Table 3.7 Two ideal-type frames of appropriateness structured following Entmann (1993, 52)

Variant	Legitimizing CE	Delegitimizing CE
Rationale	Logic of appropriateness: Decision based on what is considered appropriate with regards to a specific set of normative values	
Problem definition	Climate change presents a wrong with reference to a specific set of values	CE presents a wrong with reference to a specific set of values
Analysis of cause	Behavior conflicting with set of values leading to climate change	Behavior conflicting with set of values leading to conducting CE
Evaluation of CE	Pro	Counter
Proposed remedy	CE	Not CE

(Huttunen and Hildén 2014: 18). The reference to specific values and the evaluation of CE along criteria of appropriateness is conducted when CE is "framed as morally wrong and an expression of human hubris" (Huttunen and Hildén 2014: 18).

What remains to be seen is whether only fragments of this appropriateness frame in the present understanding can be traced in the discussion, or whether complete or, at least, partly complete, appropriateness frames can be unearthed, and whether those are dominant. To be able to do so, I operationalize appropriateness moves as frames, summarizing the considerations above: The *appropriateness* frame defines a problem in reference to some set of values. It introduces the rationale of appropriateness as the central criterion in decision-making: A measure is considered legitimized if it conforms to the relevant set of *values*. In Nisbet's frame set, the theme of appropriateness is represented by the frame called 'morality/ethics', which he describes as defining a "science-related issue […] in terms of right or wrong; respecting or crossing limits, thresholds or boundaries" (Nisbet 2010a: 46) (Table 3.7).

3.4 Summary: Framing the Scope of Analysis

This chapter has presented a theoretical construct of abstract categories of analysis which allow the researcher to empirically and systematically answer the central research question. Thus, this chapter defines, on the basis of theoretical literature, which portion of reality is considered relevant for the present study to answer this question and how this portion of reality can be conceptualized and made accessible for empirical analysis. The analysis is based on the theoretical perspective of modernist social constructivism. The general perspective, together with five more specific assumptions made on the basis of STS, framing theory and the EC approach, provide the basis for the two central claims on which rests the relevance of the present study:

(I) *The legitimacy of CE approaches, for which the early academic discussion sets the stage, can influence future decision-making concerning these approaches.*

(II) *An analysis of frames in the academic discussion on CE can give insights into the construction of the meaning of CE with regards to its influence on future decision-making.*

Further, this chapter described in detail the theoretical construct which is applied in the empirical analysis of this study. In the following chapter, the theoretical construct is illustrated through the case of the early academic discussion on nuclear technology.

It is important to note, at this point, where the limits lie of the theoretical construct presented here, which are implied in its derivation from theory. What can it not be used for? First, it is not the intention of this study to show a causal relationship between the academic discussion on CE and political decision-making concerning these approaches. Thus, such a causal relationship is not built into the design of this study as a possible finding. Instead, it resides, as an assumption, at the starting point of the analysis, which indicates why the assessment of framing in the academic discussion is of political significance. To be precise, it is not assumed that the academic discussion about emerging technologies is the only factor determining political decisions on these technologies; it is assumed to be one of many factors relevant to this. In short: An academic discussion in which a frame legitimizing the technology is dominant is not, in itself, a sufficient condition for favorable political decision-making concerning the technology.

Further, and more broadly, the theoretical construct is limited by the very assumptions that have been made to provide for it a basis of non-contradictory statements about social reality and about the nature of knowledge. If, for example, the reader was to claim that the academic discussion on an emerging technology does not influence political decision-making in any way, then, for him or her, the rest of the study loses much of its relevance for climate policy and technology governance. I have tried to make arguments in support of the assumptions made above. This does not mean, however, that they are proven to be correct. This needs to be kept in mind in order to facilitate critical reflection about the results of this study and their implications for decision-making as well as further analyses.

References

Adler E (1997) Seizing the middle ground: constructivism in world politics. European J Int Rel 3(3):319–363

Adler E (2008) The spread of security communities: communities of practice, self-restraint, and NATO's post-cold war transformation. European J Int Rel 14(2):195–230. https://doi.org/10.1177/1354066108089241

Adler E, Bernstein S (2005) Knowledge in power: the epistemic construction of global governance. In: Barnett M, Duvall R (eds) Power on global governance. Cambridge University Press, Cambridge, pp 294–318

Adler E, Haas PM (1992) Conclusion: epistemic communities, world order, and the creation of a reflective research program. Int Org 46(1):367–390

Adler E, Pouliot V (2011) International Practices. Int Theory 3(1):1–36. https://doi.org/10.1017/S175297191000031X

Anais S (2013) Objects of security/objects of research. In: Mutlu CE, Salter MB (eds) Research methods in critical security studies. Routledge, New York, pp 195–198

Anshelm J, Hansson A (2014a) Battling promethean dreams and Trojan Horses: revealing the critical discourses on geoengineering. Energy Res Soc Sci 2:135–144. https://doi.org/10.1016/j.erss.2014.04.001

Anshelm J, Hansson A (2014b) The last chance to save the planet? An analysis of the geoengineering advocacy discourse in the public debate. Environ Hum 5:101–123

Anshelm J, Hultman M (2015) Discourses of global climate change. Apocalyptic framing and political antagonisms. Routledge, London

Armeni C (2015) Global experimentalist governance, international law and climate change technologies. Int Compara Law Q 64(4):875–904

Aronson JD (2010) The nature of technology: what it is and how it evolves. Int J Commun 4:258–260

Arthur WB (2009) The nature of technology: what it is and how it evolves. Free Press, New York, NY

Asayama S (2014) Catastrophism toward 'opening up' or 'closing down'? Going beyond the apocalyptic future and geoengineering. Curr Soc 63(1):89–93. https://doi.org/10.1177/0011392114559849

Asayama S, Sugiyama M, Ishii A (2017) Ambivalent climate of opinions: tensions and dilemmas in understanding geoengineering experimentation. Geoforum 80:82–92

Austin JL (1962) How to do things with words. The William James lectures delivered at Harvard University in 1955. Clarendon Press, Oxford, UK

Balzacq T (2005) The three faces of securitization: political agency, audience and context. European J Int Rel 11(2):171–201

Balzacq T (2010) Constructivism and securitization studies. In: Cavelty MD, Mauer V (eds) The Routledge handbook of security studies. Routledge, London, pp 56–72

Balzacq T (2011) A theory of securitization: origins, core assumptions, and variants. Securitization theory: how security problems emerge and dissolve. Routledge, London, pp 1–30

Balzacq T (2015) The 'essence' of securitization: Theory, ideal type, and a sociological science of security. Int Rel 29(1):103–113

Balzacq T, Guzzini S, Williams MC, Waever O, Patomäki H (2015) What kind of theory—if any—is securitization? Int Rel 29(1):96–136. https://doi.org/10.1177/0047117814526606

Balzacq T, Léonard S, Ruzicka J (2016) 'Securitization' revisited: theory and cases. Int Rel 30(4):494–531. https://doi.org/10.1177/0047117815596590

Barberis NC (2013) Thirty years of prospect theory in economics: a review and assessment. J Econ Perspect 27(1):173–196

Barry A (2013) The translation zone: between actor-network theory and international relations. Millenium J Int Stud 41(3):413–429

Beck S (2015) Science. In: Bäckstrand K, Lövbrand E (eds) Research handbook on climate governance. Edward Elgar, Cheltenham et al., pp 286–296

Beck S, Mahony M (2018) The politics of anticipation: the IPCC and the negative emissions technologies experience. Global Sustain 1(e8):1–8. https://doi.org/10.1017/sus.2018.7

Beck U (1995) Ecological politics in an age of risk. Polity Press, Cambridge

Beck U (1998) Politics of risk society. In: Franklin J (ed) The politics of risk society. Blackwell, Oxford, pp 9–22

Behnke J, Baur N, Behnke N (2012) Empirische Methoden der Politikwissenschaft (2. aktualisierte, Auflage edn. Ferninand Schöningh, Paderborn

Bellamy R (2013) Framing geoengineering assessment. Geoengineering our climate working paper and opinion article series

Bellamy R, Chilvers J, Vaughan NE, Lenton TM (2012a). Appraising geoengineering. Tyndall working paper series, 153

Bellamy R, Chilvers J, Vaughan NE, Lenton TM (2012b) A review of climate geoengineering appraisals. WIREs Climate Change 3(6):597–615

Bellamy R, Hulme M (2011) Beyond the tipping point: understanding perceptions of abrupt climate change and their implications. Weather Clim Soc 3:48–60. https://doi.org/10.1175/2011WCAS1081.1

Belter CW, Seidel DJ (2013) A bibliometric analysis of climate engineering research. WIREs Climate Change 4(5):417–427. https://doi.org/10.1002/wcc.229

Benford RD, Snow AD (2000) Framing processes and social movements: an overview and assessment. Ann Rev Sociol 26:11–39

Bergek A, Jacobsson S, Sandén B (2008) 'Legitimation' and 'development of positive externalities': two key processes in the formation phase of technological innovation systems. Technol Anal Strateg Manag 20(5):575–592

Biermann F (ed) (2010) Global climate governance beyond 2012: architecture, agency and adaptation. Cambridge University Press, Cambridge

Bigo D (2013). International political sociology. In: Williams PD (ed) Security studies. an introduction, 2nd edn. Routledge, London and New York, NY, pp. 120–133

Bigo D (2014) The (in)securitization practices of the three universes of EU border control: Military/Navy—border guards/ police—database analysts. Sec Dial 45(3):209–225

Bijker WE (2006) Why and how technology matters. In: Goodin RE, Tilly C (eds) The Oxford handbook of contextual political analysis. Oxford University Press, New York, pp 681–706

Borgmann A (2012) The setting of the scene: technological fixes and the design of the good life. In: Preston CJ (ed) Engineering the climate: the ethics of solar radiation management. Lexington, Plymouth, UK, pp 189–200

Bowden G (2010) Rock, paper, scissors: uncertainty, innovation, and the evolving ecoloy of climate change discourse. Proc Media Ecol Ass 11:67–78

Brown MB (2015) Policizing science: conceptions of politics in science and technology studies. Soc Stud Sci 45(1):3–30. https://doi.org/10.1177/0306312714556694

Brzoska M (2009) The securitization of climate change and the power of conceptions of security. Sicherheit Frieden 27(3):137–208

Brzoska M, Oels A (2011) Versicherheitlichung des Klimawandels? Die Konstruktion des Klimawandels als Sicherheitsbedrohung und ihre politischen Folgen. In: Brzoska M, Kalinowski M, Matthies V, Meyer B (eds) Klimawandel und Konflikte. Versicherheitlichung versus präventive Friedenspolitik? Nomos, Baden-Baden

Buck HJ (2013) Climate engineering: spectacle, tragedy or solution? A content analysis of news media framing. In: Methmann C, Rothe D, Stephan B (eds) Interpretive approaches to global climate governance. (De)constructing the greenhouse. Routledge, London, pp 166–180

Bueger C (2013) Actor-network theory, methodology, and international organization. Int Political Sociol 7(3):338–342

Buzan B, Waever O, De Wilde J (1998) Security. a new framework for analysis. Lynne Rienner Publishers, Boulder, CO

Caldeira K, Keith DW (2010) The Need for climate engineering research. Iss Sci Technol Fall 2010:57–62

Carr W, Yung L, Preston CJ (2014) Swimming upstream: Engaging the American public early on climate engineering. Bull Atomic Sci 70(3):38–48. https://doi.org/10.1177/0096340214531180

Chong D, Druckman JN (2007a) Framing theory. Annu Rev Polit Sci 10:103–126. https://doi.org/10.1146/annurev.polisci.10.072805.103054

Chong D, Druckman JN (2007b) A theory of framing and opinion formation in competitive elite environments. J Commun 57(1):99–118

Cohen MJ (2011) Is the UK preparing for 'war'? Military metaphors, personal carbon allowances, and consumption rationing in historical perspective. Clim Change 104(2):199–222

Collingridge D (1980) The social control of technology. Frances Pinter, London

Connolly WE (2013) The 'new materialism' and the fragility of things. Millenium J Int Stud 41(3):399–412

Coole D (2013) Agentic capacities and capacious historical materialism: thinking with new materialisms in the political sciences. Millenium: J Int Stud 41(3):451–469

Cornelissen JP, Holt R, Zundel M (2011) The role of analogy and metaphor in the framing and legitimation of strategic change. Org Stud 32(12):1701–1716

Corner A, Pidgeon N (2010) Geoengineering the climate: the social and ethical implications. Environ Sci Policy Sustain Devel 52(1):24–37

Corner A, Pidgeon N (2015) Like artificial trees? The effect of framing by natural analogy on public perceptions of geoengineering. Clim Change 130:425–438. https://doi.org/10.1007/s10584-014-1148-6

Corry O (2012). Securitization and 'riskification': second-order security and the politics of climate change. Millenium J Int Stud 40(2):235–258

Cross MAD (2013) Rethinking epistemic communities twenty years later. Rev Int Stud 39(1):137–160. https://doi.org/10.1017/s0260210512000034

Crutzen PJ (2006) Albedo enhancement by stratospheric sulfur injections: a contribution to resolve a policy Dilemma. Clim Change 77:211–219

Cudworth E, Hobden S (2013) Of parts and wholes: international relations beyond the human. Millenium J Int Stud 41(3):430–450

Dalby S (2014) Rethinking geopolitics: climate security in the anthropocene. Global Policy 5(1):1–9

Detraz N, Betsill MM (2009) Climate change and environmental security: for whom the discourse shifts. Int Stud Perspect 10:303–320

Donk A, Metag J, Kohring M, Marcinkowski F (2012) Framing emerging technologies: risk perceptions of nanotechnology in the German press. Sci Commun 34(1):5–29. https://doi.org/10.1177/1075547011417892

Druckman JN (2001) The implications of framing effects for citizen competence. Polit Behav 23(3):225–256

Druckman JN (2004) Political prefernce formation: competition, deliberation, and the (ir)relevance of framing effects. Am Polit Sci Rev 98(4):671–686

Druckman JN (2011) What's it all about?: framing in political science. In: Keren G (ed) Perspectives on framing. Psychology Press, New York et al., pp 279–301

Druckman JN, Nelson KR (2003) Framing and deliberation: how citizens' conversations limit Elite influence. Am J Polit Sci 47(4):729–745

Dryzek JS (2005) The politics of the earth. environmental discourses, 2nd edn. Oxford University Press, Oxford

Dryzek JS (2006) Deliberative global politics. Discourse and democracy in a divided world. Polity, Cambridge

Edelman M (1993) Contestable categories and public opinion. Polit Commun 10(3):231–242

Engelkamp S, Glaab K (2015) Writing norms: constructivist norm research and the politics of ambiguity. Alternat Global Local Polit 40(3–4):201–218. https://doi.org/10.1177/0304375415612270

Entman RM (1993) Framing: toward clarification of a fractured paradigm. J Commun 43(4):51–58

Entman RM (2004) Projections of power: framing news, public opinion, and US foreign policy. Chicago University Press, Chicago

Entman RM (2010) Framing media power. In: D'Angelo P, Kuypers JA (eds) Doing news framing analysis. Routledge, New York and London, pp 331–355

Finnemore M, Sikkink K (1998) International norm dynamics and political change. Int Org 52(4):887–917

Gamson WA (2005) Book review of Robert M. Entman's projections of power. Public Opin Q 69(2):324–336

Gamson WA, Modigliani A (1989) Media discourse and public opinion on nuclear power: a constructionist approach. Am J Sociol 95(1):1–37

Gardiner SM (2013b) Geoengineering and moral schizophrenia. What is the question? In: Burns WCG, Strauss AL (eds) Climate change geoengineering. philosophical perspectives, legal issues, and governance frameworks. Cambridge University Press, New York, pp 11–38

Geels FW (2004) From sectoral systems of innovation to socio-technical systems. Insights about dynamics and change from sociology and institutional theory. Res Policy 33:897–920

Geels FW (2007) Transformations of large technical systems: a multilevel analysis of the Dutch highway system (1950–2000). Sci Technol Human Values 32(2):123–149
Geels FW (2012) A socio-technical analysis of low-carbon transitions: introducing the multi-level perspective into transport studies. J Transp Geogr 24:471–482
Geels FW (2014) Regime resistance against low-carbon transitions: introducing politics and power into the multi-level perspective. Theory Cult Soc 31(5):21–40
Giddens A (1984) The constitution of society. Outline of the theory of structuration. Polity Press, Cambridge
Gieryn TF (1983) Boundary-work and the demarcation of science from non-science: strains and interests in professional ideologies of scientists. Am Sociol Rev 48(6):781–795
Goffman E (1974) Frame analysis. An essay on the organization of experience. Penguin, Harmondsworth
Grauvogel J, Diez T (2014) Framing und Versicherheitlichung: Die diskursive Konstruktion des Klimawandels. Zeitschrift für Friedens- und Konfliktforschung 3(2):203–232
Gschwend T, Schimmelfennig F (2007) Forschungsdesign in der Politikwissenschaft: ein Dialog zwischen Theorie und Daten. In: Gschwend T, Schimmelfennig F (eds) Forschungsdesign in der Politikwissenschaft: Probleme - Strategien - Anwendungen. Campus, Frankfurt am Main, pp 13–35
Gupta A, Möller I (2018) De facto governance: how authoritative assessments construct climate engineering as an object of governance. Environ Polit. https://doi.org/10.1080/09644016.2018.1452373
Haas PM (1992) Introduction: epistemic communities and international policy coordination. Int Org 46(1):1–35
Hajer M (1997) The politics of environmental discourse. Ecological modernization and the policy process. Clarendon Press, Oxford
Hajer M (2006) Doing discourse analysis: coalitions, practices, meaning. In: Van den Brink M, Metze T (eds) Words matter in policy and planning. Discourse theory and method in the social sciences. KNAG, Utrecht, pp 65–74
Hall T (2012) Electronic voting. In: Kersting N (ed) Electronic democracy. Barbara Budrich, Opladen, pp 153–176
Hamilton C (2013) Earthmasters. The dawn of the age of climate engineering. Yale University Press, New Haven, CT
Hardin G (1968) The tragedy of the commons. Science 162:1243–1248
Hay C (2007) Why we hate politics. Polity, Cambridge, MA
Hay C (2014) Depoliticisation as process, governance as practice: what did the 'first wave' get wrong and do we need a 'second wave' to put it right? Policy Politics 42(2):293–311. https://doi.org/10.1332/030557314X13959960668217
Heidegger M (1954) Die Frage nach der Technik. In: Heidegger M (ed) Vorträge und Aufsätze. Pfullingen, Günter Neske
Heyward C, Rayner S (2013) Apocalypse Nicked! Climate geoengineering governance working paper series, 006
Hilpert U (2016) Introduction. In: Hilpert U (ed) Routledge handbook on politics and technology. Routledge, London
Hollis M, Smith S (1990) Explaining and understanding in international relations. Clarendon Press, Oxford
Hoppe R (1999) Policy analysis, science and politics: from 'speaking truth to power' to 'making sense together'. Sci Public Policy 26(3):201–210
Hoppe R (2005) Rethinking the science-policy nexus: from knowledge utilization and science technology studies to types of boundary arrangements. Poeisis Prax 3(3):199–215
Horton JB (2013) Geoengineering and the myth of unilateralism: pressures and prospects for international cooperation. In: Burns WCG, Strauss AL (eds) Climate change geoenegineering: philosophical perspectives, legal issues, and governance frameworks. Cambridge University Press, New York, USA, pp 168–181

Horton JB (2015) The emergency framing of solar geoenengineering: time for a different approach. Anthropocene Rev Prepubl. https://doi.org/10.1177/2053019615579922
Horton JB, Keith DW, Honegger M (2016) Implications of the Paris agreement for carbon dioxide removal and solar geoengineering. Harvard project on climate agreements—viewpoints
Horton JB, Reynolds J, Buck HJ, Callies DE, Schäfer S, Keith DW et al (2018) Solar geoengineering and democracy. Global Environ Politics 18(3):5–24. https://doi.org/10.1162/glep_a_00466
Hulme M (2008) The conquering of climate: discourses of fear and their dissolution. Geograph J 174(1):5–16
Hulme M (2014) Can science fix climate change? A case against climate engineering. Polity Press, Cambridge, UK
Hurd I (1999) Legitimacy and authority in international politics. Int Org 53(2):379–408
Huttunen S, Hildén M (2014) Framing the controversial: geoengineering in academic literature. Sci Commun 36(3):3–29
Jachtenfuchs M (1996a) Frames and learning. In: Jachtenfuchs M (ed) International policy-making as a learning process? The european union and the greenhouse effect. Ashgate, Aldershot et al., pp 23–43
Jachtenfuchs M (1996b) A methodology for frame analysis. In: Jachtenfuchs M (ed) International policy-making as a learning process? The european union and the greenhouse effect. Ashgate, Aldershot et al., pp 43–60
Jacobsen KL (2015) Experimentation in humanitarian locations: UNHCR and biomtric registration of Afghan refugees. Sec Dial 46(2):144–164
Jann W, Wegrich K (2007) Theories of the policy cycle. In: Fischer F, Miller GJ, Sidney MS (eds) Handbook of public policy analysis. Theory, politics, and methods. CRC Press, Boca Raton
Jasanoff S (1987) Contested boundaries in policy-relevant science. Soc Stud Sci 17(2):195–230
Jasanoff S (2004a) The idiom of co-production. In: Jasanoff S (ed) States of knowledge. The co-production of science and social order. Routledge, London and New York, pp 1–12
Jasanoff S (2004b) Ordering knowledge, ordering society. In: Jasanoff S (ed.), States of knowledge. The co-production of science and social order. Routledge, London and New York, pp 13–45
Jasanoff S (2015) Science and technology studies. In: Bäckstrand K, Lövbrand E (eds) Research handbook on climate governance. Edward Elgar, Cheltenham et al., pp 36–48
Jaworski A, Coupland N (2006) The discourse reader, 2nd edn. Routledge, Oxon
Kahneman D, Tversky A (1984) Choices, values, and frames. Am Psychol 39(4):341–350
Keith DW (2013) A case for climate engineering. Boston Review Cambridge, London
Kessler O, Daase C (2008) From insecurity to uncertainty: risk and the paradox of security politics. Alternatives 33:211–232
Kintisch E (2010) Hack the Planet. Science's best hope—or worst nightmare—for averting climate Catastrophe. Wiley, New York, NY
Kreuter J (2015) Technofix, plan B or ultima ratio? A review of the social science literature on climate engineering technologies. INSIS Occasional Papers Series, India
Kreuter J (2018) Climate engineering and political decision-making: the importance of polarity in the academic debate. INoGOV policy brief
Lakoff G (2014) The all new don't think of an elephant! Know your values and frame the debate, 2nd edn. Chelsea Green Publishing, Chelsea, VT
Lapid Y (1989) The Third debate: on the prospects of international theory in a post-positivist era. Int Stud Quart 33:235–254. https://doi.org/10.2307/2600457
Latour B (1996) On actor-network theory: a few clarifications. Soziale Welt 47(4):369–381
Launder B, Thompson JMT (2008) Preface. Philosop Trans Royal Soc A 366(1882):3841–3842. https://doi.org/10.1098/rsta.2008.0151
Lee C-J, Scheufele DA, Lewenstein BV (2005) Public attitudes toward emerging technologies. examining the interactive effects of cognitions and affect on public attitudes toward nanotechnology. Sci Commun 27(2):240–267. https://doi.org/10.1177/1075547005281474
Levy DL, Spicer A (2013) Contested imaginaries and the cultural political economy of climate change. Organization 20(5):659–678

Levy JS (2003) Applications of prospect theory to political science. Synthese 135:215–241
Linnér B-O, Wibeck V (2015) Dual high-stake emerging technologies: a review of the climate engineering research literature. WIREs Climate Change 6:255–268
Litfin K (2000) Environment, wealth, and autority: global climate change and emerging modes of Legitimation. Int Stud Rev 2(2):119–148
Lloyd ID, Oppenheimer M (2014) On the design of an international governance framework for geoengineering. Global Environ Polit 14(2):45–63
Lohmann L (2013) Interpretive openness and climate action in an age of market environmentalism. In: Methmann C, Rothe D, Stephan B (eds) Interpretive approaches to global climate governance. (De)constructing the greenhouse. Routledge, London, pp 72–87
Long J, Loy F, Morgan MG (2015) Policy: start research on climate engineering. Nature 518(7537)
Low S (2017) The futures of climate engineering. Earth's Fut 5:67–71. https://doi.org/10.1002/2016EF000442
Lucke FV, Wellmann Z, Diez T (2014) What's at stake in securitizing climate change? Towards a differentiated approach. Geopolitics 19(4):857–884
Lucke FV, Wellmann Z, Diez T (2016) Klimakämpfe: Eine komparative Studie der Versicherheitlichung von Klimawandel. Zeitschrift für Internationale Beziehungen 23(2):112–143. https://doi.org/10.5771/0946-7165-2016-2-112
Lundborg T, Vaughan-Williams N (2015) New materialisms, discourse analysis, and international relations: a radical intertextual approach. Rev Int Stud 41:3–25
Luokkanen M, Huttunen S, Hildén M (2014) Geoengineering, news media and metaphors: framing the controversial. Public Understand Sci 23(8):966–981. https://doi.org/10.1177/0963662513475966
Markusson N (2013) Tensions in framings of geoengineering: constitutive diversity and ambivalence. Climate geoengineering governance working paper series, 003
Markusson N, Ginn F, Ghaleigh NS, Scott V (2013) 'In case of emergency press here': framing geoengineering as a response to dangerous climate change. WIREs Clim Change 5(2):281–290. https://doi.org/10.1002/wcc.263
Markusson N, Wong P-H (2015) Geoengineering governance, the linear model of innovation, and the accompanying geoengineering approach. Climate geoengineering governance working paper series, 020
Masco J (2010) Bad weather: on planetary crisis. Soc Stud Sci 40(1):7–40. https://doi.org/10.1177/0306312709341598
Matzner N, Barben D (2018) Verantwortungsvoll das Klima manipulieren? Unsicherheit und Verantwortung im Diskurs um Climate Engineering. In: Janich N, Rhein L (eds) Unsicherheit als Herauforderung für die Wissenschaft. Peter Lang, Berlin et al., pp 143–179
Mayer M (2012) Chaotic climate change and security. Int Polit Soc 6(2):165–185
McDonald M (2012) The failed securitization of climate change in Australia. Australian J Polit Sci 47(4):579–592
McDonald M (2013) Discourses of climate security. Political Geogr 33:42–51
McLaren D (2013). Mirror mirror, on the wall: Fairness and justice in geo-engineering discourses. Paper presented at the Science in Public Notthingham
Mert A (2013) Discoursive interplay and co-constitution. Carbonification of environmental discourses. In: Methmann C, Rothe D, Stephan B (eds) Interpretive approaches to global climate governance. (De)constructing the greenhouse. Routledge, London, pp 23–39
Methmann C, Rothe D (2012) Politics for the day after tomorrow: the logic of apocalypse in global climate politics. Sec Dial 43(4):323–344
Methmann C, Rothe D (2013) Apocalypse now! From exception rhetoric to risk management in global climate politics. In: Methmann C, Rothe D, Stephan B (eds) Interpretive approaches in global climate governance. (De)constructing the greenhouse. Routledge, London, pp 105–121
Methmann C, Rothe D, Stephan B (2013b) Introduction. How and why to deconstruct the greenhouse. In: Methmann C, Rothe D, Stephan B (eds) Interpretive approaches to global climate governance. (De)constructing the greenhouse. Routledge, London, pp 1–22

Methmann C, Rothe D, Stephan B. (2013c). Reflections. In: Methmann C, Rothe D, Stephan B (eds) Interpretive approaches to global climate governance. (De)constructing the greenhouse. Routledge, London, pp 248–256

Miller C (2001) Hybrid management: boundary organizations, science policy, and environmental governance in the climate regime. Sci Technol Human Values 26(4):478–500

Morozov E (2014) To save everything, click here. Penguin, London

Morth U (2000) Competing frames in the European commission—the case of the defence industry and equipment issue. J European Public Policy 7(2):173–189

Mutlu CE (2013) The material turn—introduction. In: Mutlu CE, Salter MB (eds) Research methods in critical security studies. Routledge, New York, pp 385–400

Nelson TE, Oxley ZM, Clawson RA (1997) Towards a psychology of framing effects. Polit Behav 19(3):221–246

Nerlich B, Jaspal R (2012) Metaphors we die by? Geoengineering, metaphors, and the argument from catastrophe. Metaphor Symbol 27(2):131–147

Nexon DH, Pouliot V (2013) Things of networks: situating ANT in international relations. Int Political Sociol 7(3):342–345

Nisbet MC (2010a) Framing science. A new paradigm in public engagement. In: Kahlor L, Stout PA (eds) Communicating science. New agendas in communication. Routledge, New York, pp 40–67

Nisbet MC (2010b) Knowledge into action. Framing the debates over climate change and poverty. In: D'Angelo P, Kuypers JA (eds) Doing news framing analysis. Empirical and theoretical perspectives. Routledge, New York and London, pp 43–83

Nisbet MC, Brossard D, Kroepsch A (2003) The stem cell controversy in an age of press/politics. Press/Politics 8(2):36–70. https://doi.org/10.1177/1081180X02251047

Nisbet MC, Mooney C (2007) Framing science. Science 316(5821):56

Nwogugu M (2006) A further critique of cumulative prospect theory and related approaches. Appl Math Comput 179:451–465

Oels A (2012) From 'securitization' of climate change to 'climatization' of the security field: comparing three theoretical perspectives. In: Scheffran J, Brzoska M, Brauch HG, Link PM, Schilling J (eds) Climate change, human security and violent conflict. Challenges for societal stability (Hexagon series on human and environmental security and peace, vol. 8). Springer, Heidelberg

Oels A (2016) Resisting the climate security discourse. In: O'Lear S, Dalby S (eds) Reframing climate change. Construcing ecological geopolitics. Routledge, London, pp 188–202

Oels A, von Lucke F (2015) Gescheiterte Versicherheitlichung oder Sicherheit im Wandel: Hilft uns die Kopenhagener Schule beim Thema Klimawandel? Zeitschrift für Internationale Beziehungen 22(1):43–70

Oldham P, Szerszynski B, Stilgoe J, Brown C, Eacott B, Yuille A (2014) Mapping the landscape of climate engineering. Philosop Trans Royal Soc A 372(2031). https://doi.org/10.1098/rsta.2014.0065

Parsons RJ (2010) Climate change: the hottest issue in security studies? Risk HazardCrisis Public Policy 1(1):87–116

Porter KE, Hulme M (2013) The emergence of the geoengineering debate in the UK print media: a frame analysis. Geograph J 179(4):342–355. https://doi.org/10.1111/geoj.12003

Pouliot V (2010) Materials of practice: nuclear warheads, rhetorical commonplace and committee meetings in Russian-Atlantic relations. Coop Conflict 45(3):294–311

Rayner S (1991) A cultural perspective on the structure and implementataion of global environmental agreement. Eval Rev 15(1):75–102

Rayner S (1995) A conceptual map of human values for climate change decision making. In: Katama A (ed) Equity and social considerations related to climate change. Papers presented at the IPCCC working group III workshop. ICIPE Science Press, Nairobi, pp 57–73

Rein M, Schön D (1993) Reframing policy discourse. In: Fischer F, Forester J (eds) The argumentative turn in policy analysis and planning. Duke University Press, Durham et al., pp 145–166

Reynolds J (2015) A critical examination of the climate engineering moral hazard and risk compensation concern. Anthrop Rev2(2):174–191. https://doi.org/10.1177/2053019614554304

Rip A, Kemp R (1998) Technological change. In: Rayner S, Malone EL (eds) Human choice and climate change: resources and technology, vol 2. Battelle Press, Columbus, pp 327–400

Robock A (2008) 20 reasons why geoengineering might be a bad idea. Bull Atomic Sci 64(2):14–18. https://doi.org/10.2968/064002006

Rothe D (2011) Managing climate risks or risking a managerial climate: state, security and governance in the international climate regime. Int Rel 25(3):330–345. https://doi.org/10.1177/004711 7811415486

Rothe D (2015) Von weitem sieht man besser. Satelittensensoren und andere Akteure der Versicherheitlichung. Zeitschrift für Internationale Beziehungen 22(1):97–124

Ruggie JG (2014) Global governance and 'new governance theory': lessons from business and human rights. Glob Gov 20:5–17

Schlipphak B (2011) Was ist ein Frame?—Zu den Grundlagen des Frame-Ansaztes in den Sozialwissenschaften. In: Schlipphak B (ed) Framing Ideology. Die Kommunikation ideologischer Positionierungen zwischen Parteien, Wählern und Medien. Nomos, Baden-Baden, pp 21–51

Scholte S, Vasileiadou E, Petersen A (2013) Opening up the societal debate on climate engineering: how newspaper frames are changing. J Int Environ Sci 10(1):1–16

Schot J, Kanger L (2018) Deep transitions: emergence, acceleration, stabilization and directionality. Res Policy 47:1045–1059

Scott D (2012) Insurance policy or technological fix? The ethical implications of framing solar radiation management. In: Preston CJ (ed) Engineering the climate: the ethics of solar radiation management. Lexington, Plymouth, UK, pp 151–168

Scott D (2013) Philosophy of technology and geoengineering. Geoengineering our climate working paper and opinion article series

Scott SV (2008) Securitizing climate change: international legal implications and obstacles. Cambridge Rev Int Aff 21(4):603–619. https://doi.org/10.1080/09557570802452946

Shepherd J, Cox P, Haigh D, Keith DW, Launder B, Mace G et al (2009) Geoengineering the climate: science, governance and uncertainty. The Royal Society, London

Sikka T (2012) A critical discourse analysis of geoengineering advocacy. Critical Dis Stud 9(2):163–175

Smit WA (2006) Military technologies and politics. In: Goodin RE, Tilly C (eds) The Oxford handbook of contextual political analysis. Oxford University Press, Oxford (UK), pp 722–744

Sniderman PM, Theriault SM (2004) The structure of political argument and the logic of issue framing. In: Saris WE, Sniderman PM (eds) Studies in public opinion. Attidues, nonattitudes, measurement error, and change. Princeton University Press, Princeton and Oxford, pp 133–165

Spindler M, Schieder S (2010) Theorien in der Lehre von den internationalen Beziehungen. In: Schieder S, Spindler M (eds) Theorien der Internationalen Beziehungen 3, Auflage edn. Barbara Budrich, Opladen, pp 9–38

Steffek J (2003) The legitimation of international governance: a discourse approach. European J Int Rel 9(2):249–275

Stephan B (2013) How to trade 'not cutting down trees'. A governmentality perspective on the commodification of avoided deforestation. In: Methmann C, Rothe D, Stephan B (eds) Interpretive approaches to global climate governance. (De)constructing the greenhouse. Routledge, London, pp 57–71

Stern N (2006) The economics of climate change. The stern review. Cambridge University Press, Cambridge

Stirling A (2014) Transforming power: social science and the politics of energy choices. Energy Res Soc Sci 1:83–95

Szerszynski B, Galarraga M (2013) Geoengineering knowledge: interdisciplinarity and the shaping of climate engineering research. Environ Plann A 45:2817–2824

Szerszynski B, Kearnes M, Macnaghten P, Owen R, Stilgoe J (2013) Why solar radiation management geoengineering and democracy won't mix. Environ Plann A 45:2809–2816

Thakur R (2015) Nuclear weapons and international security: collected essays. Taylor and Francis, Florence

Trombetta MJ (2008) Environmental security and climate change: analysing the discourse. Cambridge Rev Int Aff 21(4):585–602

Trout BT (1975) Rhetoric revisisted. Political legitimation and the cold war. Int Stud Q 19(3):251–284

Tversky A, Kahneman D (1974) Judgment under uncertainty: heuristics and biases. Science 185(4157):1124–1131

Tversky A, Kahneman D (1981) The framing of decisions and the psychology of choice. Sci New Ser 211(4481):453–458

Tversky A, Kahneman D (1992) Advances in prospect theory: cumulative representation of uncertainty. J Risk Uncertain 5:297–323

Uther S (2014) Diskurse des Climate Engineering. Argumente, Akteure und Koalitionen in Deutschland und Großbritannien. Springer Fachmedien, Wiesbaden

Van Wijk J, Fischhendler I (2015) The construction of urgency discourse around mega-projects: the Isreali case. Policy Sci Forthcoming

Verbong G, Geels FW (2007) The ongoing energy transition: lessons from a socio-technical, multi-level analysis of the Dutch electricity system (1960–2004). Energy Policy 35:1025–1037

Waever O (1995) Securitization and desecuritization. In: Lipschutz RD (ed) On security. Columbia University Press, New York, NY, pp 46–85

Waever O (1996) The rise and fall of the inter-paradigm debate. In: Smith S, Booth K, Zalewski M (eds) International theory. Positivism and beyond. Cambridge University Press, Cambridge, pp 149–185

Wainwright J, Mann G (2012) Climate Leviathan. Antipode 45(1):1–22

Weber M (1980 [1922]) Wirtschaft und Gesellschaft. Grundriss der verstehenden Soziologie (5. Revidierte Studienausgabe ed.). Mohr, Tübingen

Weinberg AM (1966) Can technology replace social engineering? Bull Atomic Sci 22(10):4–8

Wendt A (1998) On constitution and causation in international relations. Rev Int Stud 24(Special Issue):101–117

Wendt A (1999) Scientific realism and social kinds. In: Wendt A (ed) Social theory of international politics. Cambridge University Press, Cambridge, pp 47–91

Willsher K (2017) Paris mayor unveils plan to restrict traffic and pedestrianise city centre. The Guardian

Winner L (1977) Autonomous technology. Technics-out-of-control as a theme in political thought. MIT Press, Cambridge, MA

Winner L (1980) Do artifacts have politics? Daedalus 109(1):121–136

Wood M, Flinders M (2014) Rethinking depoliticisation: beyond the governmental. Policy Politics 42(2):151–170

Zürn M (2012) Global governance as multi-level governance. In: Lewi-Faur D (ed) The Oxford handbook of governance. Oxford University Press, Oxford, pp 730–744

Zürn M, Schäfer S (2013) The paradox of climate engineering. Global Policy 4(3):266–277

Chapter 4
The Early Discussion of Nuclear Weapons Technology in the 1930s and 1940s—Illustrating the Theoretical Construct

Abstract This chapter analyzes the development of nuclear technology in the late 1930s and early 1940s as an instance in which the presence of legitimizing frames voiced by academic actors coincided with the subsequent development and eventual use of the emerging technology. Kreuter highlights similarities as well as differences between the development of nuclear technology and the current discussion of CE approaches before identifying thematic security, complexity, economy and appropriateness frames in the academic discussion on nuclear technology. Concluding by broadening the perspective, Kreuter points out a multitude of factors which bore on the eventual outcome of use of nuclear weapons technology.

In order to illustrate the theoretical construct through a concrete example and to indicate how academic framing might be influential for decision-making with regards to emerging technology, the early academic discussion on nuclear technology is outlined here. This case has, from a research perspective, one important advantage over the case of CE: It is concluded. The early academic discussion on the potential of nuclear technology in the USA, some European countries and Soviet Russia was followed up by research and development up to the actual deployment of nuclear weapons against Japanese cities in the 1940s and the use of nuclear technology for energy production, for which is it used up to this day in a number of countries.

Various studies on the potential implications of CE development and deployment draw upon the case of nuclear technology development, making a variety of arguments, for example about the influence of the respective technology on global political order (Keith 2013: XI), the (assumed) danger of unilateral action (Horton 2013: 177), the potential regulation of the technology (Reynolds 2014), the momentum of scientific development of emerging technologies (Kintisch 2010: 17–18), the scales of testing and development (Hulme 2014: 63), the uncertainties connected with the respective technologies (Huttunen et al. 2015: (2), the general approach to the human relationship to nature (Hamilton 2014: (1), or the dilemma of using risky emerging technologies (Baum 2014: (1). Lederer and Kreuter (2018) turn to the case of the

Parts of this chapter have been published, verbatim or in similar formulation, in Lederer and Kreuter (2018).

J. Kreuter, *Climate Engineering as an Instance of Politicization*,
Springer Climate, https://doi.org/10.1007/978-3-030-60340-3_4

development of nuclear technology in order to illustrate a path that CE might take if, "instead of following a legalistic rationale, climate engineering organization [was to] pursue a logic of technical feasibility, political acceptance and bureaucratic momentum" (Lederer and Kreuter 2018: 1). What this illustration does not aim to imply is that there was a mono-causal relationship between the constitution of the early academic discussion on a specific technology or set of technologies, on the one hand, and the political decision to develop and deploy these technologies, on the other hand. As pointed out above, many other factors influence political decision-making on any given issue. The claim made here is that the early academic discussion can, in specific circumstances, influence the social construction of the meaning of these technologies, which then, in turn, can influence, together with other factors, political decision-making.

In this chapter, I focus on the very first years of debate on nuclear technology in the USA, from the discovery of nuclear fission in late 1938 to the deployment of two bombs in August 1945 (Reed 2014). I do so in order to illustrate how the early and—in the beginning exclusively—academic discussion of a—at this point hypothetical—technological approach applying the frame types defined in the theoretical framework can lead to the development and, eventually, even the deployment of such a technology, even against the explicit opposition of some of the academic actors first involved in the discussion. Thus, I argue that the development of nuclear weapons can be understood as a case of discursive legitimization of a technology before it was considered feasible.

During this time between 1938 and 1945, the notion became dominant that research and eventual deployment of nuclear weapons might be considered necessary to avoid an even worse disaster (Jervis 1989; Schelling 1966), while only very few people were involved in the discussion of nuclear strategy (Brodie 1966: 9). Further, it was still an open question whether the nuclear bomb was just another weapon or whether it would completely change world politics (Kaplan 1983). There is, however, no doubt that the potential military use of nuclear weapons was already at this time dominated by a perception of threat, for example by the rather successful campaigns of Nazi forces all over Europe and the potential that Nazi Germany might build such a weapon first (Reed 2014: 5, 123, 152; Rhodes 2004: 19). In this sense, the academic discussion contributed, I claim, to the depoliticization of nuclear technology, which was subsequently considered not an issue for deliberation and dispute, but one which required decisive action. The following section develops similarities as well as differences between the two cases of CE technologies on the one hand and nuclear technology on the other. Afterwards, the presence of the four frame themes—security, complexity, economy and appropriateness—is traced in the early discussion of nuclear technology between academic experts and policy-makers in order to illustrate their relevance for constructing the meaning of nuclear technology.

4.1 Similarities and Differences

Several common properties can be identified between the two cases of the early academic discussion on CE and the phase of early research into nuclear technology.[1] First, *time* is of the essence both in beating your enemy to developing nuclear capabilities and in addressing climate change. In the former case, the urgency lies both in the development of the technological capacities to design a functioning bomb before Nazi Germany, and in securing the material necessary to build it. The first issue was all the more pressing for the fact that nuclear fission had been discovered in Nazi Germany in 1938 (Rhodes 2004: 17; Einstein 2014: 122). The second issue was brought to the attention of US President Franklin Roosevelt by economist Alexander Sachs, who argued that "with the danger of a German invasion of Belgium, it was urgent that arrangements be made with the mining firm of Union Minière du Haut-Katanga, whose head office was in Brussels [...]" (Reed 2014: 123) In the case of climate change, the projected changes in temperature and precipitation patterns and the resulting implications for the decades to come contribute to the notion of urgency to act. Thus, the later aggressive climate action starts, the lower the probability that the Paris goals can be met and dangerous climate change averted (Sanderson et al. 2016).

Second, *scientific uncertainties* and unknowns play a major role in the discussion on the development of both CE technologies and nuclear capabilities, which makes both of them cases in which the nexus between policymakers and scientists is very close. Rhodes (2004: 20) argues that, in the case of nuclear technology, it was uncertain "whether or not such weapons were inventable [sic] [...] [and] whether or not the enemy was capable of inventing them in time to affect the outcome of the war." Similarly, there are aspects, consequences and implications of CE that are not known, and some of these, especially so-called 'unknown unknowns' might not be knowable through modelling, without field testing (Michaelson 2013: 101). Accordingly, both supporters of nuclear technology in the 1930s as well as of CE today strongly favor research in order to have a clear understanding of what the actual use would entail (Hamilton 2013a: 125). However, testing in the name of research is difficult in both cases as any experiment yielding indicative results is difficult to control and limit (Hulme 2014: 63). When, in the case of nuclear weapons technology, it was shown that it could be done, the security framing had imbued the technology with such moment that deployment followed, even against the intentions of the researchers involved. In the case of CE, researchers such as David Keith and Ken Caldeira, both of whom are very visible in the CE discussion (Buck 2013: 173; Oldham et al. 2014: 8), aim to "avoid global catastrophe" (Caldeira and Keith 2010: 57). They argue that, "if SRM proves to be unworkable or to pose unacceptable environmental risk, the sooner scientists know this, the faster they can take these options off the table" (Caldeira and Keith 2010: 62). Whether it is then really in the control of these researchers to take the options 'off the table' is a different question.

[1] Some ideas expressed in this section have been published in Lederer and Kreuter (2018).

Third, some knowledge about *negative externalities* exists in both cases already during early research into the respective technologies. Accordingly, both instruments are perceived as a 'weapon of last resort' and even their proponents argue that they should only be employed when all other options have failed. The negative implications of a belligerent military demonstration of nuclear power on global politics were discussed in the Franck Report of mid-1945 which correctly predicted that the use of nuclear weapons by the USA would lead to an international arms race (Reed 2014: 370; Lanouette 2004: 77). Similarly, Brzoska et al. (2012: 191) warn that the deployment of certain CE technologies could lead to the development of new conflicts and the exacerbation of existing ones. In both cases, possible unilateral deployment of the technology in question might pose a threat to the other nations as well as the global system. On the other hand, there is the perception of potentially detrimental effects for democratic decision-making in both the case of nuclear weapons and of CE (Markusson et al. 2013: 281; Szerszynski et al. 2013: 2809; Hulme 2014: 136; Heyward and Rayner 2013: 20–21). In the case of nuclear weapons, democratic governance was restricted through the creation of the Top Policy Group which excluded legislators from decision-making (Reed 2014: 139), and by the strict application of secrecy measures to shield the Manhattan Project to develop the bomb in the USA from any and every intrusion (Lanouette 2004: 73).

Fourth, both instruments—CE and nuclear weapons—face various *governance challenges* (see, for example, Bellamy 2016). The basis of these challenges is what Rhodes calls the paradox of nuclear weapons: "[...] [T]hough nuclear weapons are the property of individual nation-states, [...] in their indiscriminate destructiveness they are a common danger to all, like an epidemic disease, and like an epidemic disease they transcend borders, disputes, and ideologies." (Rhodes 2012: 3) In the case of nuclear weapons, in 1944 there was still an undecided debate ongoing in the USA whether the weapons should come under international legal control (Reed 2014: 370). Similarly, CE technologies are indiscriminate of borders or nation-states. Whether and how potential CE deployment might be organized is still an open question (Lederer and Kreuter 2018). Whereas some claim that SRM, for example, is contrary to international law (Winter 2011), others argue that in times of a climate crisis, "the niceties of international law might be expendable" (Hamilton 2013a: 152). A need for governance instruments is commonly acknowledged, especially in the case of SRM (Stavins and Stowe 2019), but so far, no international treaty exists that covers all forms of CE and the existing mechanisms are commonly deemed insufficient to cover all governance needs (Victor 2008: 322; Banerjee 2011: 35; Humphreys 2011: 115; Wirth 2013: 413; Proelss 2012). Like in the case of nuclear weapons, however, researchers have launched calls and proposals for a dedicated international regime (Winickoff and Brown 2013: 81; Zürn and Schäfer 2013: 272; Lloyd and Oppenheimer 2014: 51–57). Very recently, these have been taken up in political circles, as Switzerland submitted a draft proposal titled "Geoengineering and its governance" at the fourth session of the UN Environment Assembly in March 2019 in Nairobi, which, however, was subsequently withdrawn (Federal Department of the Environment Switzerland 2019).

Finally, in both cases, CE and nuclear technology, *academic actors* can be assumed to play a special role in the early construction of meanings of the respective technological approach. The assumption has been, for the case of CE, developed above in detail.[2] In the case of nuclear technology, as will be outlined below, academic actors, in particular from Germany, the UK and the USA, in the 1930s worked on researching the possibilities of the technology while attempting to convince policy-makers of the necessity to support the investigation with a view, in particular, of the threat of Nazi Germany potentially possessing such a weapon. For example, in his letter of US-President Franklin Roosevelt in 1939, Albert Einstein argues that research on nuclear technology should be supported as Germany could be expected to be working on the same issue (Einstein 2014).

However, a number of important differences separate the two cases of CE and nuclear technology. First, the nature of the issue of what is aimed to be achieved through technology differs considerably (avoiding war/defeat vs. avoiding runaway climate change). In addition, in the case of nuclear weapons, the supposed threat was primarily framed in national terms, whereas the 'climate threat' is perceived to be global in nature. Also, the Manhattan Project was organized in a strictly hierarchical fashion, whereas research in CE approaches is organized in a much more 'bottom-up' direction, mainly concentrated around various research clusters (e.g. through financing by the Royal Society in the UK or the DFG in Germany) (see, e.g., Hamilton 2013a: 125). In addition, public engagement is sought actively in CE research (Bellamy and Lezaun 2017, 2015; Corner and Pidgeon 2010), while this was definitely not the case during the early research into nuclear technology. However, it is important to note that, because of comparatively little public knowledge of CE, "designers and facilitators of public engagement events had to construct the topic of geoengineering from scratch, so to speak, for participants who were largely ignorant of the subject" (Bellamy and Lezaun 2017: 413), which, again, stresses the relevance of framing in the context of CE decision-making. This, in turn, points to a further difference: The groups of people involved in the discussion of each of these technological approaches differ widely from each other. In the case of CE, it contains mainly researchers and a very few industrialists. Currently small, but continually increasing public awareness will, in all probability, play a role in this discussion. In the case of nuclear technology, this group included, in addition to the relevant research community, military and political planners early on in the debate. An involvement of military planners in CE development is, at this stage, difficult to imagine, as technologies other than CE approaches might be considered of larger strategic importance (e.g. cyber espionage or artificial intelligence). It is important to note, however, that military attempts to manipulate the weather—if not the climate—are not unprecedented (see Fleming 2010: 179). Finally, secrecy was an important feature of research into nuclear weapons (Groueff 2004: 34; Norris 2004: 65; Reed 2014: 6), which is arguably not the case in CE research.

[2] See Chap. 3.

4.2 The Development of Nuclear Technology—A Very Short Historical Overview

On the basis of the similarities and differences between CE today and nuclear technology in the USA in the 1930s and 1940s, I argue that the illustration of the theoretical construct through the latter case can be instructive as to how the four thematic sets can be relevant in the discussion of an emerging technology, and in particular, the four respective ideal-type frames legitimizing technology to address a problem. In this case, the security frame was of prime importance, but traces of the other three types can be ascertained as well.

The pertinent time period begins,[3] arguably, with the discovery that "it was possible […] to force [the nucleus] artificially to release some of its energy in artificial decay" (Rhodes 2012: 203) by French scientists Irène and Frédéric Joliot-Curie. The theoretical possibility of making use of the phenomenon for both peaceful energy generation and for producing devastating explosions was conceived of as early as 1935 by physicist Léo Szilard, who filed several patent applications (Rhodes 2012: 214). In 1938 and 1939, scientists from Germany, the UK and USA, aware of each other's progress, made relevant discoveries pertinent to the possibility of producing vast amounts of energy from fission. In December 1938, Otto Hahn and Fritz Strassmann in Germany discovered Uranium fission, and in March 1939, Enrico Fermi and Szilard found that neutrons necessary to produce a chain reaction are the product of fission (Rhodes 2012: 267, 291). Nazi German authorities became aware of the ominous potential of nuclear technology, and in September 1939 "the German atomic bomb project was well begun" (Rhodes 2012: 296). In October of the same year, US President Roosevelt was informed of the scientific discoveries and shortly after, the Advisory Committee on Uranium was founded. Further discoveries in 1940, in particular in the UK, which point more and more strongly towards the possibility of constructing a bomb based on nuclear fission, were accompanied by the German military invading several European countries in turn, first Poland in 1939, then of Norway, the Netherlands and Belgium as well as France in 1940. The USA developed its research governance infrastructure and liaised with respective institutions in the UK. In October 1941, the Top Policy Group was established by Roosevelt with the aim of establishing policy pertinent to the nuclear bomb potential. Rhodes argues:

> The United States was not yet committed to building an atomic bomb. But it was committed to exploring whether or not an atomic bomb could be built. One man, Franklin Roosevelt, decided that commitment—secretly, without consulting Congress or courts. It seemed to be a military decision and he was Commander in Chief. (Rhodes 2012: 379).

Shortly afterwards, on the basis of the third NAS report on the matter, "the United States had finally set its wheels to the bomb track […] Acceleration overcoming inertia, it now began to roll" (Rhodes 2012: 387).

[3]While I will try to give a rough overview over relevant developments on nuclear science between 1934 and 1945, the interested reader may find an excellent comprehensive history in Rhodes (2012).

According to Rhodes, there is no "smoking gun" to be found in the archives, no document which shows the "fateful decision to expedite research toward an atomic bomb" except for a short note by the President from November 27, 1941 on the third NAS report (Rhodes 2012: 387). One day after the attack on Pearl Harbor, on December 8, 1941, the US Congress agreed on a joint declaration of war "not only on Japan but Germany and Italy as well" (Rhodes 2012: 392). In December 1941, the S-1 program was started to ascertain how an atomic bomb might be assembled. In August 1942, S-1 was named the 'Manhattan Engineer District'—the Manhattan Project was born. The development work went on from 1942. In January 1945, Oak Ridge Laboratory produced enough enriched uranium to build a bomb. On April 12, 1945, President Roosevelt died of cerebral hemorrhage (Rhodes 2012: 613) and was succeeded in office by President Harry Truman. Truman is informed of the secret bomb project "within twenty-four hours of Franklin Roosevelt's death" (Rhodes 2012: 617). In May 1945, Germany surrendered to the Allied forces. In the summer of 1945, at meetings of the Interim Committee, the Target Committee and the scientific advisory panel to the Interim Committee, the questions of whether and, if so, where the bomb should be belligerently demonstrated were discussed. On July 15, the Potsdam conference commenced, bringing together Truman, Joseph Stalin and Winston Churchill in order to make decisions on how to manage defeated Germany. On July 16, Trinity test—the first successful detonation of an atomic bomb—was conducted. On July 26, the USA, the UK and China released the Potsdam declaration, calling for the surrender of Imperial Japan, threatening, in case of rejection, the "'prompt and utter destruction'" of Japan (Rhodes 2012: 692). Japan declined. On August 6 and 9, respectively, atomic bombs were dropped on the Japanese cities of Hiroshima and Nagasaki. Japan surrendered on August 13.

4.3 Thematic Frames in the Early Academic Discussion on Nuclear Technology

For this period between 1934 and 1945, the secondary literature indicates the presence of all four thematic sets of frames from the theoretical construct in the academic discussion about the research, development and, eventually, the deployment of nuclear weapons technology.[4] In accordance with a security framing of nuclear technology, a number of existential threats were pronounced to which the new weapon technology was presented as a potential protection—or even the only possible protection. For example, the threat posed by Nazi Germany or by Imperial Japan to US security was considered. However, the security framing was not the only theme in the discourse legitimizing the development—and later the use—of the bomb: Another theme brought forth arguments of world peace through nuclear stalemate and of saving (soldiers') lives and ending a bloody war early by shocking the enemy into

[4]While an analysis of the primary literature would be more satisfying at this point, for reasons of practicability I chose to make use of the extensive secondary literature instead.

surrender through the atomic bomb (Rhodes 2012: 308, 678, 696) and can be linked to the thematic set of appropriateness. A third theme pertained to the philosophy of science culturally influential in the US at the time which considered science a guarantor of human progress and presented the development of the bomb as the 'simpler' measure compared to political difficulties, which is in line with the theme of the 'techno-fix'. The theme of economy is, likewise, represented in the secondary literature on the early discussion of the technology, for example in arguments by researchers presenting the construction of nuclear plants negligible when compared to the cost of a war (Rhodes 2012: 323).

With regards to *security* frames, chronologically, the first threat identified to which the research into nuclear weapons technology was presented as a protection and which set off the discourse to begin with, were the successful campaigns of the Nazi forces all over Europe during 1939 and 1940 (see, for example, Rhodes 2012: 344) and the possibility that Nazi Germany might build such a weapon first. Interestingly, the first push towards the study of nuclear weapons and the first presentation of the threat of Nazi Germany or other enemy powers came from scientists, not militarists or government (see, for example, Rhodes 2012: 293). For example, when Szilard successfully offered his patents for safekeeping to the British Admiralty in 1935 with the help of fellow-physicist Frederick Lindemann, the latter enclosed a letter expressing concern about "the disasters which could be caused by their use on the part of certain Powers which might attack this country" (letter by Lindemann cited in Rhodes 2012: 225). The focus of nuclear scientists on Nazi Germany as a threat is not surprising on two counts: First, a number of influential European-born nuclear scientists had fled from Fascist persecution in the late 1930s and early 1940s to the US or Great Britain—among them Leo Szilard, Edward Teller, Enrico Fermi, Hans Bethe, Rudolf Peierls and Otto Frisch (Rhodes 2012: 168–198). Second, the ability of Uranium to fission and thus release energy—the physical mechanism that renders possible the atomic uranium bomb—was first discovered in Nazi Germany by Otto Hahn and Fritz Strassmann in late 1938 and published in January 1939 (Rhodes 2012: 266).[5] Accordingly, it was conceivable that research in this direction would progress quickly in Germany.

Arguably, it was a group of émigré physicists around Leo Szilard, Edward Teller and Eugene Wigner who first introduced the security framing to politics by asking Albert Einstein in July 1939 to address a letter to the US government and to alert it to the danger of a Nazi nuclear program (Rhodes 2012: 305; S. Klein 1963: 227).

[5]And really, in April 1939, a secret conference was held in Berlin and a secret research program established. Rhodes argues that, in September 1939, "the German atomic bomb project was well begun." (Rhodes 2012: 312) The extensive military successes of Nazi Forces in Germany fed the danger further, as Germany, in 1940, had access to industrial and technological assets relevant to building a bomb (e.g. a heavy-water factory, stores of uranium and a cyclotron) in Norway, Belgium and France (Rhodes 2012: 344). However, the project never seemed to have yielded any practical results. As the ALSOS intelligence mission, sent from the US to Germany in September 1944 to gather information on the German nuclear program, found: "The evidence at hand proved definitely that Germany had no atom bomb and was not likely to have one in any reasonable form." (cit. in Rhodes 2012: 607).

Szilard considered it "possible" as early as 1934 to "construct atomic bombs" (Weart and Weiss Szilard 1978: 17; cit. in Rhodes 2012: 28) "What [the discovery of fission] meant to Szilard [in February 1939], Teller remembers, was 'Hitler's success could depend on it.'" (Rhodes 2012: 281) In Einstein's letter, the issue is presented as a threat and the development of a US nuclear bomb as a way to counter this threat: The famous physicist explains that "it is conceivable […] that extremely powerful bombs of a new type may thus be constructed" (Einstein 2014: 122) and recommends that research into nuclear fission as well as the securing of uranium ore—which is needed for one type of chain reaction necessary for a nuclear bomb—should be pursued by the US government. Similarly, the Frisch-Peierls memorandum, written in March 1940 by two German scientists who had fled to Great Britain, argues that.

> [a]s a weapon, the super-bomb would be practically irresistible. There is no material or structure that would be expected to resist the force of the explosion" and that "it is quite conceivable that Germany is, in fact, developing this weapon. (Frisch and Peierls 1940; cit. in Rhodes 2012: 324f.)

Thus, the security framing presented, from its outset, features of both the legitimizing and the delegitimizing variant, which were tightly linked to each other: On the one hand, the discovery of the weapons potential of fission was considered a threat, but only if came into the wrong hands. On the other hand, the research and development of such a weapon by the right agents was considered to provide protection against enemy powers.

The legitimizing variant can be argued to increase in strength over time, however, as it was adopted increasingly by nuclear scientists between 1939 and 1941 in order to overcome government inertia and to alert the attention of both the US and the British government to the threat. Government needed to be convinced. This is apparent, for example, from the first report on nuclear fission by the NAS published in 1941 on orders of the National Defense Research Council (NDRC): "It would seem to us unlikely that the use of nuclear fission can become of military importance within less than two years" (First NAS Report, cit. in. Rhodes 2012: 365). Rhodes finds that "the American program was on danger for its life that summer, [member of the NDRC and physicist Karl] Compton thought: 'The government's responsible representatives were … very close to dropping fission studies from the war program.'" (Rhodes 2012: 368). It was saved, argues Rhodes, in part by a report prepared by the British Military Application of Uranium Detonation (MAUD) Committee in July 1941 (Rhodes 2012: 377). They argued: "The committee considers that the scheme for a Uranium bomb is practicable and likely to lead to decisive results in the war.[…] It recommends that this work be continued on the highest priority […]" (Thomson et al. 1941: 3). The writers of the report calculated that a plant producing three atomic bombs per month would cost approximately 5,000,000 £. However: "In spite of this very large expenditure we consider that the destructive effect both material and moral is so great that every effort should be made to produce bombs of this kind" (Thomson et al. 1941: 1). The threat of utter destruction is used to legitimize the development of the expensive technology. As Rhodes describes, after the MAUD Report had reached the President on October 9, 1941, "the United States had finally set its wheels to the bomb track.[…] Acceleration overcoming inertia, it now began to roll" (Rhodes 2012: 377). Not before June 1940, however, with the foundation of the NDRC in the USA, was the issue discussed seriously by government (Rhodes 2012: 337).[6] From then on out, the security framing was carried by actors from both science and politics, partly in common committees.[7] In late 1941, with the official beginning of the so-called S-1 program, military joined in the discussion.

Meanwhile, American as well as immigrant physicists continued to argue for the research of the nuclear weapon: "Patriotism contributed to many decisions, but a deeper motive among the physicists, by the measure of their statements, was fear—fear of German triumph, fear of a thousand-year Reich made invulnerable with atomic bombs" (Rhodes 2012: 368). Rhodes argues that it was "their primary reason for urging an American bomb" (Rhodes 2012: 405). This argument was used, for example, by Danish physics Nobel Prize laureate Niels Bohr when he visited Los Alamos in early 1943 to motivate the scientists working on the project, as Oppenheimer recounts in a lecture after the war:

> Bohr spoke with contempt of Hitler, who with a few hundred tanks and planes had tried to enslave Europe for a millennium. He said nothing like that would ever happen again; and his own high hope that the outcome would be good […] all of us wanted very much to believe. (Oppenheimer, cit. in Rhodes 2012: 524).

With Nazi Germany beginning to lose the Second World War in mid-1943, scientists started to "sense the country's desperation" and experienced "great apprehension" of the Nazi threat (Rhodes 2012: 512): Hans Bethe and Edward Teller wrote a memorandum to Oppenheimer in August 1943 to alert him to the possibility that Germany might already have a nuclear bomb if it had done research on it since the discovery of fission:

> Recent reports both through the newspapers and through secret service have given indications that the Germans may be in possession of a powerful new weapon which is expected to be ready between November and January. There seems to be considerable probability that this new weapon is tubealloy [i.e., uranium]. It is not necessary to describe the probably consequences which would result if this proves to be the case […]. (Teller and Bethe, cit. in Rhodes 2012: 511f.)

The physicists go on to propose a crash program to accelerate bomb development. "Nothing seems to have come from the Bethe-Teller proposal—Hitler's secret weapon proved to be the V-1 and V-2 rockets […]—but it captured the mid-war mood" (Rhodes 2012: 512) and the strength of the presentation of Nazi Germany as an existential threat.

[6]The Advisory Committee on Uranium that was founded in October 1939 started a discussion of the issue, but according to historian Richard Rhodes, "a bomb of uranium seemed a remote possibility at best. However intense their sympathies, the war was still a European war." (Rhodes 2012: 317).

[7]For example the Interim Committee which was appointed in May 1945 under the leadership of Secretary of War Henry L. Stimson to discuss the political and social effects of the bomb, which had a Scientific Panel (Hewlett and Anderson 1962: 356) which comprised Arthur Compton, Ernest Lawrence, J. Robert Oppenheimer and Enrico Fermi (Rhodes 2012: 630).

Later on, when the threat of Nazi Germany receded, the existential threat was partly perceived to emanate from other powers. After the attack on Pearl Harbor, for example, the threat of Japan rose in importance in the discussion as both Japanese soldiers and civilians were considered never to surrender and to "fight to the bitter end" (speech by US Ambassador to Japan Joseph C. Grew in 1942 cited in Rhodes 2012: 520). A more general security framing did not consider any specific nation as an existential threat to U.S. security, but the bomb itself as a threat that would come no matter what. As early as 1936, chemist Francis William Aston argued in a lecture:

> Personally, I think there is no doubt that sub-atomic energy is available all around us, and that one day man will release and control its almost infinite power. We cannot prevent him from doing so and can only hope that he will not use it exclusively to blow up his neighbor. (Cornford et al. 1936: 114; cit. in Rhodes 2012: 141).

This notion of fatalism was echoed by the scientists who urged the U.S. government to take up bomb research in 1941 and 1942: "[…] Deeper than fear was fatalism. The bomb was latent in nature as a genome is latent in flesh. Any nation might learn to command its expression. The race was […] against time." (Rhodes 2012: 379) British researcher James Chadwick came to a similar conclusion:

> I remember the spring of 1941 to this day. I realized then that the nuclear bomb was not only possible—it was inevitable. Sooner or later ideas could not be peculiar to us. Everybody would thing about them before long, and some country would put them into action. (cit. in Rhodes 2012: 356).

Similarly, Ernest Lawrence battled the lack of commitment of the authors of the first two NAS Reports of 1941:

> It will be a calamity if, when we get the answers to the uranium problem, they turn out negative from the military point of view, but if the answers are fantastically positive and we fail to get them, first, the results for our country may well be a tragic disaster. I feel strongly, therefore, that anyone who hesitates on a vigorous, all-out effort on uranium assumes a grave responsibility. (cit. in Rhodes 2012: 383).

This notion was picked up in the third NAS Report which was published in November 1941:

> The possibility must be seriously considered that within a few years the use of bombs such as described here, or something similar using uranium fission, may determine military superiority. Adequate care for our national defense seems to demand urgent development of this program. (cit. in Rhodes 2012: 386f.)

The legitimation of the bomb through a threat is summarized by a number of historians. Reed argues that also the scientists working on the Manhattan Project were "[m]otivated by the existential threat of the war and the prospect that German scientists were likely thinking along the same lines […]" in decisions made during the development process (Reed 2014: 5). Similarly, Rhodes argues:

> Why would these men of good will, who believed themselves to be members of a peaceful international community of scientists, want to build a weapon of mass destruction? Always and everywhere in that first round of nuclear proliferation the same reason repeats: because possession of such a weapon seemed the only defense against an enemy similarly armed. (Rhodes 2004: 19).

Thus, the unthinkable—the development of a bomb with unprecedented destructive power and numerous long-term effects—became legitimized, partly through the application of security framing in discussing its possibilities.

Beyond the security rationale, the academic discussion on nuclear weapons technology presented elements in accordance with a theme of *appropriateness*: While a specific set of values is not mentioned, Rhodes argues that, for many of the researchers involved, the potential of the bomb represented the possibility of achieving a specific ideal: life in which "'war as an institution [was] abolished'" (Carl-Friedrich von Weizsäcker, cited in Rhodes 2012: 312): For example, Eugene Wigner, at the time of Einstein's letter, saw a benefit to humankind which might emerge from the existence of the weapon:

> We realized that, should atomic bombs be developed, no two nations would be able to live in peace with each other unless their military forces were controlled by a common higher authority. We expected that these controls, if they were effective enough to abolish atomic warfare, would be effective enough to abolish all forms of war. (Wigner cited in Rhodes 2012: 308).

In this way, researching the bomb might have been represented as the 'right thing to do' in order to achieve world peace.

Connected to this notion was the feeling that the bomb might, inadvertently, provide a solution to the *complexity* of the political problem of providing peace:

> Wars waged with atom bombs as regularly recurring events, that is to say, nuclear wars as institutions, do not seem reconcilable with the survival of the participating nations. But the atom bomb exists. It exists in the minds of some men. According to the historically known logic of armaments and power systems, it will soon make its physical appearance. If that is so, then the participating nations and ultimately humankind itself can only survive if war as an institution is abolished. (Carl-Friedrich von Weizsäcker, cited in Rhodes 2012: 312).

As Rhodes argues: "By necessity, commonly understood, to avoid triggering a nuclear holocaust, the *deus ex machina* would have accomplished then what men had been unable to accomplish by negotiation or by conquest: the abolition of major war." (Rhodes 2012: 533) Thus, a comparatively simple technological fix was perceived to be able to provide a solution to a century-old social problem of human kind. This theme was picked up by Winston Churchill after the war, when he argued:

> To avert a vast, indefinite butchery, to bring the war to an end, to give peace to the world, to lay healing hands upon its tortured peoples by a manifestation of overwhelming power at the cost of a few explosions, seemed, after all our toils and perils, a miracle of deliverance. (Churchill 1954: 553; cit. in Rhodes 2012: 697).

This notion of a comparatively easily achieved technological solution to human suffering was bolstered by technological progressivism, which supported the scientists active in the project in their conviction that their research was for the greater good: Oppenheimer, when recruiting scientists for Los Alamos in late 1942, encountered this philosophy implicitly in the answers of the colleagues he approached:

> Almost everyone knew that it was an unparalleled opportunity to bring to bear the basic knowledge and art of science for the benefit of his country. Almost everyone knew that this job, if it were achieved, would be part of history. This sense of excitement, of devotion and of patriotism in the end prevailed. Most of those with whom I talked came to Los Alamos. (Oppenheimer in Commission 1954: 13; also cit. in Rhodes 2012: 452).

In this view, science and technology might prevent the age-old human threat of war, a problem that politics had hitherto been unable to solve. The theme of *economy* was also connected to these considerations to some degree: Rudolf Peierls, one of the authors of the Frisch-Peierls Memorandum, for example, illustrated the low cost of a nuclear plant by arguing: "The cost of such a plant [….] would be insignificant compared with the cost of the war" (Rhodes 2012: 323).

4.4 A Multitude of Factors Influencing the Decision

To sum up the illustration, instances of elements of all four thematic sets of frames can be identified in the early academic discourse on nuclear technology in the United States, albeit to different degrees. A causal relation between the presence of these elements and the events that followed—namely the increasing secrecy of the research and development of nuclear technology, followed by the eventual decision of belligerently demonstrating the power of the new bomb—cannot be drawn on the basis of these considerations. We can conclude, however, that, in the case of nuclear weapons technology, the presence of legitimizing frames in the early academic discussion coincides with its development and use. A multitude of factors bore on this outcome.

Other authors have, in tremendous detail, addressed the steps which preceded the eventual detonation of the two atomic bombs in the cities of Hiroshima and Nagasaki as well as their impact on subsequent politics (Rhodes 2012; Alperovitz 1965; Groueff 2004; Lanouette 2004; Norris 2004; Fehner and Gosling 2012; Brodie 1966; Schelling 1966; Jervis 1989; Jungk 1979; Kaplan 1983; Bernstein 1990; Churchill 1954; Groves 1965; Stimson 1947). What remains, here, to be mentioned are a few features of these developments concerning the researchers who, arguably, initiated the discussion on nuclear technology in the 1930s.

First, *secrecy* became a feature of the nuclear development program in the USA early on and was tightened subsequently. Szilard would later find that "the most powerful weapon to result from the Manhattan Project was not the A-bomb but the 'SECRET' stamp." (Lanouette 2004: 73) Science was not the only field in which deliberation was restricted. Through the establishment of the 'Top Policy Group' in October 1941, a few months after he was informed about the MAUD Report,

Roosevelt excluded legislators from decisions concerning nuclear weapons policy, most of which had no idea that the project even existed (Reed 2014: 148). He restricted decision-making instead to.

> a group comprising himself, the Vice-President, Secretary of War Henry Stimson, Army Chief of Staff General George C. Marshall, and [Director of the National Research Council Vannevar] Bush and [James B.] Conant [of the National Defense Research Committee] (Reed 2014: 139).

The decision to explore the possibilities was made exclusively by the President: "One man, Franklin Roosevelt, decided that commitment—secretly, without consulting Congress or courts" (Rhodes 2012: 379). In October 1942, decision-making moved away even from the executive to a degree, when the S-1 program was assigned to the army's Corps of Engineers under the direction of Major General Leslie R. Groves (Reed 2014: 148). The Major General was, in many respects, under no effective oversight through the executive or parliament in his dealings with the Project. Groueff (2004: 35) argues that "[t]he General was never given the extraordinary powers he used; he simply took them." According to Groueff, Groves himself conceded that the decision made were only possible in a state of exception: "He firmly believed that many urgent decisions he had to take could not be made by committees, or by consent, or by asking permission every time" (Groueff 2004: 34). Under the leadership of the military, whole Manhattan Project installations were kept secret, such as the famous Los Alamos Laboratory:

> Established in the spring of 1943 in the high desert of Northern New Mexico and directed by physicist J. Robert Oppenheimer of the University of California, the task of this secret installation was to design the weapons that would be powered by the uranium and plutonium being produced in Tennessee and Washington. (Reed 2014: 10).[8]

Secrecy was upheld by Groves, who had every scientist reviewed who was to work on the secret project (Norris 2004: 65).

While the management of the project to develop the bomb lay in military hands starting in 1942, both specific *scientific and political actors* still played a pivotal role in nuclear policy, and also in the discussion about whether or not the bomb should be belligerently demonstrated. The President himself, but also his immediate advisors as well as a number of scientists belonged to the select few who made the decisions concerning the use bomb in 1945. Two committees were of particular importance: The Target Committee, which met under Groves' leadership for the first time in late April 1945 to decide on the target of the bomb in Japan (Rhodes 2012: 626); and the Interim Committee, which was founded in May 1945 on orders of the newly instated President Truman (Rhodes 2012: 628). The two committees each met several times in April and May 1945. The Target Committee comprised three military men—two Air Force officers as well as Grove's deputy—and five scientists "including John von Neumann and British physicist William G. Penney" (Rhodes 2012: 626). In later meetings, they would be joined by Robert Oppenheimer, Hans

[8]For an overview over the institutional structure of the Manhattan Project, see Reed (2014).

Bethe and some others (Rhodes 2012: 630). The committee's aim was to decide on the possible targets of a nuclear attack against Japan. Groves set the guidelines: "I had set as the governing factor that the targets chosen should be places the bombing of which would most adversely affect the will of the Japanese people to continue the war" (see also Groves 1965: 270; Groves, cit. in Rhodes 2012: 627).[9] Rhodes maintains that "the committee completed its review by abandoning any pretension that its objectives [in Japan] were military [...]" (Rhodes 2012: 639).

The aim of the Interim Committee was to make policy proposals on dealing with the newly acquired nuclear power. Open questions concerned the control of the weapon during war-time as well as publicity "for a time when secrecy was no longer in effect" (Reed 2014: 375) and post-war research and governance. The Committee consisted of representatives from the executive branch of government as well as a number of scientists (Reed 2014: 375; Stimson 1947: 100). Truman agreed to put the committee in place, but "significantly did not even bother to name his own man to the list" (Rhodes 2012: 628) of participants: no-one there represented the President personally. The importance that the President accorded the committee can be questioned in these grounds. During the first meeting, it was suggested that leading scientists of the Manhattan Project should be invited to serve on the Committee as well, to "present their views on international relations in the context of the bomb" (Reed 2014: 375). Accordingly, a Scientific Panel was appointed (Rhodes 2012: 630). The Panel was invited to advise "not only on technical matters, 'but also to present to the Committee their views concerning the political aspects of the problem" (Reed 2014: 375). A common meeting of the Interim Committee and the Scientific Panel took place on May 31, 1945. The main question discussed at the meeting was whether the bomb should be demonstrated peacefully or used belligerently against Japan. "[It] was pivotal in the sense of arriving at a 'decision' as to how the bomb would be used. The entire committee plus the Scientific Panel was present, as were Generals Grove and Marshall as invited guests." (Reed 2014: 375) The Committee came to a decision on the use of the bomb against Japan that met, according to the minutes of the meeting as quoted by Reed, with widespread approval:

> After much discussion concerning various types of targets and the effects to be produced, *the Secretary expressed the conclusion, on which there was general agreement, that we could not give the Japanese any warning; that we could not concentrate on a civilian area; but that we should seek to make a profound psychological impression on as many of the inhabitants as possible. At the suggestion of Dr. Conant the Secretary agreed that the most desirable target would be a vital war plant employing a large number of workers and closely surrounded by worker's houses.* (minutes of the Interim Committee meeting of May 31, 1945, as cited in Reed 2014: 376).

Thus, the scientists present at the meeting did not protest against belligerent use of the weapon. On the contrary, Oppenheimer "as the head of the [Scientific] panel, [...] argued that there was no practical alternative to military use of the weapon against

[9]Kyoto, Hiroshima and Niigata were chosen, but Kyoto was taken off the list by Secretary of War Stimson (Rhodes 2012: 640; Groves 1965: 276–278). Nagasaki would be added to the list as late as July 25, probably on recommendation of Air Force staff (Rhodes 2012: 689).

Japan." (Herken 2004: 70; Rhodes 2012: 647) However, Oppenheimer proposed that the United States should "offer the world free interchange of information with particular emphasis on the development of peace-time uses" (Reed 2014: 376) in the event of a successful demonstration. Also, Oppenheimer would later, in 1949, "oppose development of the hydrogen bomb on both practical and ethical grounds"[10] (Herken 2004: 69). The reasons for the use of the bomb against Hiroshima and Nagasaki discussed in the literature include the wish to end the war Japan quickly, faster than possible with conventional means of warfare—such as an invasion—but also the idea the threatening influence of the Soviet Union in global politics needed to be curbed (Alperovitz 1965: 240; Hershberg 2004: 51; Rhodes 2004: 26; 2012: 697).

However, while the whole process had been set in motion by concerned scientists in 1939, and the discourse included until 1945 voices from the academic sphere, between 1938 and 1945, again and again a number of scientists involved in the Manhattan Project rose up in protest, trying to counter the dominant discourse legitimizing the development and then also the use of the bomb. Szilard was one of the main figures of protest. As early as 1934, "he recalled later, 'there was little doubt in my mind that the world was headed for grief'" (Rhodes 2012: 292). In 1942, when S-1 was already well under way, Szilard published a memorandum in which he strongly criticized the lack of deliberation between all parts of the project: "Stated in abstract form, the trouble at Chicago [Metallurgy Laboratory] arises out of the fact that the work is organized along somewhat […] authoritarian rather than democratic lines" (Weart and Weiss Szilard 1978: 156; cit. in Rhodes 2012: 413). In the year of the first successful test, 1945, he acted on his premonitions and recommended instead of belligerent use a peaceful demonstration to prove its capacities to Japan and thus convince it to accept the United States' terms of surrender.

> […] in June [of 1945] Szilard helped draft the Franck Committee report in which Manhattan Project scientists proposed demonstrating the bomb [instead of using it]. This was also ignored in Washington, leaving Szilard a last, symbolic effort—a scientist's petition to President Truman. Szilard […] sent it up the Army chain-of-command on July 17th, the day after the A-bomb was first tested. To undermine the petition, Groves' staff polled the scientists but discovered that 83% favored a demonstration. In all, 155 Manhattan Project scientists signed versions of the petition, which urged the president to weigh moral and diplomatic issues before using the bomb. (Lanouette 2004: 77).

The effort was to no avail. The bomb had gained such relevance in the political discourse that its necessity was taken for granted. "[O]n its way to the White House Szilard's petition was pigeonholed by Groves' staff until after A-bombs destroyed Hiroshima and Nagasaki in August 1945" (Lanouette 2004: 77). Similarly, Niels Bohr's attempts to start international cooperation on the control of the bomb before

[10] Oppenheimer was, after the end of WW II, accused of espionage for the Soviet Union. The transcripts of his 1954 hearings, conducted by the Atomic Energy Commission, were published subsequently by the US Atomic Energy Commission (Commission 1954). The complete transcript including sections deleted in this earlier version have been published by the US Department of Energy in 2014 (US Department of Energy 2014).

it would ever be used were rejected (Rhodes 2012: 534). This development indicates that the decision which researchers were to be heard in the Interrim Committee actually made a difference on the position that they took: If, for example, Szilard would have formed part of it instead of Oppenheimer, the stance might have been different.

The first successful detonation of the bomb took place on July 16, 1945 at Almogoro Air Base in New Mexico. At this point, according to historian Gar Alperovitz, in the decision-making on the bomb, its use was not questioned any longer in the executive: "It was *assumed*—not decided—that the bomb would be used" (Alperovitz 1965: 236). According to Alperovitz, President Truman and his advisors did "not seem to have shared Eisenhower's 'grave misgivings' [...]. [T]hey simply assumed that they would use the bomb, never really giving serious consideration to not using it" (Alperovitz 1965: 237). Reed agrees with Alperovitz' assessment when he states: "President Truman did not make a 'hard decision' to use the bomb so much as he elected not to alter a chain of events that was already far along when he inherited the Presidency" (Reed 2014: 383). Thus, a mere three weeks after the first successful test, the bomb was used in combat, first in Hiroshima, on August 6, and then in Nagasaki, on August 9.

4.5 Summary

These developments give a first indication of the relevance that specific legitimizing framings might have in the construction of the meaning of a new, or emerging, technology: For instance, the early introduction of the issue into political circles by researchers like Szilard, Teller and Wigner in 1939 and Peierls and Frisch in 1940 presented nuclear technology partly through security framing. The legitimizing security frame is subsequently present throughout the discussions of the technology from then onwards, through its stages of research and development until the moment when the bomb is actually available for use. The frame proves resilient to changes in political conditions when it survives the recession of the Nazi German threat, shifting its focus onto threats posed by Imperial Japan and Soviet Russia. And it is carried on by political and military actors who use it to argue, not for the research into the potential of the bomb, but its belligerent use, even against the opposition of those researchers who first used the frame to draw attention to the issue. This interpretation of events, while it cannot prove the relevance of this framing, can render it plausible by presenting an account of how the frame influences decision-making at various junctures. Furthermore, it indicates that science does not always speak with one unified voice: Had a different set of scientists been chosen to "present their views on international relations in the context of the bomb" (Reed 2014: 375) to the Interrim Committee, the stance taken by 'science' with regards to the potential use of the bomb might have been different. This argument supports the claim that the diversity of framing in the academic discussion of CE should be preserved when decision-making is conducted.

References

Alperovitz G (1965) Atomic diplomacy: hiroshima and potsdam. Simon and Schuster, New York

Banerjee B (2011) The limitations of geoengineering governance in a world of uncertainty. Stanford J Law Sci Policy 4(1):15–36

Baum SD (2014) The great downside dilemma for risky emerging technologies. Phys Scr 89:1–10. https://doi.org/10.1088/0031-8949/89/12/128004

Bellamy R (2016) A sociotechnical framework for governing climate engineering. Sci Technol Human Values 41(2):135–162. https://doi.org/10.1177/0162243915591855

Bellamy R, Lezaun J (2015) A review of deliberative public engagements with climate engineering. In: Climate geoengineering governance working paper series, 019

Bellamy R, Lezaun J (2017) Crafting a public for geoengineering. Public Underst Sci 26(4):402–417. https://doi.org/10.1177/0963662515600965

Bernstein BJ (1990) An analysis of 'two cultures': writing about the making and the using of the atomic bombs. Public Historian 12(2):83–107

Brodie B (1966) Escalation and the nuclear option. Princeton University Press, Princeton, NJ

Brzoska M, Link PM, Neuneck G (2012) Geoengineering—Möglichkeiten und Risiken. Sicherheit and Frieden 30(4):185–193

Buck HJ (2013) Climate engineering: spectacle, tragedy or solution? a content analysis of news media framing. In: Methmann C, Rothe D, Stephan B (eds), Interpretive approaches to global climate governance: (De)constructing the greenhouse, London, Routledge, pp 166–180

Caldeira K, Keith DW (2010) The need for climate engineering research. Issues Sci Technol Fall 2010:57–62

Churchill W (1954) In: Triumph and tragedy: the second world war, vol 6. London et al., Cassell and Co. Ltd

Commission USAE (1954) In the matter of J. Robert Oppenheimer. Transcript of hearning before personnel security board April 12, 1954, through May 6, 1954. Washington D.C, Government Printing Office

Corner A, Pidgeon N (2010) Geoengineering the climate: the social and ethical implications. Environ Sci Policy Sustain Develop 52(1):24–37

Cornford FM, Dampier W, Rutherford, Bragg WL, Aston FW, Eddington AS et al (1936) In: Background to modern science. Ten lectures at cambridge arranged by the history of science committee (The Cambridge Library of Modern Science) Cambridge, Cambridge University Press

Einstein A (2014) Letter to President Roosevelt from August 2, 1939. In: Reed BC (ed) The history and science of the manhattan project. Springer, Heidelberg, pp 121–123

Federal Department of the Environment Switzerland (2019) Concept note, accompanying note to draft resultion for the 4th session of the united nations environment assembly of UNEP, working title: geoengineering and its governance. United Nations Office at Nairobi, Nairobi

Fehner TR, Gosling FG (2012) The manhattan project. U.S, Department of Energy

Fleming JR (2010) Fixing the sky: the checkered history of weather and climate control. Columbia University Press, New York, NY

Frisch OR, Peierls R (1940) Memorandum on the properties of a radioactive 'super-bomb'

Groueff S (2004) The Manhattan project: an extraordinary achievement of the 'american way'. In: Kelly CC (ed) Remembering the manhattan project: perspectives on the making of the atomic bomb, New Jersey, World Scientific, pp 31–38

Groves LR (1965) Jetzt darf ich sprechen. Kiepenheuer and Witsch, Köln et al

Hamilton C (2013) Earthmasters: the dawn of the age of climate engineering. Yale University Press, New Haven, CT

Hamilton C (2014) Geoengineering and the politics of science. Bull Atom Sci 70(3):17–26. https://doi.org/10.1177/0096340214531173

Herken G (2004) Science in the service of the state: the cautionary tale of Robert Oppenheimer. In: Kelly CC (ed) Remembering the manhattan project: perspectives on the making of the atomic bomb and its legacy, New Jersey, World Scientific, pp 69–71

Hershberg JG (2004) A footnote on hiroshima and atomic morality: conant, niebuhr, and an 'emotional' clergyman 1945–46. In: Kelly CC (ed) Remembering the manhattan project: perspectives on the making of the atomic bomb and its legacy. New Jersey, World Scientific

Hewlett RG, Anderson OEJ (1962) The new world, 1939/1946, A History of the United States Atomic Energy Commission, vol 1. University Park (PA), The Pennsylvania State University Press

Heyward C, Rayner S (2013) In: Apocalypse nicked! Climate geoengineering governance working paper series, 006

Horton JB (2013) Geoengineering and the myth of unilateralism: pressures and prospects for international cooperation. In: Burns WCG, Strauss AL (eds) Climate change geoenegineering: philosophical perspectives legal issues and governance frameworks. Cambridge University Press, New York, USA, pp 168–181

Hulme M (2014) Can science fix climate change? a case against climate engineering. Polity Press, Cambridge, UK

Humphreys D (2011) Smoke and mirrors: some reflections on the science and politics of geoengineering. J Environ Develop 20(2):99–120. https://doi.org/10.1177/1070496511405302

Huttunen S, Skytén E, Hildén M (2015) Emerging policy perspectives on geoengineering: an international comparison. Anthropocene Rev 2(1):14–32. https://doi.org/10.1177/2053019614557958

Jervis R (1989) The meaning of the nuclear revolution: statecraft and the prospect of armageddon. Cornell University Press, Ithaca

Jungk R (1979) The nuclear state. John Calder, London

Kaplan FM (1983) The wizards of armageddon. Stanford University Press, Stanford, CA

Keith DW (2013) A case for climate engineering. Boston Review, Cambridge/London

Kintisch E (2010) Hack the planet. science's best hope—or worst nightmare—for averting climate catastrophe. Wiley, New York, NY

Klein S (1963) Review: the new world, 1939/1946 by Hewlett RG, Anderson OE. Technol Culture 4(2):227–231

Lanouette W (2004) Leo szilard: baiting brass hats. In: Kelly CC (ed), Remembering the manhattan project: perspectives on the making of the atomic bomb and its legacy , New Jersey, World Scientific, pp 73–78

Lederer M, Kreuter J (2018) Organising the unthinkable in times of crises: will climate engineering become the weapon of last resort in the Anthropocene? Organization 25(4). https://doi.org/10.1177/1350508418759186

Lloyd ID, Oppenheimer M (2014) On the design of an international governance framework for geoengineering. Global Environ Politics 14(2):45–63

Markusson N, Ginn F, Ghaleigh NS, Scott V (2013) In case of emergency press here': framing geoengineering as a response to dangerous climate change. WIREs climate change 5(2):281–290. https://doi.org/10.1002/wcc.263

Michaelson J (2013) Geoengineering and climate management. from marginality to inevitability. In: Burns WCG, Strauss AL (eds) Climate change geoengineering: philosophical perspectives legal issues and governance frameworks, New York, USA, Cambridge University Press, pp 81–114

Norris RS (2004) General Leslie R. Groves and the Scientists. In Kelly CC (ed), Remembering the manhattan project: perspectives on the making of the atomic bomb and its legacy. New Jersey, World Scientific, pp 63–73

Oldham P, Szerszynski B, Stilgoe J, Brown C, Eacott B, Yuille A (2014) Mapping the landscape of climate engineering. Philos Trans Royal Soc A 372(2031).https://doi.org/10.1098/rsta.2014.0065

Proelss A (2012) Geoengineering and international law. Sicherheit and Frieden 30(4):205–211

Reed BC (2014) The history and science of the manhattan project. Springer, Heidelberg

Reynolds J (2014) The international regulation of climate engineering: lessons from nuclear power. J Environ Law 26(2):269–289. https://doi.org/10.1093/jel/equ006

Rhodes R (2004) The atomic bomb in the second world war. In: Kelly CC (ed) Remembering the manhattan project: perspectives on the making of the atomic bomb and its legacy. New Jersey, World Scientific, pp 17–30
Rhodes R (2012) In: The making of the atomic bomb (25th Anniversary Edition ed.) London et al., Simon and Schuster
Sanderson BM, O'Neill BC, Tebaldi C (2016) What would it take to achieve the Paris temperature targets? Geophys Res Lett 437133–437142. https://doi.org/10.1002/2016GL069563
Schelling TC (1966) Arms and influence. Yale University Press, New Haven
Stavins RN, Stowe RC (eds) (2019) Governance of the deployment of solar geoengineering. Harvard Project on Climate Agreements, Cambridge
Stimson HL (1947) The decision to use the atomic bomb. Harper's Mag 194(1161):97–107
Szerszynski B, Kearnes M, Macnaghten P, Owen R, Stilgoe J (2013) Why solar radiation management geoengineering and democracy won't mix. Environ Plan A 45:2809–2816
Thomson SGP, Oliphant M, Blackett P, Chadwick J, Moon P, Cockcroft J (1941) Report by the MAUD committee on the use of uranium for a bomb. Ministry of Aircraft Production, London
US Department of Energy (2014) In: Robert Oppenheimer J (ed) personnel hearning transcripts, U. D. o. Energy, Washington D.C., US Department of Energy
Victor DG (2008) On the regulation of geoengineering. Oxford Rev Econ Policy 24(2):322–336
Weart SR, Weiss Szilard G (1978) Leo szilard: his version of the facts. MIT Press, Cambridge (MA)
Winickoff DE, Brown MB (2013) Time for a government advisory committee for geoengineering research. Issues Sci Technol 29:79–85
Winter G (2011) Climate engineering and international law: last resort or the end of humanity?. Rev Eur Commun Int Environ Law 20(3):277–289. https://doi.org/10.1111/j.1467-9388.2012.00730.x
Wirth DA (2013) Engineering the climate: geoengineering as a challenge to international governance. Boston College Environ Aff Law Rev 40(2):413–438
Zürn M, Schäfer S (2013) The paradox of climate engineering. Global Policy 4(3):266–277

Chapter 5
The Tools for Empirical Analysis—The Method of Qualitative Content Analysis

Abstract Outlining the technical steps of empirical analysis, this chapter operationalizes content analysis as a tool to assess meaning-construction of emerging technologies through framing. Kreuter explores methodological implications of assessing peer-reviewed journal articles from various disciplines addressing CE approaches. Focusing on sampling and manual text analysis, Kreuter also points out possibilities for follow-up studies based on automated content analysis. The chapter concludes with an overview of eight ideal-type legitimizing or delegitimizing frames of security, complexity, economy and appropriateness.

The following chapter shortly outlines the method applied for empirical data generation and data analysis to assess the presence and frequency of the ideal-type frames in the case of the academic discussion of CE. The methodical approach in its entirety was informed by qualitative content analysis following Mayring and Früh. Data was generated by creating a corpus of academic journal articles addressing CE approaches according to a systematic search term, drawing a sample from this corpus according to specific rules and then coding the texts in the sample in accordance with the codebook. Data analysis was conducted through qualitative and quantitative analysis steps Table 5.1.

5.1 Content Analysis as Data Generation and Analysis Method

The method used in this thesis to answer the research question is qualitative content analysis following Mayring (2010, 2015, 2014) and Früh (2011). Following Früh, content analysis is here defined as.

> an empirical method for the systematic and intersubjectively reproducible description of formal as well as content features of messages, most often with the aim of drawing an

J. Kreuter, *Climate Engineering as an Instance of Politicization*,
Springer Climate, https://doi.org/10.1007/978-3-030-60340-3_5

Table 5.1 Important terms in analysis

Term	Definition
Code	Individual designator which can be applied to a text segment to categorize the content of that segment (e.g. 1.2 s or 3.1 SRM)
Coding	Instance of application of a code to a segment of text/document
Coded segment	Part of a document in the sample which has been tagged with a code (here: paragraph, due to choice of unit of coding = paragraph, see Codebook 5.0)
Set of coded segments	Entirety of coded segments in all documents in the sample
Document	A text in the sample
Sample	Set of texts coded in accordance with the codebook
Document variable	A variable which takes up one value for each document, e.g. 'Author' or 'Disciplinary Focus'; if codes are transferred into variables, the variable takes up an integer as value representing the number of codings for said code in the text
Text data	Information in the text itself
Meta/formal data	Information about the text, taken from outside of the text

> interpretative inference on the basis of this to circumstances external to the message.[1] (Früh 2011: 133).

Empirical methods, according to Früh, aim at translating a theoretical construct into a data model. He differentiates three levels on which empirical researchers operate: The *object* level contains the portion of reality that is of interest to the study. In the case of this thesis, that is the academic discussion on CE. The *theory* level contains a theoretical construct, a "structured image of perception"[2] (Früh 2011: 22), which is created to represent all features and relations of the portion of reality which are relevant to address the research question. In the case of this study, this would be the construct containing the four ideal-type frames. The *data* level, finally, contains a data model which represents the theoretical construct in the form of empirical data established through empirical method. In this study, the data model contains all codings applied to all texts studied. Thus, it represents the portion of reality in question—albeit seen through the perspective of the coders, who subjectively make decisions on coding specific text sections with specific codes. While the subjectivity of the coders has been lowered through intra—and inter-coder reliability testing procedures, social reality, according to the social constructivist perspective adopted in this study, can only ever be constructed. In the case of this study, the data level contains all codings in the texts studied in the empirical analysis. Content analysis is a tool that can be used to move from the second to the third level (see Image 5.1).

Mayring provides "basic principles" (Mayring 2014: 39) for content analysis: The definition above implies that the analysis of texts done through content analysis

[1]Original quote: "[...] als eine empirische Methode zur systematischen und intersubjektiv nachvollziehbaren Beschreibung inhaltlicher und formaler Merkmale von Mitteilungen, meist mit dem Ziel

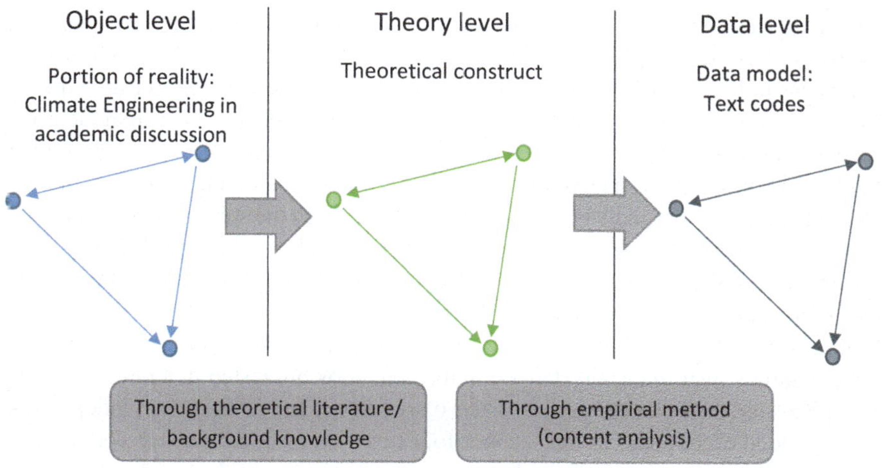

Image 5.1 Model of empirical research on the basis of Früh (2011: 23)

should be guided by a *systematic procedure* (1) which discloses every single step that is taken during analysis (Mayring 2015: 51). This procedure has to be determined in advance and made transparent to the reader. Mayring calls this the "procedural model" of analysis (Mayring 2014: 54; 2015: 62). The quality of the systematic procedure has to comply with various *quality criteria* (2), such as "objectivity, reliability and validity" (Mayring 2014: 42). The drawing of an *inference* (3) from the findings to circumstances external to the text is not a necessary feature of content analysis is commonly done. Mayring claims that "[t]he text is always interpreted within its context, i.e. the material is examined with regards to its origin and effect." (Mayring 2014: 39) As I assess the academic discussion of CE in order to understand how it constructs the legitimacy of these approaches, an inference to circumstances outside of the text, namely its effect on the audience, are drawn tentatively in the conclusion of this study. The required communication model (Mayring 2015: 50) is provided, in the present study, through frame theory.

The term 'empirical' in the definition above introduces, for Früh, the use of quantitative as well as qualitative steps in the analysis (Früh 2011: 36). While Mayring calls his method 'qualitative', he also calls for the integration of *quantitative steps* (4) if the particular research design requires these. (Mayring 2014: 40) In the case of this study, coding is a qualitative step as it comprehends the close reading of the relevant passages of each text in the sample. Quantitative analysis is required in order to be able to make a statement about the frequency of frames and frame elements (Früh 2011: 36). *Categories* (5) are central for content analysis (Früh 2011: 82, 133; Mayring 2014: 40). In this thesis, deductive categories have been constructed, on the

einer darauf gestützten interpretativen Inferenz auf mitteilungsexterne Sachverhalte." (Früh 2011: 133) translated by JK.

[2]Original quote: "strukturiertes Vorstellungsbild " (Früh 2011: 22), translated by JK.

one hand, from existing political and sociological theory, and on the other hand, on the basis of existing exploratory empirical studies on the framing of CE. These categories are in line with the individual elements of frames according to frame theory. This *theory-guidedness* (6) "means that in all procedural decisions systematic reference is made to the latest research on the particular subject and on comparable subject fields." (Mayring 2014: 41) While the notion of theory-guidedness ensures that the content analysis is based firmly in existing knowledge on the subject matter, the analysis shall also take its empirical object into account at all stages: The method is not a formalistic application of theoretical concepts to an empirical issue, but rather entails interpretive effort by the researcher (Mayring 2014: 40; see also Mayring 2015: 603; Früh 2011: 86). This *object reference* (7) is accounted for when the specific category system and its instruments as defined in the codebook are tested in a *pilot study* (8). The pilot study, in this thesis, consisted of two tests. In the first, two independent coders coded eight texts. In the second, four texts were coded by each coder. After each test, the codebook was revised to prevent coding errors in the main coding phase Image 5.2.

In order to ensure that the procedure of analysis is systematic, intersubjectively reproducible and transparent, numerous decisions have to be made and the content of these decisions revealed in a "standardized research procedure" (Früh 2011: 97; see also Mayring 2015: 61) concerning the research strategy and method. In the present study, the *research strategy* is guided by the research question: *How is the meaning of CE approaches constructed in the academic discussion through frames which legitimize or delegitimize the use of these approaches to address anthropogenic climate change?* This question is operationalized further to guide the qualitative and quantitative steps of the empirical analysis. The relevance of this research is based on the assumptions that CE is discussed mainly in academic circles, the academic discussion can affect decision-making concerning CE, and framing is done in the communication of content on CE in the academic discussion. From these assumptions, it can be inferred that, if a dominant frame exists in the mainly academic discussion on CE, this might be picked up by decision-makers in a situation characterized by uncertainty and urgency, thus bolstering a decision that is not based on the 'best-available science' but rather on a specific legitimizing or delegitimizing framing of the issue, for example either a complete moratorium of CE or an all-out development of one or more of these approaches. The theoretical construct is operationalized in this study by disassembling each of the ideal type frames identified in the theoretical chapter, namely Security, Economy, Complexity and Appropriateness, into four distinct frame elements, Problem definition, causal analysis, normative evaluation and suggested remedy, which serve as categories for the analysis. For each of the eight ideal-type frames, the categories take up specific values. Qualitative content analysis allows for the systematic coding of the values for each of the categories for all frame types.

Content analysis as a *method* provides specific individual steps that have to be taken, one by one, in the analysis (Mayring 2014: 54; see also Mayring 2015: 62). Content analysis is, to some degree, an interpretive method: It does not provide clear-cut instructions which can be applied to any and every research object in a formalistic fashion. Categories and other instruments might be tailored for each

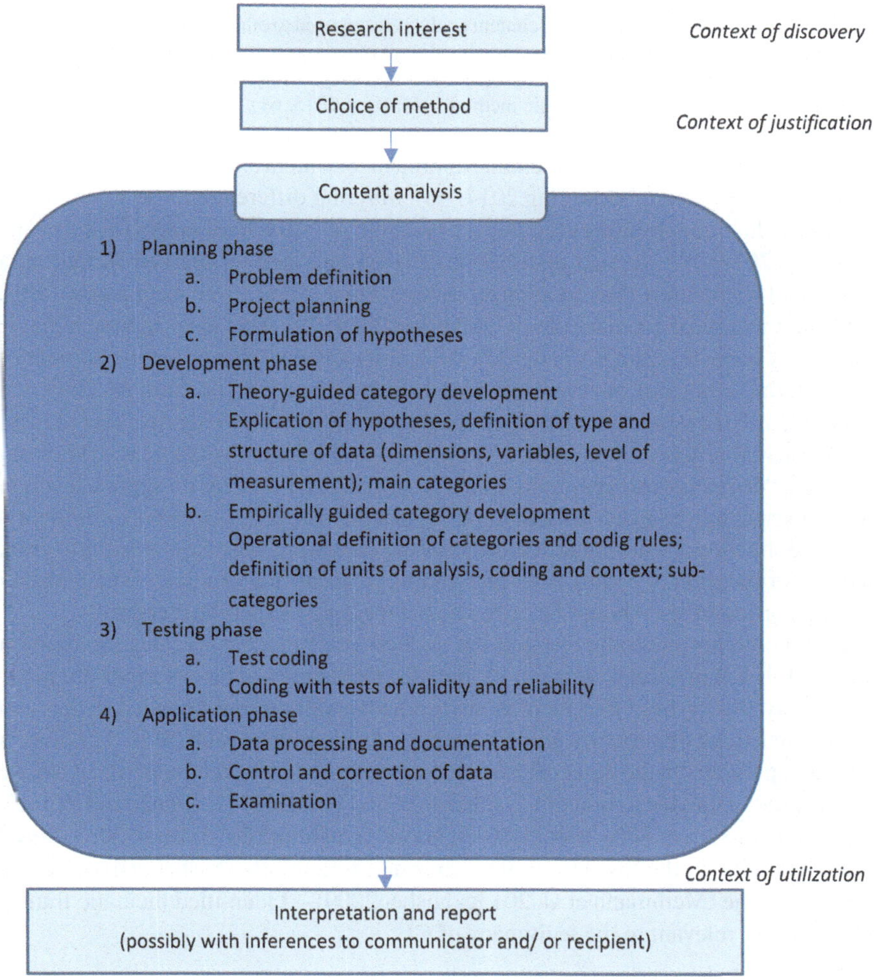

Image 5.2 Standardized procedure according to Früh (2011: 102), translated JK

research object anew. However, the method is still designed to ensure the highest possible standard of objectivity and replicability, as opposed, for example, to methods of 'free interpretation' (Mayring 2015: 51). To make this possible, the procedural model provides instructions according to which every single analytical step has to be taken and rules have to be defined which the actual coding of the text material has to follow (Mayring 2015: 51). While the instructions can be formulated in a general way, the rules for the actual coding have to be tailored to each research object specifically. These rules are then put down in the code book, according to which the empirical material is structured. Mayring argues that this is the specific advantage of content analysis in comparison to other text interpretation methods:

> Here lies the strength of qualitative content analysis as opposed to other interpretive methods: that the analysis is disassembled into individual steps of interpretation, which are determined in advance. Thus, it can be traced by others and intersubjectively reproduced, [...] usable for others, thus it becomes scientific method.[3] (Mayring 2015: 61).

The most important step of a systematic approach to content analysis is the "definition of the category system" (Mayring 2014: 96). Mayring differentiates between inductive and deductive approaches to category creation: In "inductive category formation" (Mayring 2014: 79), the text material itself is summarized further and further up a gradient of abstraction, thus creating categories which come "from the material itself, not from theoretical considerations" (Mayring 2014: 79). This is an approach that is commonly used in explorative studies which aim to assess what a specific amount of text material is focusing on. For the academic literature on CE, for example, Huttunen and Hildén (2014) study which frames are used by inductively analyzing a set of most cited research articles on the topic. As opposed to the inductive approach, Mayring describes "deductive category assignment" (Mayring 2014: 96) as an approach which aims to structure the material according to specific categories derived from theory. This is done through the clear definition of the categories based on existing theory, and the formulation of coding rules and finding of anchoring examples which clarify when a portion of text should be coded as belonging to a specific category.

Früh combines the inductive and the deductive approaches to category creation (Früh 2011: 73) in order to account for the idea that empirical observations are never completely theory-free, that there is always some background knowledge that can be used to extend empirically guided category development (Früh 2011: 72). I use a similar approach: On the basis of existing empirical studies which inductively isolate frames used in the discussion of CE (Huttunen and Hildén 2014; Linnér and Wibeck 2015; Bellamy 2014; Scholte et al. 2013; Porter and Hulme 2013; Buck 2013)—and more generally, in the discussion of science and technology (Nisbet 2010a) and of climate change (Methmann et al. 2013c; Nisbet 2009)—I identified thematic frames which can be relevant to the legitimacy of CE.

5.2 Operationalizing the Method of Content Analysis for the Present Study

The empirical analysis of this study aims to translate the theoretical construct into a data model which can then be interpreted to answer the research question. In operationalizing content analysis methodology for my study, I follow closely the instructions in Früh's in standardized procedure and Mayring's procedural model.

[3]Original quote: "Eben darin besteht die Stärke der qualitativen Inhaltsanalyse gegenüber anderen Interpretationsverfahren, dass die Analyse in einzelne Interpretationsschritte zerlegt wird, die vorher festgelegt werden. Dadurch wird sie für andere nachvollziehbar und intersubjektiv nachprüfbar, [...] für andere benutzbar, dadurch wird sie zur wissenschaftlichen Methode.“ (Mayring 2015: 61) Quote translated by JK.

5.2.1 Research Interest and Choice of Material

The *empirical starting point* of this study consists of a peculiar imbalance with regards to CE technologies concerning, on the one hand, their potential relevance for climate governance and as well as their potential impact on our climate, and, on the other hand, the current limitation of political deliberation and decision-making on the questions as to whether and how we want these technologies to enter our lives. In Früh's differentiation of types of *cognitive interest* (Früh 2011: 44) driving content analysis, this study pursues a descriptive approach: It aims to describe how CE approaches are framed in the academic discussion. The relevance of this pursuit is based on the assumption that text has an effect on the audience, namely on decision-makers' perceptions of the legitimacy of CE approaches to address climate change. Frame theory thus provides a communication model (step 3 in Mayring's procedural model) with which to contextualize the texts studied, and which serves as a basis for inference from text to receiver (Entman 1993: 52–53).

Drawing on the theoretical framework, the following *hypotheses* have been formulated in order to facilitate the development of categories for the codebook[4]:

Group 1: Security

H1a: If CE approaches are presented as a way to protect a valued referent object against an existential threat, their legitimacy increases.

H1b: If CE approaches are presented as an existential threat to a valued referent object, their legitimacy decreases.

Group 2: Economy

H2a: If CE approaches are presented as a cheap tool to address a challenge, their legitimacy increases.

H2b: If CE approaches are presented as an expensive tool to address a challenge, their legitimacy decreases.

Group 3: Complexity

H3a: If CE approaches are presented as a simple technological fix, their legitimacy increases.

H3b: If CE approaches are presented as technologies inside a complex socio-technical system with many physical as well as social effects (some unknown), their legitimacy decreases.

Group 4: Appropriateness

H4a: If CE approaches are presented as the appropriate thing to do, their legitimacy increases.

[4]Note: These hypotheses are not tested in the present thesis, but developed only in order to provide a basis for the systematization of the text material.

H4b: If CE approaches are presented as an inappropriate thing to do, their legitimacy decreases.

These hypotheses are explained and justified in more detail in the dimensional analysis below. After research interest and research question as well as preliminary hypotheses have been formulated, what remains is to define the material to be analyzed (Mayring 2015: 55) and to choose the units of analysis (Früh 2011: 40).

The *material* is differentiated into corpus and sample. The corpus of texts that addressed here are all English-language articles published in academic journals between 1971 and 2016 on the topic of CE. As Reppen and Simpson (2002: 96) argue: "The composition of the corpus should represent the anticipated research goals." As I am trying to make an argument about the constitution of the academic discussion on CE, I produced a corpus of the academic literature on CE in the period of interest. To define the corpus of academic articles, I follow an approach similar to Rössler (2010: 54): Concerning date of publication, I analyze academic journal articles on CE from between 1971 and 2016. 1971 is the earliest year for which academic journal articles can be found using the search term. Concerning regional space, I do not exclude any publication on the basis of its region or country of origin. However, in order to render the content analysis coherent without having to deal with translation error, I exclusively analyze English-language publications.[5] Concerning the media type, I analyzed academic journal articles which are available online. I did not pick specific journals, but analyzed all articles that can be found under these criteria. This means I do not exclude any academic discipline. Also, all types of publication have been considered, meaning research articles, but also comments, reviews etc. In order to collect the corpus, which needs conform to the criteria of representativeness and balance (Reppen and Simpson 2002: 93), I used the online database Web of Science. In order to be able to identify as many texts on CE as possible—some of which do not refer to the umbrella term 'climate engineering' (or 'geoengineering'), but focus on a specific approach—I used a complex search term (see Belter and Seidel 2013: 418; Oldham et al. 2014: 3):

i. (geoengineer* OR "geo-engineer*" OR "climate engineer*" OR "radiation management" OR ("carbon dioxide removal" AND "climate change"))
ii. OR (((("ocean fertili*ation" OR "ocean iron fertili*ation" OR "iron fertili*ation") OR ("enhanced weathering" OR "artificial weathering" OR "accelerated weathering") OR ("artificial upwelling" OR "ocean upwelling") OR ((atmosph* OR air) NEAR/5 capture)) AND (((carbon OR "CO(2)" OR "CO2") AND sequest*) OR ((carbon OR "CO(2)" OR "CO2") AND remov*)))
iii. OR ((biochar* AND ("climate change" OR "global warm*" OR "negative emission*"))).
iv. OR ((("cloud brightening" OR "cloud seeding") OR ("aerosol injection" OR "stratospheric aero-sol" OR "aerosol in the stratosphere") OR "space mirror")

[5] I am aware that this potentially excludes important voices from my analysis. Linguistically more broad analyses would need to be conducted in future research in order to review my findings.

AND ("albedo enhanc*" OR "en-hanc* albedo" OR "albedo increas*" OR "increase* albedo" OR "albedo modif*" OR "mod-if* albedo"))

v. AND ("climate change" OR "global warm*" OR "negative emission*"))

The search term comprises a number of specific CE approaches by name, as many publications which deal specifically with one individual approach—commonly found in the natural sciences—do not make mention the terms 'geoengineering' or 'climate engineering' specifically. With this search term, I aim to cover all CE approaches discussed in various comprehensive reports on the issue (NAS 2015b; Rickels et al. 2011b).[6] The final search term yielded 1650 hits.

Subsequently, the corpus was adjusted by close reading of title and abstract of each article in order to assess whether they actually did address CE or if they contained only noise that had not been filtered out by the search term. After several steps of cleaning and comparison to the corpus created by Nils Matzner for the project CE-SciPol (Matzner and Barben 2018), the final corpus comprises 1101 articles. From this corpus, a sample was drawn on which the actual empirical analysis was conducted. To do so, the following criteria were applied:

1. The sample is to contain at least 10% of the corpus.
2. The most highly cited articles of each year are to appear in the sample.
   This means that not the most highly cited articles of the entirety of the corpus were chosen, but the corpus was broken down by years and for each year, the most highly cited articles were chosen. Thus, an overemphasis of older articles—had a longer exposure time and thus could be cited more without actually being of higher impact—in the sample can be avoided.
3. The number of articles of each year which are added to the sample is the fraction of the corpus which originated in said year.

For example, if I find that 17% of texts in the corpus were published in 2012:

$$17*110 = 12.7$$
$$\text{rounded to integers} = 13.$$

In this case, 13 texts would be added to the sample from 2012.

This is to ensure that I do not add articles to the sample with a very low citation number on the basis of the year of publication having yielded only a very limited number of articles. In concrete terms, this means that, if the corpus contained 130 articles from the year 2012, for example, then 13 articles from 2012 (10% of 130) are added to the sample.

4. At least one article of each year in which any article on CE was published is to be added to be corpus, notwithstanding the fraction that articles of said year present of the corpus.

[6] Approaches afforestation and BECCS were included in this study even though their inclusion under the umbrella term ‚climate engineering ' is contested. In order to ensure that the texts are pertinent to the research question, abstracts of all texts were checked through close reading against their relevance with regards to the CE discussion.

This is to ensure that articles from early years, in which little publishing activity on CE is evident, nevertheless are added to the sample, as these early articles can be expected to have large impact on the research coming later.

I combine two approaches of representative sampling namely sampling by quota and sampling of typical cases (Rössler 2010: 62), in order to be able to draw general and generalizable conclusions: "If one seeks to achieve representative, i.e. generalizable statements, then the chosen unit has be represent a structurally identical, minimized copy of the main unit about which a statement is to be made."[7] (Rössler 2010: 58) Further, I deviate slightly from these two approaches, favoring publications from the early years of the discussion. This is to ensure that these publications, which can be influential as they first set the stage of the discussion, are not excluded from the sample on the basis of low publication numbers in these early years.

Coding rules define the way in which coding proceeds during content analysis. The unit of coding defines for which amount of text a category may be coded. For example, the unit of coding could be one word, or it could be one sentence. In the former case, one sentence can be coded numerous times. In the latter, it can be coded only once (Früh 2011: 96). The unit of context indicates which amount of text may be used to understand the coding unit properly in case explanation becomes necessary. This could be, for example, the entirety of the text which came before the passage in question. The unit of analysis defines the unit "about which a statement is to be made in the study (the 'case' in analysis jargon)."[8] (Früh 2011: 95) The unit of analysis can be a sentence, if a statement is to be made about individual sentences in a text. It could also be an entire article. In this case, the codes of each article need to be summarized in order to infer from this the tendency of each article.

The units of analysis vary for two types of analysis conducted here (Mayring 2010: 605): For meta data analysis, several types of data were collected on the level of the individual document as unit of analysis: author names, disciplinary focus, publication types, year of publication and journal name. While author names, year of publication, publication type and journal name could be derived from the Web of Science database, it is more difficult to define the disciplinary focus of each article. To do so, I gathered information from the title, abstract and affiliations of the authors of each text through close reading. As a clear attribution to a specific academic discipline is hardly possible on the basis of this information, I categorized the articles roughly into 'Social Sciences and Humanities', 'Natural Sciences and Engineering', 'unclear' and 'both'. The decision of categorization was made in each case, i.e. for each text, on the basis of the following question: Is the objective of the article to provide information on natural—i.e. physical, chemical, biological, geological—or on social—i.e. economic, sociological, political, linguistic, communication, ethical or legal—phenomena? Thus, I differentiate into the groups of disciplines by focusing

---

[7]Original quote: "Strebt man repräsentative, d.h. verallgemeinerbare Aussagen an, so muss die Auswahleinheit ein strukturgleiches, verkleinertes Abbild der Grundgesamtheit darstellen, über die eine Aussage getroffen werden soll." (Rössler 2010: 58), translated by JK.

[8]Original quote: "[...] über die in der Studie eine Aussage getroffen werden soll (den ‚Fall ' im Analysejargon)." (Früh 2011: 95), translated by JK.

on the object of interest of the article in question, and not, as has been proposed, on the method of study or the original motivation of research (Klein and Rietschel 2007: 18). Articles studying both natural and social phenomena were categorized as 'both'. Those articles in which a decision could not be made concerning the question outlined above were categorized as 'unclear'. In addition, a list of disciplines was created, differetiating academic disciplines into the categories above. Only if the categorization on the basis of the question and the categorization on the basis of the list were consistent, was the discipline coded. In case they did not, the discipline was coded as 'unclear'.

In the coding of these meta data points, the unit of coding is the article: For each article, one code was given for author names, publication year, name of journal, publication type and discipline. The unit of analysis is the article as well. The unit of context is the entire corpus: To understand the attribution of one article to a discipline, the entire corpus can be drawn upon for additional information. In the coding of frames in text analysis, the unit of coding is the paragraph. The coding by sentence would be easier to automate, as sentences are usually formally defined by the full stop. However, coding frame elements by sentence would present a high frequency of frames in texts by authors who favor a set of short sentences over a single long sentence. Also, many frames and even frame elements are outlined not in one sentence, but in a string of sentences. Choosing the argument as unit of coding is attractive. Specific linguistic indicators often introduce a new argument, such as "moreover" or "in addition". These were used as guidelines in the coding process. Also, a paragraph break often indicates a new argument. The inter-coder reliability test revealed, however, that coding units were chosen very differently between the two coders when coding arguments. Thus, the paragraph was chosen as unit of coding in the application. The unit of context as well as the unit of analysis for the coding of frame elements is the article or document.

5.2.2 Definition of Categories and Formulation of Codebook

It is a special feature of frame analyses that frames can be broken down into elements and that all four frame elements (problem definition, causal analysis, proposed remedy and moral evaluation) can be coded individually (Gamson and Modigliani 1989: 11; Matthes and Kohring 2008: 263; Donk et al. 2012: 12; Scholte et al. 2013: 6). The aim is to increase reliability by hiding from the coder, at least to some degree, which element belongs to which frame: "During the content analysis, the affiliation of the frame elements to a specific frame remains unknown. This increases the reliability of the analysis and enables the identification of new and modified frames." (Donk et al. 2012: 12) In this study, the four frame elements provide the main categories for coding. Table 5.2 gives an overview over the coding scheme. In order to assess whether there is frame dominance in the academic discussion, this study assessed the frequency of each individual frame/frame element as well as the competition

Table 5.2 Coding scheme

Type of data/type of analysis	Qualitative analysis	Quantitative analysis
Text data	Legitimization/delegitimization frames • Problem definition • Diagnosis of causes • Moral evaluation • Suggestion of remedies	Frequency Dominance/competition
Meta data	Author Disciplinary focus of article (nat or soc or none or both) Type of publication Publication date	Frequency

between various frames. This is in line with Entman's criterion of internal consistency presented above. The criterion of external acceptance has been ignored as the receiver has not been analyzed and the effect of the individual frames is assumed on the basis of theoretical considerations. Table 5.3 presents an overview over the categories and sub-categories of the codebook as well as the values that the sub-categories can take. These values have been derived from the formulation of the hypotheses.

In a pilot study, the codebook was tested to ensure a high degree of reliability. Inter-coder reliability was tested by two coders independently coding several texts from the sample. This is based on the procedure recommended by Mayring:

> Inter-coder reliability (or rather objectivity of analysis) is tested by presenting at least a part of the material to a second coder. The rules here are not as strict as those of quantitative content analysis: The second coders are comprehensively instructed in the use of the regime. In the case of non-consistency, the text sections in question are discussed in a coder conference. It only counts as mis-coding if second coders can prove faulty codings to primary coders, who usually command more background knowledge about the material. (Mayring 2010: 604).

This inter-coder reliability test was conducted twice and the codebook adjusted based on the findings of these tests.

5.3 Coding of Sample

It is important to note that the results presented below are based on a qualitative and, to some degree, subjective coding process. It was attempted to reduce both the subjectivity of the process and the potential of mis-codings as much as possible: Subjectivity was reduced through intercoder reliability testing and subsequent amendment of the codebook and through the coding of individual frame elements instead of complete frames. Mis-codings were detected through cross-review of all codings after the process of coding (see Fig. 5.3) as well as through qualitative analysis of as many

Table 5.3 Overview over categories and subcategories

Category	Sub-categories	Scaling of values	Unit of coding	Values = codes
Problem definition	What is the main challenge mentioned in the text?	Nominal	Paragraph	Climate change CE Both Other Unclear
	How is the problem characterized?	Nominal	Paragraph	Threat/danger Cost Complexity Appropriateness Other None
Causal analysis	Are there one or more social actors presented as causally responsible for climate change?	Nominal	Paragraph	Political actors Economic actors Science Society Media Other None Unclear
Evaluation of CE	Which set of approaches is assessed?	Nominal	Paragraph	CDR SRM Both Unclear
	Which criterion is applied to assess advantages and disadvantages?	Nominal	Paragraph	Threat versus Safety Cost versus Benefit Simplicity versus complexity Appropriateness Regulation Other Unclear
	Which degree of activity is assessed?	Nominal	Paragraph	Research and development Deployment Several Unclear
	What is the outcome of the evaluation of CE?	Ordinal/Nominal	Paragraph	Pro Carefully pro Undecided/dilemma Carefully con Con Open Unclear
Suggested remedy	Is a remedy to climate change identified?	Nominal	Paragraph	Exclusively CE CE in portfolio CE Other than CE Undecided

(continued)

Table 5.3 (continued)

Category	Sub-categories	Scaling of values	Unit of coding	Values = codes
	Is a specific avenue of activity proposed for the suggested remedy to the problem climate change?	Nominal	Paragraph	R&D Deployment Regulation Unclear Several

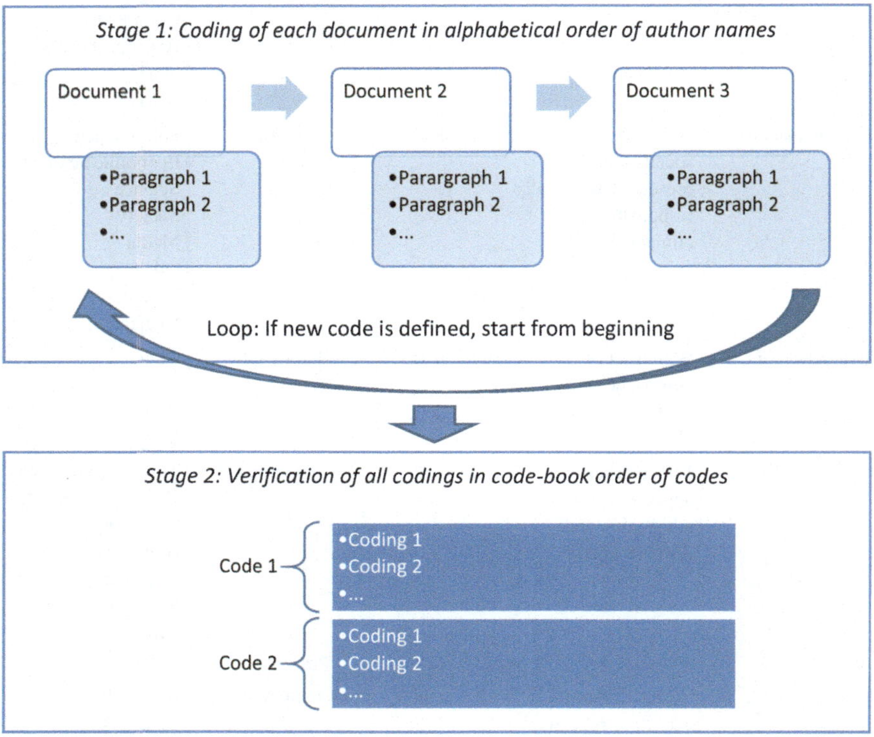

Image 5.3 Two-stage process of coding in application phase

conspicuous findings in quantitative analysis as possible. However, a completely objective assessment or complete error-free analysis is not possible. The first is due to the nature of the coding process as an interpretive approach. The second is due to a residual amount of human failure which cannot be completely excluded. This should be kept in mind when considering the results.

The following procedure was applied to code the sample in accordance with the codebook: In a first stage of coding, each paragraph of each text was coded in order of appearance, moving through the sample alphabetically by last name of the first author of each text. During this stage, whenever a new code was created inductively on the basis of the empirical data, the entire material coded up to that point was run through again, in order to re-code for the newly created code. Thus, the material was coded in loops. In a second stage, each coding was verified by going through all codings for each specific code.

During coding, new, **inductive codes** were generated whenever the text material provided content which presented as a frame element as defined by the code book, but could not be covered by any of the codes present in the code book. This means that no new codes were created during coding outside of the structure of frame elements. All in all, 6031 codings were applied for 79 codes over the entire sample during the coding process. After application of the codebook to the sample was completed, all codings were cross-reviewed against each other in order to weed out mis-codings. In this process, a number of codings were deleted and codings for other codes were applied where necessary. This process modified the distribution of codings, in some cases significantly. For example: Before cross-review, 571 codings were applied to 125 documents for code '1.1 CC', meaning that 571 segments of text were found in which climate change was defined as the central problem of the text. After cross-review, 483 codings were left in 122 documents for this code. The rest of the codings for '1.1 CC' were deleted and re-coded for '1.1 Other' or '1.1 CE' or '1.1 None'. The reason for this was that many passages originally coded '1.1 CC' could, during cross-review, not be found to be focused on climate change as the central problem, but rath≥er on CE or a completely different problem.

5.4 Pre-analysis Processing Steps of Findings from Sample

After coding was completed in the manner described above, the data thus created encompassed 6906 codings for 103 codes and subcodes (both meta data and text data) applied in 1321 coded segments (paragraphs) in 136 documents. The first data processing step consists of the adjustment of the number of coded segments: 74 segments needed to be subtracted from this number as they referred to direct or indirect quotations and are not part of the sample, bringing the number of coded segments in the sample to 1247. 331 segments (26.56% of all segments in the sample) in 113 documents bore no codes from the codebook. The second pre-processing step consists of the generation of document variables from formal data assessed during the coding process: The formal data collected during coding for the categories 'disciplinary focus', 'year', 'approach' and 'type of text' was stored in MaxQDA in the form of codings as this information was derived from the text. Afterwards, this information was transferred into document variables as this information came from the text, but was applicable not only for the specific section coded, but for the entirety of the text.

The third pre-analysis processing step consists of the re-assembly of frame elements and frames. In this study, the frame elements derived from frame theory were applied as main categories into the codebook and then subdivided into subcategories conforming to the ideal-type frames. The coding of individual frame elements or sub-elements instead of entire frames is a protection mechanism to prevent coders

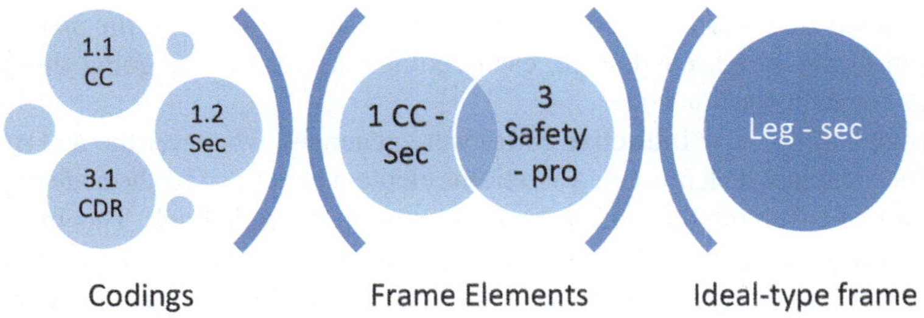

Image 5.4 Illustration of process of re-assembling ideal-type frames in the documents of the sample (example)

from creating artifacts. This means, however, that, in order assess whether frame elements or even entire frames in accordance with the ideal-type frames exist in the material, i.e. in order to be able to answer the research question, the codes applied to code these subcategories now need to be assembled in order to present as complete frame elements Image 5.4.

Thus, a frame element can be represented by a specific combination of codes in a text segment or document. At this point, the theoretical considerations which preceded the empirical analysis are helpful to distinguish relevant information from noise: Based on the ideal-type legitimizing and delegitimizing frames developed in the theoretical construct, the following code combinations are particularly interesting as they represent elements of the ideal-type frames. Table 5.4 presents the values that would need to be found in the text material in order to conform with the individual ideal-type frames from the theoretical construct.

The set of inductively developed frames consists of two frames legitimizing and two delegitimizing CE. Of these four, two (one legitimizing and one delegitimizing) frames present themes of characterization and criteria for evaluation other than the ideal-type frames ('Leg—Oth' and 'Deleg—Oth') and two (again, one legitimizing and one delegitimizing) present no characterization of the main problem at all ('Leg—None' and 'Deleg—None'). One frame neither legitimizing nor delegitimizing completes the set ('None—x'). In this inductively developed frame, no problem is identified or characterized. Any criteria can be applied for the evaluation of CE, but no clear outcome of evaluation or proposed remedy is presented. Accordingly, this 'frame' could be described as a non-frame, indicating the absence of frames as defined in this study Table 5.5.

In order to unveil the shared occurrence of codes relevant for each of the deductive ideal-type and inductively developed frames and thus assess whether complete frame elements or even frames exist and how close to completion incomplete frames are, first, frame elements were assembled from codings, and then, complete or incomplete deductively and inductively developed frames were assembled from frame elements. The values for the various categories were chosen either on the basis of the ideal-type frames or the inductively developed frames (see tables above). The number

Table 5.4 Strong and weak variants of ideal-type frames

Assumed effect on decision-making	Ideal-type frame	1 Problem definition (problem identification—characterization)	3 Evaluation of CE (Criteria—outcome of evaluation)	4 Proposed remedy (remedy—avenue of activity)
Legitimizing CE	Security (Leg—Sec)	CC/other—threat	Safety—pro/cpro	CE/Port/ ExCE—use (strong variant) CE/Port/ ExCE—research/development (weak variant)
	Economy (Leg—Cost)	CC/other—expensive	Low cost—pro/cpro	
	Complexity (Leg—Comp)	CC/other—complex	Simplicity—pro/cpro	
	Appropriateness (Leg—Norm)	CC/Other—inappropriate	Appropriateness—pro/cpro	
Delegitimizing CE	Security (Deleg—Sec)	CE—threat	Threat—con/ccon	Other—use (strong variant) Other—research/development (weak variant)
	Economy (Deleg—Cost)	CE—expensive	High cost—con/ccon	
	Complexity (Deleg—Comp)	CE—complex	Complexity—con/ccon	
	Appropriateness (Deleg—Norm)	CE—inappropriate	Inappropriateness—con/ccon	

Frame element (relevant subcategories in codebook)

The abbreviations in the table and all following discussions read a follows: Ideal-type frames: Leg = legitimizing frame variant; deleg = delegitimizing frame variant; Sec = security; Econ = economy; Comp = complexity; Norm = appropriateness; Problem definition: CC = climate change; CE = Climate Engineering; Evaluation of CE: pro = positive, cpro = carefully positive; con = negative; ccon = carefully negative; Proposed remedy: CE = Climate Engineering; Port = portfolio of measures including some sort of CE; ExCE = exclusively CE

Table 5.5 Frame element values for inductively developed frames

Frame Element (relevant subcategories in codebook)				
Assumed effect on decision-making	Inductively developed frame	1 Problem definition (problem identification—characterization)	3 Evaluation of CE (Criteria—outcome of evaluation)	4 Proposed remedy (remedy—avenue of activity)
Legitimizing CE	Other (Leg—Oth)	CC/Other—other	Pol—pro/cpro Other—pro/cpro	CE/Portfolio/ ExCE—use/development/ research
	'None' (Leg—None)	CC/Other—none	Other—pro/cpro	
Delegitimizing CE	Regulation (Deleg—Oth)	CE—other	Pol—con/ccon Other—con/ccon	Other—use/development/ research
	'None' (Deleg—None)	CE—none	Other—con/ccon	
None [a]	'None' (None—x)	CC/Other/ CE—none	Sec/Cost/ Comp/Norm/ Pol/Other—Open/Undecided	–

[a]The 'None—x' frame is not actually a frame, as it is not defined by a specific proposed remedy. It is used here to serve as residual category

Table 5.6 Overview over values for each frame element for the eight ideal-type frames of the construct

Legitimizing or delegitimizing CE	Frame	Problem definition	Causal analysis (actor responsible)	Evaluation	Suggested remedy
Legitimizing	Security	CC = threat	?	Pro CE	CE
	Economy	CC = expensive	?		
	Complexity	CC = complex socio-technical challenge	?		
	Appropriateness	CC = inappropriate	?		
delegitimizing	Security	CE = threat	?	Counter CE	Not CE
	Economy	CE = expensive	?		
	Complexity	CE = complex socio-technical challenge	?		
	Appropriateness	CE = inappropriate	?		

of codings after the generation of the new codes for co-occurrences of the codings presented above is 8.676 codings for 136 documents.

5.5 Summary

Table 5.6 presents an overview over the definitions of each frame element of the eight ideal-type frames of the theoretical construct on the basis of Matthes and Kohring (2008: 267).[9] This means that this is what I would have to find if the frame existed in the discussion in its idealized form.

The quantitative interpretation of the results of the coding of meta data allows me to address the following questions: Which codes/frame elements are dominant in the corpus? Through a combination of the coding of frames and of metadata, I am able ascertain: Is there a frame dominant during a specific time period/in a specific journal/in the texts by a specific author/in a specific document group? Image 5.5 provides an overview of the operationalization of Mayring's procedural model as applied to the present study Table 5.7

[9]Please note that these frames are ideal types. It remains to be seen if they exist in this idealized form in the discussion.

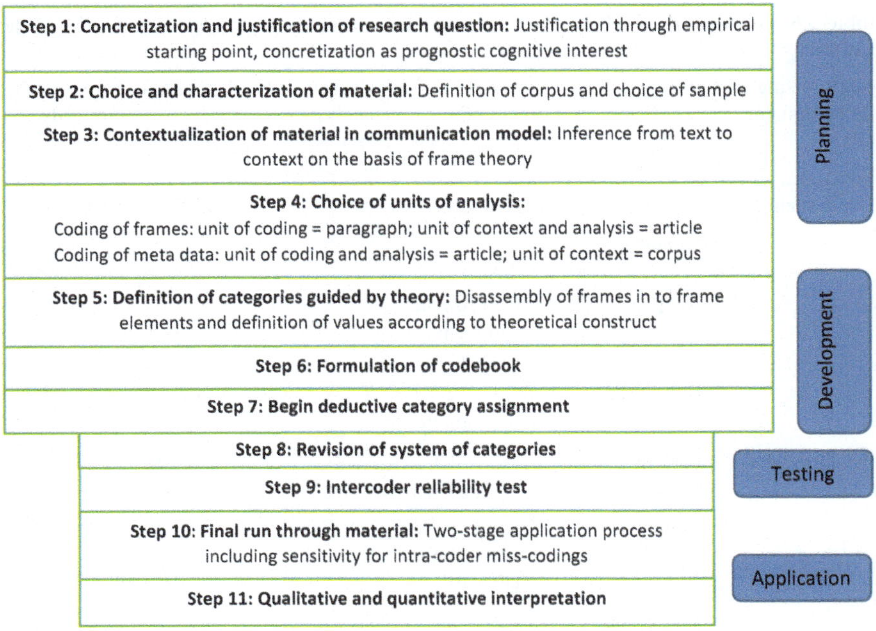

Image 5.5 Operationalization (non-bold in green boxes) of procedural model of deductive category assignment based on Mayring (2010: 605) (bold green boxes), with phases following to Früh (2011: 102) (blue boxes)

Table 5.7 Overview over units of analysis and level of aggregation of data

Unit of analysis	Segment	Document	
Meta data	–	Variable values	
Text data: levels of aggregation	Codings	Codings	Code profiles
	Frame elements across sample and by document group (based on metadata values)	Frame elements across sample	
	Frames	Frames	

Both text data and meta data were analyzed for the sample. The analysis of meta-data took place on the level of the individual documents, as the meta data variables—'year of publication', 'approach addressed', 'disciplinary focus' and 'type of article'—applied for the entire text. The analysis of text data in the sample took place on two levels: One, on the level of coded segments, i.e. paragraphs; and two, on the level of coded documents, i.e. individual articles. The analysis on the level of segments allows a quantitative and comparative analysis of frames and frame elements in the entire set of coded segments, notwithstanding from which text in the sample they originate, for example across years of publication or across disciplinary boundaries. As the research question aims at assessing potentially dominant frames in the entire sample, the comparison on the level of segments is indicated. The analysis on the level of the document was done in order to review the findings from the segment-level analysis: As each text has a very different length, more codings are derived from longer texts than from shorter ones, thus adding more coded segments to the sample which might lead to misapprehensions of the results: For example, a frame element might appear often in the sample, but this might be due to the fact that an author using this frame element might have written longer articles than authors who do not, and thus does not necessarily prove this frame element to be particularly relevant to the sample. The unit of analysis is indicated in the text below for each step individually.

References

Bellamy R (2014) Safety first! framing and governing climate geoengineering experimentation. In: Climate geoengineering governance working paper series, 014

Belter CW, Seidel DJ (2013) A bibliometric analysis of climate engineering research. WIREs Climate Change 4(5):417–427. https://doi.org/10.1002/wcc.229

Buck HJ (2013) Climate engineering: spectacle, tragedy or solution? a content analysis of news media framing. In: Methmann C, Rothe D, Stephan B (eds) interpretive approaches to global climate governance: (De)constructing the greenhouse. London, Routledge, pp 166–180

Donk A, Metag J, Kohring M, Marcinkowski F (2012) Framing emerging technologies: risk perceptions of nanotechnology in the german press. Sci Commun 34(1):5–29. https://doi.org/10.1177/1075547011417892

Entman RM (1993) Framing: toward clarification of a fractured paradigm. J Commun 43(4):51–58

Früh W (2011) Inhaltsanalyse: theorie und Praxis. UKV, Konstanz

Gamson WA, Modigliani A (1989) Media discourse and public opinion on nuclear power: a constructionist approach. Am J Sociol 95(1):1–37

Huttunen S, Hildén M (2014) Framing the controversial: geoengineering in academic literature. Sci Commun 36(3):3–29

Klein M, Rietschel E-T (2007) Schnittstellen zwischen Geistes—und Naturwissenschaften. APuZ, 46/2007, pp 15-20

Linnér B-O, Wibeck V (2015) Dual high-stake emerging technologies: a review of the climate engineering research literature. WIREs Climate Change 6:255–268

Matthes J, Kohring M (2008) The content analysis of media frames: toward improving reliablity and validity. J Commun 58:258–279

Matzner N, Barben D (2018) Verantwortungsvoll das Klima manipulieren? Unsicherheit und Verantwortung im Diskurs um Climate Engineering. In: Janich N, Rhein L (eds) Unsicherheit als Herauforderung für die Wissenschaft. Peter Lang, Berlin et al., pp 143–179

Mayring P (2010) Qualitative inhaltsanalyse. In: Mey G, Mruck K (eds) Handbuch qualitative forschung in der psychologie. Springer, Wiesbaden, pp 601–613

Mayring P (2014) Qualitative content analysis: theoretical foundation basic procedures and software solution. GESIS, Klagenfurth

Mayring P (2015) Qualitative Inhaltsanalyse (12th revised ed.), Weinheim et al, Beltz

Methmann C, Rothe D, Stephan B (2013c) Reflections. In: Methmann C, Rothe D, Stephan B (eds) Interpretive approaches to global climate governance: (De)constructing the greenhouse, London, Routledge, pp 248–256

NAS (2015b) In: Climate intervention: carbon dioxide removal and reliable sequestration. reflecting sunlight to cool earth. Washington, DC, National Academy of Sciences

Nisbet MC (2009) Communicating climate change: why frames matter for public engagement. Environ Sci Policy Sustain Develop 51(2):12–23

Nisbet MC (2010a) Framing science. a new paradigm in public engagement. In: Kahlor L, Stout PA (eds) Communicating science. new agendas in communication, New York, Routledge, pp 40–67

Oldham P, Szerszynski B, Stilgoe J, Brown C, Eacott B, Yuille A (2014) Mapping the landscape of climate engineering. Philos Trans Royal Soc A 372(2031). https://doi.org/10.1098/rsta.2014.0065

Porter KE, Hulme M (2013) The emergence of the geoengineering debate in the UK print media: a frame analysis. Geographical J 179(4):342–355. https://doi.org/10.1111/geoj.12003

Reppen R, Simpson R (2002): Corpus linguistics. In: Schmitt N (ed) An introduction to applied linguistics. London, Arnold

Rickels W, Klepper G, Dovern J, Betz G, Brachatzek N, Cacean S et al (2011b) Large-scale intentional interventions into the climate system? assessing the climate engineering debate. Scoping report conducted on behalf of the german federal ministry of education and research (BMBF). Kiel, Kiel Earth Institute

Rössler P (2010) Inhaltsanalyse, 2nd edn. UKV, Konstanz

Scholte S, Vasileiadou E, Petersen A (2013) Opening up the societal debate on climate engineering: how newspaper frames are changing. J Integr Environ Sci 10(1):1–16

Chapter 6
The Academic Framing of Climate Engineering—A Diverse Field Presenting Interesting Landmarks

Kreuter offers a timely analysis of non-strategic framing of Climate Engineering in the academic discussion. The chapter present the results of an empirical analysis of a sample of peer-reviewed academic journal articles from all disciplines. While framing is found to be rare across the sample, frames appearing in the sample are highly diverse. Furthermore, Kreuter highlights that many documents present a combination of counteracting frames which provide a curious tension inside the individual document. Kreuter concludes the chapter by discussing the roles of interdisciplinarity and the discussion across various CE approaches for academic framing of CE.

The results of the empirical analysis show the academic discussion of CE, as represented in academic journal articles, as a diverse field when it comes to framing. Framing is, as expected, rather rare across the sample. Further, there is no dominant frame present in the sample. Many documents present either no complete frames—while individual frame elements or frame element fragments are present in every document of the sample—or a combination of counteracting frames which provides for a curious tension inside the individual document. Further, an economic characterization of both the problem of climate change and the problem of CE was the first to emerge. However, later on, climate change is more commonly characterized as a security problem, and CE as a complex challenge. Also, specific patterns appear when documents are grouped according to the specific approach in question, and according to their disciplinary focus. Finally, positive evaluations of CE are very rare across the sample. In this chapter, a few interesting insights are highlighted based on metadata analysis and subsequently, central findings from the text analysis are presented in order of increasing aggregation of data.

J. Kreuter, *Climate Engineering as an Instance of Politicization*,
Springer Climate, https://doi.org/10.1007/978-3-030-60340-3_6

6.1 Metadata Analysis[1]

The comparison of sample and corpus give some interesting indications concerning the distribution of texts addressing specific CE approaches and of texts from different academic discipline groups. Due to the sampling rules applied here, the sample represents the texts with the highest number of citations for each year. Because of this, divergences between distributions of variable values in the sample and the corpus are interesting when considered under the assumption that articles with higher citation numbers have larger impact, not only on the academic community, but beyond it, and thus potentially influence decision-making concerning CE. The findings show that DAC, a CDR approach, and SAI, an SRM approach, are most prevalent in the sample. Further, natural science texts make up a larger proportion of texts in the sample than the corpus. As could be assumed on the basis of sampling by year, the corpus and the sample have a very similar structure with regards to years of publication.

Metadata for variables 'Year' and 'Type of Article' were applied from the information collected during the generation of the corpus. In the corpus, metadata for variables 'Approach' and 'Disciplinary Focus' was collected through close reading of each document's abstract.[2]

Values for the variable 'Approach' were coded on the basis of definitions from NAS (2015a, 2015b), Rickels et al. (2011). Image 6.1 shows the distribution of texts addressing various approaches in (a) the sample and (b) the corpus. It means that, for example, 4.4% of all texts in the sample were understood, during coding, to cover mainly Afforestation. The graphs indicate a very mixed distribution of frequencies of approaches in both corpus and sample. Of the sample, 68 texts focus on CDR approaches (50%) and 46 texts on SRM approaches (33.82%). Quite a number of texts cover more than one CE approach. Those cases are presented in the graph by values 'CDR—several' and 'SRM—several', respectively. To those cases which cover not only various approaches, but approaches belonging to categories CDR and SRM, the value 'both—several' was applied. 'Unclear' is a residual category for those documents for which no other value could be clearly assigned. This was the case much more often in the corpus than in the sample, which is not surprising given that a smaller amount of text from each document was used as the basis for assigning values in the corpus than in the sample. An interesting finding is that the approaches CDR—DAC and SRM—SAI are particularly frequent in the sample. While SAI is the SRM approach most frequently focused on in the documents of both corpus and sample, this does not apply for DAC: While it is the most frequent approach in the sample, OF is more frequently the object of the text in the corpus. From this we can

[1]Numbers and percentages in this section were rounded to one digit after point.

[2]Please note that the values assigned to the texts during the generation of the corpus do not, in all cases, entirely conform to the values assigned to the texts in the sample during coding. This is due, at least in part, to the divergence in the unit of reference: During the generation of the corpus, only the abstract of each text was read to categorize each document for each variable. During coding, the values for the variables mentioned above were applied on the basis of the entire coded text, i.e. abstract, introduction and conclusion. Thus, the distribution of variable values in the corpus can serve only as qualitative illustration, not basis for a quantitative comparison.

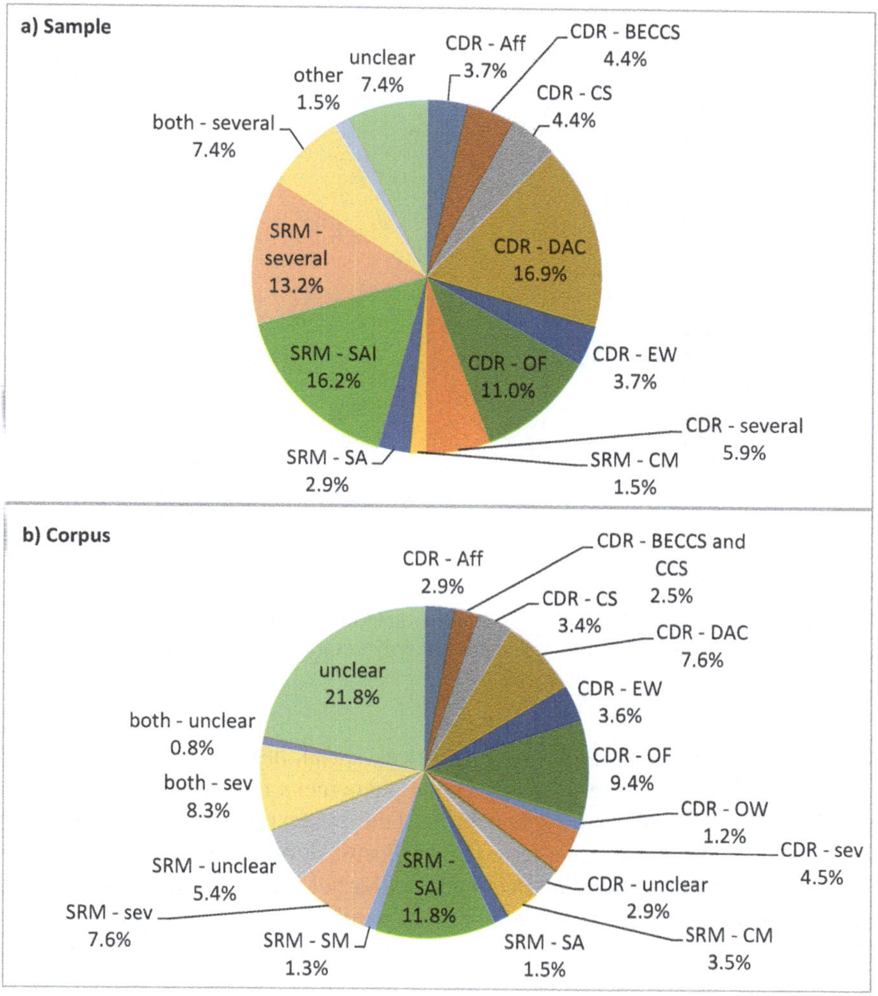

Image 6.1 Relative distribution of values for variable 'Approach' in **a** sample (136 documents) and **b** corpus (1101 documents). Please note: As units of reference were different in the sample and the corpus, the values for the corpus can serve only as qualitative comparison

infer, on the basis of the rules for the generation of the sample, that DAC texts are more highly cited than OF texts.

Differences between sample and corpus are present also concerning the variable 'Disciplinary Focus'. This variable was coded according to the following coding rule: Title, abstract, and, where necessary, the rest of the material was closely read in order to assess the main research focus. Texts focusing on processes or phenomena in their relatability to laws of nature, i.e. physical, chemical, biological or otherwise 'natural' phenomena were coded 'natural'. Texts assessing processes/phenomena in

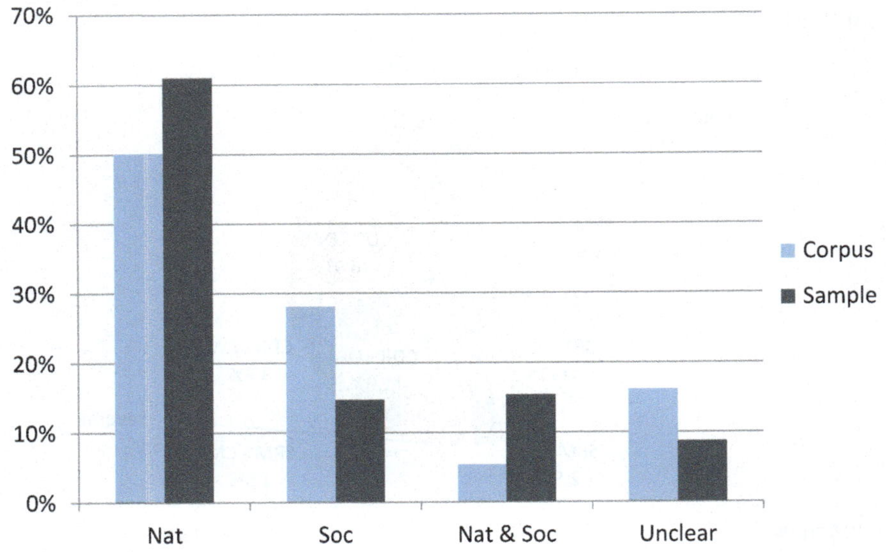

Image 6.2 Relative distribution of values for variable 'Disciplinary Focus' in corpus (1101 documents) and sample (136 documents). Please note: As units of reference were different in the sample and the corpus, the values for the corpus can serve only as qualitative comparison. Nat = natural sciences; Soc = social sciences and humanities

their relatability to social issues—e.g. political, social, historical, and economic—were coded 'social'. Those texts addressing phenomena in both perspectives were coded 'natural & social'.[3] Nearly two thirds of the 136 texts in the sample were given the value 'natural' for variable 'Disciplinary Focus' during coding (see Image 6.2). In the corpus, only a little more than half of all articles (50.2%) were assigned the value 'natural'. Similarly, the value 'social and natural' was assigned in the sample to 15.4%, while the corpus garnered only 5.5% of texts addressing questions concerning both social and natural phenomena. Conversely, the values 'unclear' and 'social' were more prevalent in the corpus than in the sample. From this, it can be inferred that CE articles addressing questions concerning natural phenomena are, on average, cited more often than those addressing social phenomena, meaning they made up a larger share of the sample on the basis of the sampling rules.

Concerning variable 'Year', the frequencies present in both corpus and sample are very closely correlated. This is hardly surprising when considering that the rules for the generation of the sample stipulate that a number of texts for each year had to be added to the sample depending on the share of texts from that year present in the corpus. Earlier texts were privileged over later texts in order to ensure that early discussions of CE were present in the sample.

[3] I refer to this category, in the course of the text, either as 'natural & social' or as 'both'.

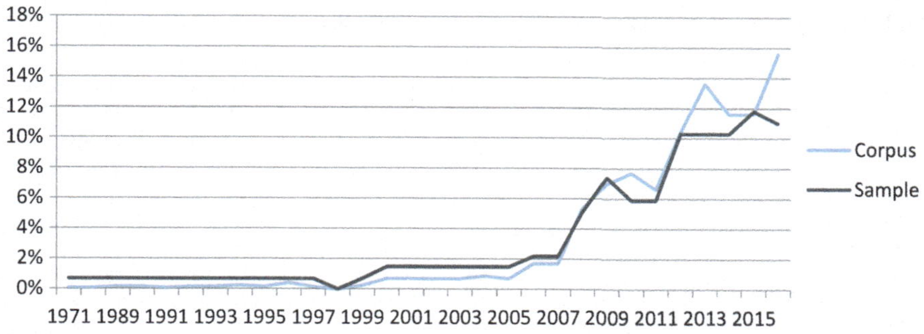

Image 6.3 Relative distribution of values for variable 'Year' in corpus (1101 documents) and sample (136 documents). Please note: Between years 1971 and 1977, not documents were found for the corpus. The same applies for the time frame between 1977 and 1989

6.2 Text Analysis

Text data was analyzed for both the segment as well as the document as the unit of analysis on three levels: the individual codes, frame elements assembled from individual codes as described above, and complete frames.[4] The following sections highlight the most interesting findings in increasing order of data aggregation: from the analysis of individual codings, to frame elements, to entire frames.

6.2.1 Codings

The lowest level of aggregation of data for this study is the level of individual codes. No aggregation has been applied to this level of data, as it is taken directly from the results of the coding process. This level already indicates some very interesting results. Firstly, with regards to all frame elements, most text segments do not carry any codings, meaning that there are no frame elements present. Secondly, of all problems identified in the sample, climate change is the most common. Thirdly, for all problems taken together (climate change, CE and others), problem characterizations took place most commonly along the lines of complexity and security. And fourthly, only very few texts identify a causally responsible actor for the problems identified.

Table 6.1 presents the frequencies of codings for codes for frame elements problem definition (category 1) and causal analysis (category 2). Category 1.1 indicates what is identified as the central problem in the text segment. Different values for category 1.2 present different themes according to which the central problem is characterized. Category 2 takes on different values concerning the actor that is identified as causally responsible for the central problem. These categories were coded for each segment

[4] Numbers and percentages in this section were rounded to two digits after point.

Table 6.1 Frequencies of codings for categories 1 (problem definition) and 2 (causal analysis)

Value	Code	Number of codings	% of codings in all documents (100% = 8676)	Number of documents	% of coded segments (100% = 1247) (%)	% documents in sample (100% = 136) (%)
Category 1.1: problem identification						
Climate change	1.1 CC	486	5.59	123	38.79	90.44
Climate Engineering	1.1 CE	187	2.16	60	15.00	44.12
None	1.1 None	77	0.89	51	6.17	37.50
Other	1.1 Other	185	2.13	74	14.84	54.41
Category 1.2: problem characterization						
Threat/Danger	1.2 Sec	99	1.14	51	7.94	37.50
Cost	1.2 Cost	54	0.62	33	4.33	24.26
Complexity	1.2 Comp	100	1.15	48	8.02	35.29
Appropriate-ness	1.2 Norm	35	0.40	13	2.81	9.56
Other	1.2 Oth	42	0.48	24	3.37	17.65
None	1.2 None	626	7.21	131	50.20	96.32
Category 2: causal analysis						
Political actors	2 Pol	38	0.44	23	3.05	16.91
Economic actors	2 Econ	12	0.14	11	0.96	8.09
Science	2 Sci	13	0.15	8	1.04	5.88
Society	2 Soc	17	0.20	13	1.36	9.56
Media	2 Media	1	0.01	1	0.08	0.74
Other	2 Oth	0	0.00	0	0.00	0.00
None	2 None	841	9.69	135	67.44	99.26
Unclear	2 Uncl	8	0.09	7	0.64	5.15
No frame elements						
	No frame elements	260	3.00	105	20.85	77.21

Grey areas in all tables in this chapter highlight particularly high frequencies of coding for the respective code

in the sample. This means that it becomes visible which value the category takes on for each category, but also if the category does not take on any value.

Concerning the problem identified in each segment of text coded, Image 6.4 shows that climate change (CC) is, of all potential problems outlined in the codebook, by far the one which is presented most frequently across all texts: 486 codings for this code across the entire sample were applied in 123 documents, which means that this code is present in 38.79% of all coded segments in 90.44% of all documents. Conversely,

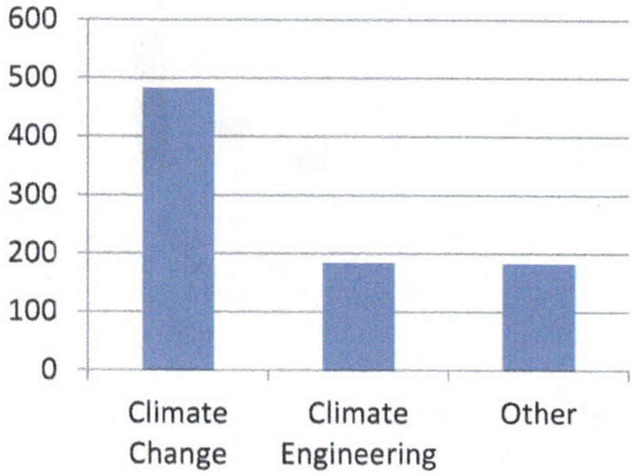

Image 6.4 Absolute frequencies of codings for codes in category 1.1: identification of problem

however, I find that nearly one tenth of all documents do not discuss climate change as a central problem in any segment. CE is problematized in at least one segment in 60 documents from the sample, which represents a share of 44.12% of all documents. 15% of all coded segments in the sample address CE as a central problem. Altogether, 337 segments were found in the sample where no central problem has been identified, representing 27.02% of all coded segments in the sample. The finding that climate change is, in a majority of documents, referred to as the central problem in texts about CE is not surprising when common definitions of the term are considered such as the one included in the 2009 Royal Society Report: CE or "geoengineering" is defined here as "the *deliberate large-scale manipulation of the planetary environment to counteract anthropogenic climate change*" (Shepherd et al. 2009). However, it is interesting that, while a majority of texts discuss climate change as a central problem, only half of them problematize CE in the same way. At this level of aggregation, this does not yet indicate any dominance of legitimizing or delegitimizing frames, however. Inferences concerning these can be made only on the level of frames, not on the level of codings. Other central problems that climate change and CE addressed in the sample are, for example, population growth, agricultural productivity or the 'Anthropocene' as social challenges, or issues of research practice such as science communication.

Concerning the characterization of the central problem of the text (category 1.2), which problem that might be in the concrete case of each segment cannot be ascertained on the level of codings alone, but will be assessed on the level of frame

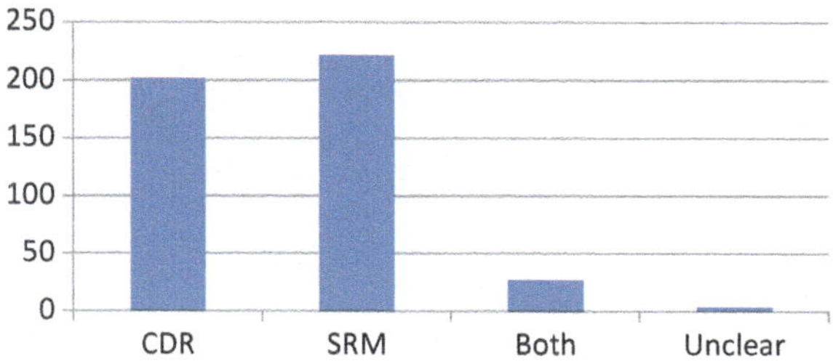

Image 6.5 Absolute frequencies of codings for approaches evaluated (numbers of codings)

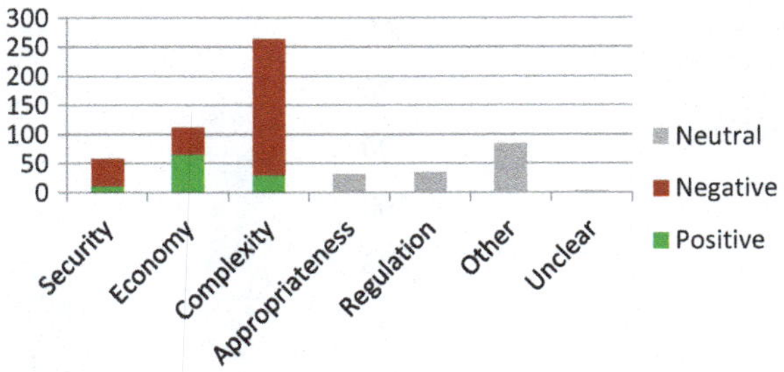

Image 6.6 Absolute frequencies of codings of criteria of evaluation (numbers of codings)

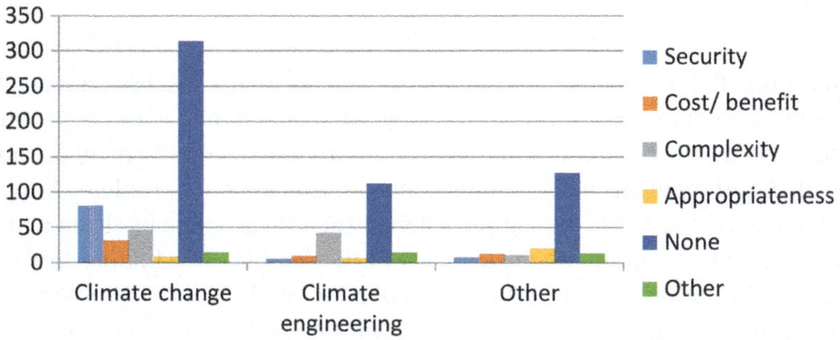

Image 6.7 Absolute frequencies of values for frame element 1: Problem Definition (number of codings)

Image 6.8 Absolute frequencies of values for frame element 3 (Evaluation of CE) (number of codings)

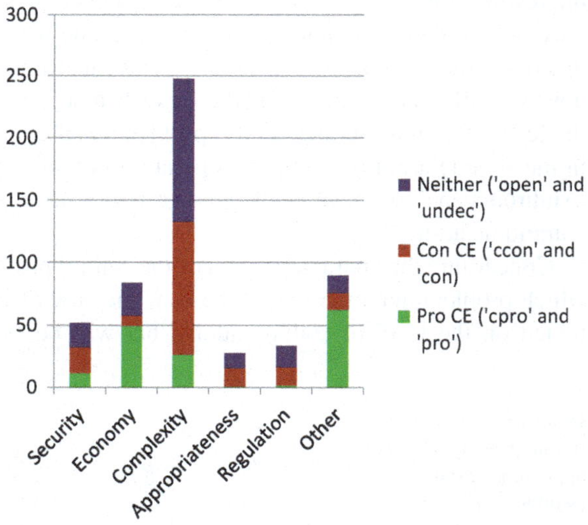

elements. Of those characterizations present, most frequent are the representation of any main problem either as one of high complexity (100 instances in 48 documents) or as a threat/danger to a valued referent object (99 instances in 51 documents). Unknown unknowns, unpredictability, uncertainty, messiness and a wealth of alternatives are concepts mentioned regularly when the central problem is characterized as one of complexity. For example:

> In practice, even independent of uncertainties in the ability to achieve the desired climate system changes, there are substantial uncertainties in both the radiative forcing exerted by a change in insolation and the climate response to that radiative forcing [...]. (Kravitz et al. 2015: 1639).

Urgency, risk, danger, fight, emergency, disaster, catastrophe, crisis and combat are common concepts related to the security characterization. For example:

> Moreover, the climatic response to elevated CO_2 concentration is uncertain, so a small risk of catastrophic impacts exists even today, and that risk grows monotonically as emissions continue to drive up the atmospheric CO_2 burden. (Keith 2009: 1654).

With a view to the frame element Causal Analysis (category 2), the results show that very few text segments across the entire sample identify causally responsible actors for the central problem identified in the text: When causal analysis is understood as the identification of a party responsible for the problem, this frame element is existent only in a tiny fraction of texts across the entire sample. In those few texts that do present some causal analysis in this manner, most commonly responsibility is posited as lying with political actors. Lack of political will, ineffectual international political cooperation and inadequate climate policies are common concepts with relation to this code. For example, Crutzen, in his oft-cited contribution from 2006, argues:

> Given the grossly disappointing international political response to the required greenhouse gas emissions, and further considering some drastic results of recent studies [...], research on the feasibility and environmental consequences of climate engineering of the kind presented in this paper, which might need to be deployed in the future, should not be tabooed. (Crutzen 2006: 214).

Concerning the evaluation of CE (category 3) and the proposed remedy (category 4), individual codings need to be connected to form frame elements in order to show these results. Only first indications can be drawn from this level.[5] For instance, the criterion of complexity was coded most often of all thematic criteria to evaluate CE approaches (234 codings, represented in 18.77% of all coded segments and 64.71% of all documents). Concepts such as side effects, unknowns, uncertainty or political complexity are named in relation to this value for category 3.2. For example:

[5]Categories 3 (Evaluation of CE) and 4 (Proposed Remedy) were not coded for every text segment in the sample. Instead, codes for these categories were applied only where the following coding rule was fulfilled: CE is assessed concerning advantages and disadvantages; arguments are listed for and/or against CE. Thus, no 'none' values exist for these two categories. Value 'unclear' serves as a residual value.

> The interactions and feedback of the suggested geoengineering methods with the natural variability of the atmospheric and surface reflectivity are not yet fully explored. The atmosphere is a complex system and any deliberate interventions should be treated with great care as they may have unanticipated adverse effects. (Bais et al. 2015: 39).

The contrasting criterion of simplicity was coded only 31 times in the entire sample. However, these codings are present in 14.71% of all documents, which means that, while few codings were made, these are spread across a large number of texts. Other evaluation criteria which are present in large fractions of the sample are existential threat, benefits or low costs and technological effectiveness. Notions of threat were coded 47 times across the sample, being present in 24.26% of all texts. Most commonly, the concept of risk is invoked in sections with this code:

> Initial studies suggest that the known risks are small, but unanticipated risks remain a serious underlying concern. If the world relies solely on SRM to limit warming, these problems will eventually pose risks as large as those from uncontrolled emissions. (Keith et al. 2010: 426).

Beyond this, concepts such as danger, drought, geopolitical conflict, irreversible damage, death and catastrophe are mentioned. Notions of safety in evaluating CE were comparatively rare in the sample, with 12 codings across all texts.

The criterion of benefit or relatively low cost was coded 67 times across the sample, being present in 28.68% of all documents. Relative arguments make comparisons to various objects under this category, for example with other, similar CE approaches or with GHG mitigation efforts. Absolute arguments present the CE approach in question as cost-effective, providing potential benefits or co-benefits. For example, Dumbrell et al. (2016: 30) make an argument about agricultural productivity as a co-benefit of carbon sequestration in soils. With 46 codings, slightly fewer texts were coded with the contrasting value of high costs as a criterion for evaluation of CE. 25.74% of all texts in the sample contain segments coded with the inductively created code of effectiveness as an evaluation criterion. Segments coded thus contain evaluations of various CE approaches referring to their effectiveness in actually counteracting effects of anthropogenic climate change, e.g.: "Some shortwave geoengineering measures, most promisingly stratospheric aerosol injections, have the potential to roughly cancel mitigated CO_2 radiative forcing." (Lenton and Vaughan 2009: 5556).

Concerning the outcome of an evaluation of CE (category 3.4), the results indicate that, across the entire sample and without differentiation concerning the specific CE approach and the type of activity (i.e. research, development or deployment), positions carefully supportive of CE and those carefully against it are in nearly perfect balance: 97 codings were applied for the former and 85 for the latter, with carefully pro-codings present in 38.24% of all documents and carefully con-codings in 35.29%. More frequent, however, are evaluations where the outcome remains open: 136 codings were applied for this value in 47.79% of documents in the sample. When adding all codings for any of the values for category 3.4 together, we can see that only in 465 coded segments, 37.28% of the entire set, is an evaluation presented at all.

Concerning remedies proposed in the texts (category 4), it needs to be noted that only in very few texts in the sample is any remedy brought forth as a potential solution to the central problem—only in 13.47% of all segments coded. This means that, in most segments, this frame element is not present at all. Concerning category 4.1 (proposed remedy), first and foremost, the results show that in none of the texts in the sample did I find the proposal to turn to CE exclusively to address whatever central problem was outlined in the text (4.1 ExCE). 100 codings were applied proposing some kind of activity with regards CE—either research or development or use in a portfolio, the data at this unaggregated level do not inform us which—as the remedy proposed to address the problem at hand.

6.2.2 Frame Element Values Across Entire Sample[6]

In order to assess whether elements of either ideal-type frames or inductively developed frames are present in the sample, and if so, whether specific frame elements are dominant, frame elements were re-assembled from codings for specific codes. These analysis steps consider the sample in its entirety. This analysis step shows, firstly, again, that most text segments identify any central problem (for instance, climate change or CE) without any further characterization of that problem. Secondly, if it is characterized, climate change is presented most commonly as a complex challenge or a security threat. Where CE is characterized as a central problem, it is most commonly presented as a complex challenge. Thirdly, evaluation of CE approaches takes place most commonly according to criteria of complexity or simplicity. If positive evaluation outcomes are found, very commonly those are derived on the basis of economic criteria.

Table 6.2 presents the frequencies of codings for the values for frame element 1 (Problem Definition). It indicates that in 64.81% of all text segments in which climate change was identified as a central problem, this was not characterized according to any of the ideal-type frames outlined above. For example:

> Despite 20 years of effort to curb emissions, greenhouse gas (GHG) emissions grew faster during the 2000s than in the 1990s, which represents a major challenge for meeting the international goal of limiting warming to $<2°$ relative to the preindustrial era (Smith 2016: 1315).

Climate change is mentioned as a problem or a challenge, but not further qualified along the themes of the ideal-type frames. Of all values for this frame element corresponding to any of the ideal-type frames, segments representing climate change as a threat or as a complex problem are most frequent in the sample (82 and 46 codings, respectively). 33.82% of all documents in the sample carry segments which describe climate change as a threat in some way, corresponding to 6.58% of all coded segments, such as:

[6]Numbers in this section have been rounded to two digits after point.

Table 6.2 Frequencies of values for frame element 1 (problem definition)

Combination from…							
1.1 (Problem identification)	1.2 (Problem characterization)	Number of codings[a]	% of entirety of segments coded (100% = 1247) (%)	Number of documents	% of entirety of documents in the sample (100% = 136) (%)	Value of frame element	Element of CE frame…
1.1 CC	1.2 Sec	82	6.58	46	33.82	1 CC—Sec	Security, legitimizing
	1.2 Cost	32	2.57	17	12.50	1 CC—Cost	Economy, legitimizing
	1.2 Comp	46	3.69	29	21.32	1 CC—Comp	Complexity, legitimizing
	1.2 Norm	9	0.72	6	4.41	1 CC—Norm	Appropriateness, legitimizing
	1.2 None	315	25.26	115	84.56	1 CC—none	Legitimizing without theme
	1.2 Oth	15	1.20	14	10.29	1 CC—Oth	Legitimizing, other theme
1.1 CE	1.2 Sec	6	0.48	6	4.41	1 CE—Sec	Security, delegitimizing
	1.2 Cost	10	0.80	9	6.62	1 CE—Cost	Economy, delegitimizing
	1.2 Comp	42	3.37	23	16.91	1 CE—Comp	Complexity, delegitimizing
	1.2 Norm	7	0.56	6	4.41	1 CE—Norm	Appropriateness, delegitimizing

(continued)

Table 6.2 (continued)

Combination from…							
1.1 (Problem identification)	1.2 (Problem characterization)	Number of codings[a]	% of entirety of segments coded (100% = 1247) (%)	Number of documents	% of entirety of documents in the sample (100% = 136) (%)	Value of frame element	Element of CE frame…
	1.2 None	113	9.06	47	34.56	1 CE—None	Delegitimizing without theme
	1.2 Oth	15	1.20	8	5.88	1 CE—Oth	Delegitimizing, other theme
1.1 Other	1.2 Sec	8	0.64	6	4.41	1 Oth—Sec	Security, delegitimizing
	1.2 Cost	13	1.04	11	8.09	1 Oth—Cost	Economy, delegitimizing
	1.2 Comp	11	0.88	6	4.41	1 Oth—Comp	Complexity, delegitimizing
	1.2 Norm	21	1.68	5	3.68	1 Oth—Norm	Appropriateness, delegitimizing
	1.2 None	128	10.26	68	50.00	1 Oth—None	Legitimizing without theme
	1.2 Oth	14	1.12	9	6.62	1 Oth—Oth	Legitimizing, other theme
1.1 None	1.2 None	73	5.85	51	37.50	1 None—none	None
No frame elements		260	20.85	105	77.21	No frame elements	None

[a]The numbers of all codings for a combination of several codes do not necessarily add up to the same sum as the first code in the combination alone. For example, the number of codings for a combination of '1.1 CC' and any value for category 1.2 do not add up to equal the number of codings for '1.1 CC', as some text segments coded '1.1 CC' were coded for more than one code for 1.2.
Inductively developed frames are presented in grey type

Table 6.3 Frequencies of values for frame element 3 (Evaluation of CE)

Combination from…							
3.2 Evaluation criteria	3.4 Outcome of evaluation	Number of codings	% of entirety of segments coded (100% = 1247) (%)	Number of documents	% of entirety of documents in the sample (100% = 136) (%)	Value of frame element	Element of CE frame…
3.2 Threat	3.4 Con	12	0.96	9	6.62	3 Threat—con	Security, delegitimizing
	3.4 CCon	9	0.72	8	5.88	3 Threat—ccon	
	3.4 Open	16	1.28	13	9.56	3 Threat—open	Security, open
	3.4 Undec	3	0.24	3	2.21	3 Threat—undec	
3.2 Safety	3.4 Pro	9	0.72	6	4.41	3 Safety—pro	Security, legitimizing
	3.4. CPro	3	0.24	3	2.21	3 Safety –cpro	
	3.4 Open	0	0.00	0	0.00	3 Safety—open	Security, open
	3.4 Undec	0	0.00	0	0.00	3 Safety—undec	
3.2 Cost high	3.4 Con	4	0.32	3	2.21	3 Cost—con	Economy, delegitimizing
	3.4 CCon	4	0.32	4	2.94	3 Cost—ccon	
	3.4 Open	13	1.04	10	7.35	3 Cost high—open	Economy, open
	3.4 Undec	0	0.00	0	0.00	3 Cost high—undec	
3.2 Cost low	3.4 Pro	27	2.17	18	13.24	3 Cost—pro	Economy, legitimizing
	3.4 Cpro	23	1.84	18	13.24	3 Cost—cpro	
	3.4 Open	11	0.88	11	8.09	3 Cost low—open	Economy, open
	3.4 Undec	2	0.16	2	1.47	3 Cost low—undec	

(continued)

Table 6.3 (continued)

Combination from…							
3.2 Evaluation criteria	3.4 Outcome of evaluation	Number of codings	% of entirety of segments coded (100% = 1247) (%)	Number of documents	% of entirety of documents in the sample (100% = 136) (%)	Value of frame element	Element of CE frame…
3.2 Complexity	3.4 Con	38	3.05	23	16.91	3 Comp—con	Complexity, delegitimizing
	3.4 CCon	68	5.45	38	27.94	3 Comp—Ccon	
	3.4 Open	99	7.94	51	37.50	3 Comp—open	Complexity, open
	3.4 Undec	12	0.96	11	8.09	3 Comp—undec	
3.2 Simplicity	3.4 Pro	13	1.04	11	8.09	3 Simp—pro	Complexity, legitimizing
	3.4 CPro	14	1.12	12	8.82	3 Simp—cpro	
	3.4 Open	4	0.32	4	2.94	3 Simp—open	Complexity, open
	3.4 Undec	0	0.00	0	0.00	3 simp—undec	
3.2 Inappropriateness	3.4 Con	5	0.40	3	2.21	3 Norm—con	Appropriateness, delegitimizing
	3.4 CCon	10	0.80	8	5.88	3 Norm—ccon	
	3.4 Open	11	0.88	10	7.35	3 Norm con—open	Appropriateness, open
	3.4 Undec	1	0.08	1	0.74	3 Norm con—undec	
3.2 Appropriateness	3.4 Pro	1	0.08	1	0.74	3 Norm—pro	Appropriateness, legitimizing
	3.4 Cpro	0	0.00	0	0.00	3 Norm—cpro	
	3.4 Open	0	0.00	0	0.00	3 Norm pro—open	Appropriateness, open

(continued)

Table 6.3 (continued)

Combination from…							
3.2 Evaluation criteria	3.4 Outcome of evaluation	Number of codings	% of entirety of segments coded (100% = 1247) (%)	Number of documents	% of entirety of documents in the sample (100% = 136) (%)	Value of frame element	Element of CE frame…
	3.4 Undec	0	0.00	0	0.00	3 Norm pro—undec	
3.2 Regulation	3.4 Con	4	0.32	3	2.21	3 Pol—Con	Other, delegitimizing
	3.4 CCon	11	0.88	10	7.35	3 Pol—Ccon	
	3.4 Open	16	1.28	11	8.09	3 Pol—Open	Other, open
	3.4 Undec	1	0.08	1	0.74	3 Pol—undec	
	3.4 Pro	1	0.08	1	0.74	3 Pol—pro	Other, legitimizing
	3.4 Cpro	1	0.08	1	0.74	3 Pol—Cpro	
3.2 Other	3.4 Con	5	0.40	5	3.68	3 Oth—con	Other, delegitimizing
	3.4 CCon	8	0.64	6	4.41	3 Oth—Ccon	
	3.4 Open	13	1.04	12	8.82	3 Oth—open	Other, open
	3.4 Undec	1	0.08	1	0.74	3 Oth—undec	
	3.4 Pro	23	1.84	10	7.35	3 oth—pro	Other, legitimizing
	3.4 Cpro	40	3.21	27	19.85	3 oth—cpro	

[a]Caveat: For 3, if several CDR/SRM approaches appeared in one paragraph, these were not subdivided and coded individually, as this would have gone counter the coding rule that the unit of coding is to be the paragraph.
Inductively developed frames are presented in grey type

Table 6.4 Frequencies of values for frame element 4 (proposed remedy)

Combination from…							
4.1 Proposed Remedy	4.2 Avenue of Activity	Number of codings	% of entirety of segments coded (100% = 1247) (%)	Number of documents	% of entirety of documents in the sample (100% = 136) (%)	Value of frame element	Ascribed to frame variants
4.1 CE	4.2 Use	12	0.96	9	6.62	4 CE—use	Legitizing
4.1 Port		14	1.12	11	8.09	4 Port—use	
4.1 ExCE		0	0.00	0	0.00	4 ExCE—use	
4.1 Other		25	2.00	18	13.24	4 Oth—use	Delegitimizing
4.1 CE	4.2 Research	73	5.85	46	33.82	4 CE—res	Legitimizing
4.1 Port		14	1.12	12	8.82	4 Por—res	
4.1 ExCE		0	0.00	0	0.00	4 ExCE—res	
4.1 Other		7	0.56	5	3.68	4 Oth—res	Delegitimizing
4.1 CE	4.2 Development	9	0.72	9	6.62	4 CE—dev	Legitimizing
4.1 Port		3	0.24	3	2.21	4 Port—dev	
4.1 ExCE		0	0.00	0	0.00	4 ExCE—dev	
4.1 Other		1	0.08	1	0.74	4 Oth—dev	Delegitimizing

> The burning of our diminishing fossil fuel reserves is accompanied by a large anthropogenic CO_2 release, which is outpacing nature's CO_2 recycling capacity, causing significant environmental harm. Much is being done to avert this crisis, including more efficient technology for usage, savings and replacing carbon fuel […]. (Goeppert et al. 2012: 7833).

Text segments representing climate change as a threat describe the phenomenon according to dangers or risks to, for example, ecological, social or economic systems. Concepts of 'fight' or 'combat' in relation to addressing the challenge are formulated. Similarly, notions of a potential climate emergency or crisis or of urgency in addressing the challenge are mentioned. Segments of 21.32% of documents in the sample present climate change as a complex issue. These amount to 3.69% of all coded segments, such as:

> The time period and method of temporal averaging, in combination with interannual variability, will lead to constant insinuations that 1.5/2 °C has been exceeded. Together with a potential peak and decline in temperatures after carbon dioxide removal (CDR), it may not be known for many decades if 1.5/2 °C has been exceeded or successfully avoided. (Peters 2016: 646).

Arguments are made about scientific uncertainty, as well as about manifold physical as well as social and political factors relevant for the magnitude, development and effects of climate change and for decisions concerning approaches to address this challenge:

> Recognition of the complexity of human-caused changes in climate does not, however, weaken the importance of actions that would seek to minimize our disturbance of the Earth's environmental system and that would reduce societal and ecological vulnerability to environmental change and variability. (Marland et al. 2003: 150).

Only 9 codings were applied to representations of climate change as a normative problem, present in 4.41% of documents and 0.72% of all coded segments in the sample. For example, Steffen et al. describe:

> The pathways of development followed by today's wealthy countries after the Second World War – built on plentiful, cheap fossil fuel energy resources, an abundance of other material resources, and large expanses of productive land to be developed – cannot be followed by the 75–80% of the human population who are now at various stages of their trajectories out of poverty, and are beginning to compete with today's wealthy countries for increasingly scarce resources. (Steffen et al. 2011b: 739).

All frames including climate change as a problem are considered to be legitimizing CE to some degree, as the acceptance of climate change as a serious problem might strengthen the position of CE as a remedy. If climate change was not a serious problem, CE would be obsolete.

Conversely, CE as a central problem in the text is, in this study, considered to be, generally, delegimizing CE. 60.43% of all codings of CE as a central problem do not characterize these further along the themes identified here. For example, Keith outlines the challenges in assessing solar engineering in the following way:

> One cannot meaningfully evaluate solar geoengineering without a scenario for its implementation.[…] Yet the balance between risks and benefits depends at least as strongly on how SRM is deployed (for example on technology choice, timing and magnitude of the induced radiative forcing) as it depends on the climate's response to a specified SRM scenario. (Keith and MacMartin 2015: 201).

Of those values for frame element 1 corresponding with the ideal-type frames, a characterization of CE as a complex issue is most frequent: 42 codings in 16.91% of all documents, but only 3.37% of all coded segments in the sample conform with this representation. For example, Keith argues: "Clear language is an essential tool for analyzing this messy topic" (Keith 2015, 1). Rayner et al. state: "Research in many disciplines, including climate science, engineering, law, economics, politics, and ethics, will be necessary to understand and deal with the challenges of developing geoengineering technologies" (Rayner et al. 2013: 2). Concerning CE as the central problem, security and appropriateness characterizations are rarest: Only 7 and 6 codings, respectively, were applied for these values, corresponding with 0.56 and 0.48% of all coded segments in the sample. Applying an argument of non-appropriateness, Macnaghten and Szerszynski argue that solar CE was inappropriate when assessed according to the norm of democratic decision-making:

> Democracy, in its various forms, depends on the articulation, negotiation and accommodation of plural views and interests. It relies on an evolving and partially flexible relationship between citizens and governance institutions. Solar radiation management by contrast exists as a planetary technology. For these reasons, one important role of bringing publics into the democratic governance of solar radiation management may be, ironically, as a method of articulating the anti-democratic constitution of the technology and its incompatibility with liberal democracy. (Macnaghten and Szerszynski 2013: 472).

Keith et al. outline an argument of possible rogue use of CE as a risk when describing:

> […] a nation might grow frustrated at the pace of international cooperation and establish a national programme of gradually expanding research and field tests.[…] In this case, one nation – which need not be a large and rich industrialized country – would effectively seize the initiative on global climate, making it extremely difficult for other powers to restrain it. (Keith et al. 2010: 427).

Concerning problems other than climate change and CE identified in the text segments, such as energy production, innovation governance or population increase with 21 codings (11.35% of all codings for '1.1 other'), a normative characterization is the most frequent. It needs to be noted, however, that this value is present in only 3.68% of all documents of the sample, in the characterization of issues such as innovation governance and the Anthropocene.

Concerning frame element 3 (Evaluation of CE),[7] the analysis of the co-occurrence of codings for Evaluation Criteria (category 3.2) and Outcome of Evaluation (category 3.4 shows that the evaluation of CE according to the criterion of security is more frequent through a representation of CE as a threat and a corresponding negative evaluation outcome ('3.4 ccon' or '3.4 con'; 21 codings) than of

[7]It is important to note again that the co-occurrence of 3.2 and 3.4 was assessed here. This is due to small numbers of codings, which are difficult to further differentiate.

CE as a (comparatively) safe measure ('3.4 cpro' or '3.4 pro'; 12 codings). Of the text segments evaluating CE according to the criterion of security and coming to a negative outcome, 3 were applied to segments evaluating CDR and 16 were applied to segments evaluating SRM approaches, for example:

> Creating a risk of widespread drought and reduced freshwater resources for the world to cut down on global warming does not seem like an appropriate fix. Our results suggest that considerable caution should be used regarding any human intervention into the climate system that we do not fully understand. (Trenberth and Dai 2007: 4).

However, both representations are fairly rare in the sample, being represented in only 1.68% and 0.96% of all coded segments, respectively. The corresponding value 'open' was applied to 16 text segments and present in 1.28% of coded segments in the sample. Many documents in the sample did not apply the criterion of security at all when evaluating CE (see above: Analysis of codings).

The majority of evaluations of CE applying the criterion of costs and benefits come to a positive assessment of CE approaches: 23 codings were applied to a carefully positive evaluation outcome and 27 codings to a positive evaluation outcome of CE for criterion economy. 34 positive evaluation outcomes were coded in text segments addressing CDR approaches, 12 in those addressing SRM approaches. In comparison, only 4 codings were applied for a carefully negative and 4 for a negative evaluation outcome. Carefully positive and positive evaluation outcomes are each represented in 13.24% of the documents in the sample. For example, Nordhaus, when comparing the costs of various policies to address climate change in 1992, states:

> A final policy, geoengineering, would introduce a hypothetical technology that provides costless mitigation of climate change. This could occur, for example, if one of the geoengineering options proved technically feasible and environmentally benign. Two interesting proposals include shooting smart mirrors into space with 16-inch naval rifles or seeding the oceans with iron to accelerate carbon sequestration. [Several geoengineering solutions have extremely low economic costs compared to conventional mitigation techniques and can therefore be treated as costless; […]] (Nordhaus 1992: 1317).

However, also for this criterion, a substantial proportion of the documents in the sample bear segments in which CE is evaluated according to this criterion, but no definite outcome of the assessment is forthcoming ('3.4 open').

The criterion of complexity/simplicity is coded most frequently as an evaluation criterion of CE. Of the text segments bearing this code, the majority does not provide a clear outcome of evaluation ('3.4 open': 99 codings; '3.4 undecided': 12 codings). For example, Tilmes et al. argue:

> SRM has been suggested as a mechanism to stabilize global temperatures. The continuous increase in positive forcing from increasing greenhouse gases is countered by a continuous increase in negative forcing from SRM produced by an increase in the Earth's albedo. This may be achieved by adjusting the amount of aerosol loading in the stratosphere […]. However, such experiments will alter the flow of energy throughout the climate system and reduce surface downwelling shortwave and latent heat fluxes, as studied in many different model […]. Some geoengineering studies have shown a significant decrease of precipitation, for example, over the Indian monsoon, as a result of geoengineering […]. […] However,

> different models and scenarios do not always agree in the sign of the change of monsoonal precipitation in response to geoengineering […]. Therefore, the impact of SRM in the context of a high CO_2 environment on the global and regional precipitation across different climate models has yet to be quantified. (Tilmes et al. 2013: 11,037).

However, 29.05% of all segments in which CE is evaluated according to complexity present a carefully negative evaluation of CE (68 codings) and 16.24% (38 codings) present a negative evaluation. For the criteria regulation and appropriateness, no clear evaluation outcome is the most common configuration, presenting dominance of neither positive nor negative evaluation outcomes of CE. Inductively coded 'other' evaluation criteria, which were found to be criteria of effectiveness or feasibility, present mostly positive evaluation outcomes in the sample: 23 codings were applied to segments which present a positive, 40 codings to segments which present a carefully positive evaluation outcome.

Concerning frame element 4 (Proposed Remedy), it needs to be noted that proposals of how to address the central problem are made very rarely in the texts in the sample, in general: Only in 13.47% of all segments coded is any coding applied for this frame element (see above). The coding produced here allows for a differentiation between text segments which talk of CE in general ('CE') and those which explicitly discuss CE as part of a portfolio of approaches ('Port'). Where the deployment of CE approaches is considered as a potential remedy (12 codings in 9 documents for CE in general, 14 codings in 11 documents for the portfolio approach), the specific approaches discussed are, either, general CDR or NETs, or, where the texts are more specific, reforestation endeavors or the modification of surfaces in order to increase albedo (SRM approach SA), but also SAI. This means that none of the texts in the sample propose to undertake anything like large-scale SAI or OF efforts as an exclusive strategy of addressing climate change. However, Wigley, for example, in a discussion of application of SAI, argues:

> Projected anthropogenic warming and increases in CO_2 concentration present a twofold threat, both from climate changes and from CO_2 directly through increasing the acidity of the oceans. Future climate change may be reduced through mitigation (reductions in greenhouse gas emissions) or through geoengineering. Most geoengineering approaches, however, do not address the problem of increasing ocean acidity. A combined mitigation/geoengineering strategy could remove this deficiency. Here we consider the deliberate injection of sulfate aerosol precursors into the stratosphere. This action could substantially offset future warming and provide additional time to reduce human dependence on fossil fuels and stabilize CO_2 concentrations cost-effectively at an acceptable level. (Wigley 2006: 452).

Aside from this, the focus of the majority of texts in the sample lies on research (73 codings in 46 documents) or development (9 codings in 9 documents) endeavors, not deployment. Other proposed remedies cover, for example, intensified climate change research in order to increase understanding of the phenomenon, energy transition efforts or mitigation in general. Codings for these proposed remedies were applied in 25 cases to 18 documents.

6.2.3 Frame Element Values Across Document Groups[8]

In order to ascertain whether specific frame element values are more common in texts with a clear focus on questions concerning natural phenomena or those focusing on social phenomena, or whether differences exist between texts addressing different CE approaches or across year of publication, frame element distributions were assessed across document groups differentiated according to their values for variables 'disciplinary focus', 'year of publication' and 'approach'. The main findings of this analysis step were the following: Firstly, CE is more commonly presented as the central problem in segments from social science or humanities texts than in those from natural science texts. Secondly, social science texts present both CE and climate change as security problems more commonly than natural science texts. Thirdly, the evaluation of CE is more often positive or carefully positive in natural science texts or those addressing issues from both natural and social science perspectives than in text addressing social science or humanities questions alone.

6.2.3.1 Differentiation by Disciplinary Focus

Using the variable 'Disciplinary Focus', documents are grouped according to the disciplinary focus of the research questions pursued in each document (Natural and engineering sciences (Nat), social sciences and humanities (Soc), both (Soc & Nat) or Unclear). Concerning frame element 1 (Problem Definition), it is interesting to compare the document groups concerning the central problem addressed. While the following graphs show the distribution of codings in each of these groups relative to the sum of codings, it is important to remember that the groups are not equally represented in the sample: Nearly two thirds of all documents belong to group 'Nat'. Texts focusing on social phenomena and texts addressing both kinds of questions are equally frequent in the sample; texts in the residual category 'unclear' cover less than 10% of the sample. Image 6.9 indicates that the problematization of CE is more common in 'Soc' texts than in 'Nat' texts.

Zooming in on codings for specific problems identified in the text, Image 6.10 shows the relative distribution of values for text segments identifying climate change as the central problem. A representation of coding frequencies relative to the sum of codings for climate change as the central problem for each of the document groups (Nat: 245, Soc: 80, Soc & Nat: 111, Unclear: 65) was chosen in order to render the numbers more intelligible.

The most common characterization in the group of documents ascribed to social sciences and humanities describes climate change as a problem of security, rather than one of costs and benefits or of complexity or appropriateness. In contrast, texts addressing both social and natural phenomena present the characterization of climate change as a security problem almost as frequently as they characterize it as a cost problem or a complex issue. Diagram part b shows the relative distribution of values

[8]Numbers in this section have been rounded to two digits after point.

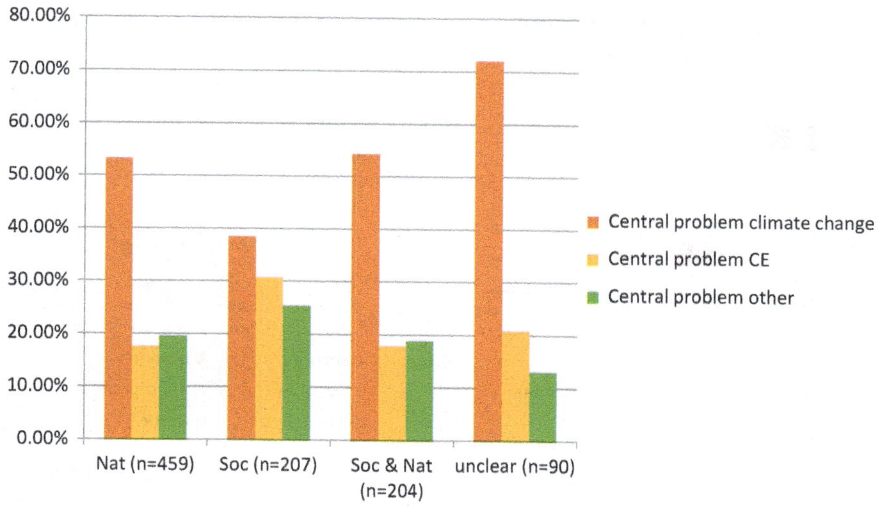

Image 6.9 Comparison of distribution of central problems addressed in texts with different disciplinary foci relative to sum of codings for frame element 1 (Nat: 459; Soc: 207; Soc & Nat: 204; Unclear: 90)

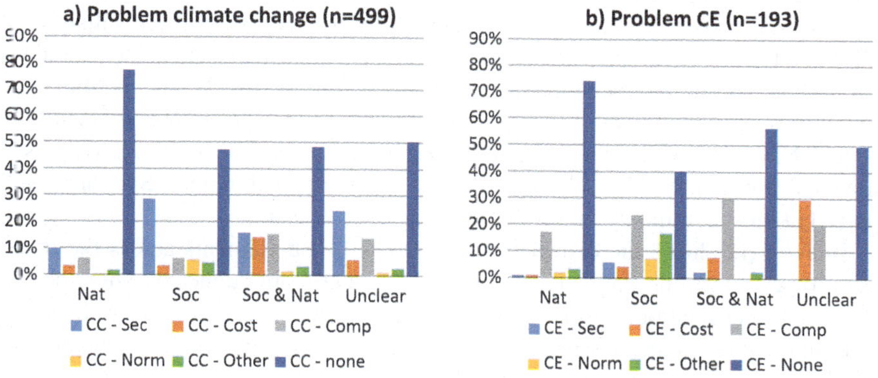

Image 6.10 Comparison of distribution of characterization of problem climate change (**a**) and climate engineering (**b**) across disciplinary foci relative to sum of codings for the two values for category 1.1: problem identification, **a** Nat: 245; Soc: 80; Soc & Nat: 111; unclear: 65; **b** Nat: 82; Soc: 64; Soc & Nat: 37; unclear: 10

for text segments identifying CE as the central problem. The graph shows that the non-characterization of CE is most frequent in all document groups, just as in the case of the problem of climate change. However, the representation of CE as a complex problem is common across all document groups. Documents addressing both natural and social phenomena have the highest share of representations of CE as a complex problem.

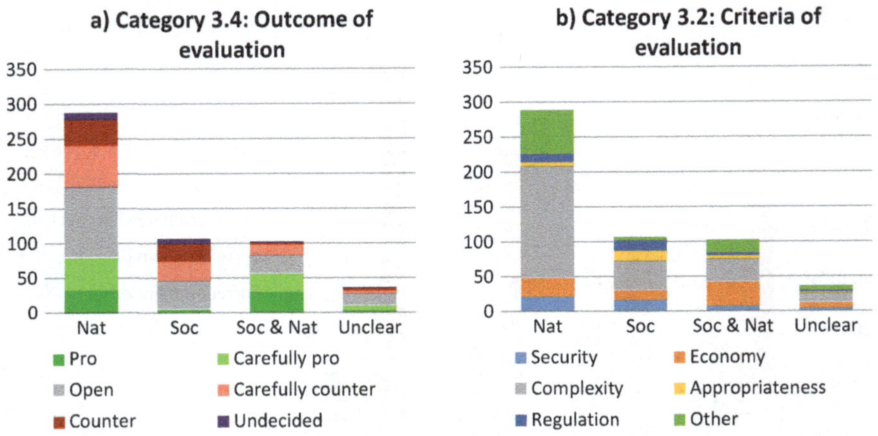

Image 6.11 Comparison of frequencies of codings for various **a** outcomes of evaluation **b** criteria of evaluation across document groups (numbers of codings)

Concerning the evaluation of CE (frame element 3), part a) of Image 6.11 shows the frequencies of codings of the various outcomes of evaluation (category 3.4) differentiated across the document groups.[9] The results indicate that segments in 'Nat' or 'Soc & Nat' texts more often come to a carefully positive or positive evaluation of CE than those in 'Soc' texts. Part b of the graph presents the frequencies of codings for various criteria of evaluation across the document groups. While complexity is common across all document groups, security criteria are applied less frequently in 'Soc & Nat'. 'Other' criteria—commonly arguments of feasibility or effectiveness—are most common in 'Nat'. Criteria of economy are applied in both 'Soc' and 'Nat' texts, but most commonly in texts from group 'Soc & Nat'.

Image 6.12 shows the distribution of various frame elements across the document types oncerning how specific evaluation criteria are connected with specific outcomes of evaluation. It is important to remember, however, that some of these are much more frequent in the sample than others: Negative or open-ended evaluations of CE using complexity criteria are particularly frequent. The graphs show that codings for text segments presenting a positive outcome of a security evaluation of CE are more common in 'Nat' texts and do not exist in 'Soc' texts, while a negative outcome as well as a lack of specific outcome are shared nearly equally between the two document groups. Positive evaluations according to economic criteria are most common in texts from the group 'Both'. Evaluation according to criteria of complexity are very common generally, but 'Nat' texts cover a larger share of codings for this criterion than the other document groups. A larger share of codings for the criteria

[9]Note: As indicated above, for frame element 3, no ‚none ' codings were applied. Segments of text in which this category did not appear were not coded for this category. The same applies to frame element 4.

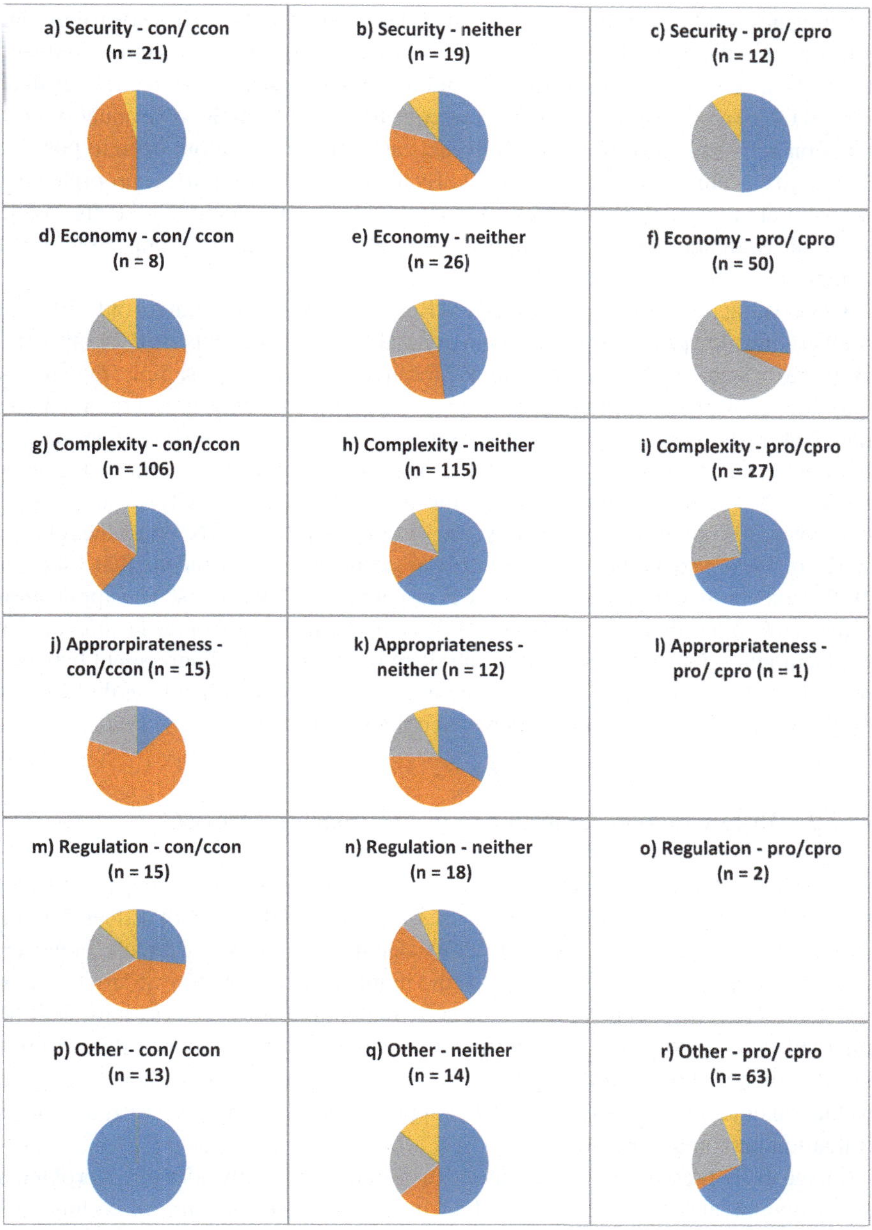

Image 6.12 (This page and page before this): Comparison of distribution of absolute number of codings for frame element 3 (Evaluation of CE) (number of codings) (blue: Nat; red: Soc; green: Soc & Nat, purple: Unclear)

Effectiveness and Feasibility is covered by 'Nat' texts as well.[10] These distributions, considered in the light of the shares each document group has in the sample, indicate that texts in group 'Soc', on average, put more emphasis on negative security evaluations of CE or reserve judgement on security evaluations. Complexity evaluations are common across all groups, while 'Nat' texts are, on average, more open to positive complexity evaluations than 'Soc' texts. Evaluations according to Appropriateness or potential of Regulation are more common in 'Soc' texts than in 'Nat' texts. While economic evaluation criteria are applied across all groups, texts of group 'Both' most commonly evaluate CE positively on these grounds.

Concerning frame element 4 (Proposed Remedy), 'Nat' texts contain only 39.91% of all codings for proposed remedies in general, while 'Soc' texts bear 24.68% and 'Both' texts bear 25.95%. As 'Nat' texts present two thirds of the sample, this means that 'Soc' or 'Both' texts more commonly present any kind of proposal for a remedy than 'Nat' texts. The values 'CE—Research' and 'Other—Use' are most dominant in the sample, i.e. the research of CE approaches and the application of measures other than CE. 'Nat' texts contain 26 codings for the remedy of further CE research. 'Soc' texts cover 20 codings for this remedy, texts from group 'Both' 16. When the relative shares of these text groups in the sample are considered, this means that 'Soc' or 'Both' texts more commonly propose CE research than 'Nat' texts. The application of measures other than CE ('Other—Use') is distributed in a similar fashion across the document groups. This indicates that there is a high level of agreement between the text groups concerning the best remedy for the problem at hand, while 'Soc' or 'Both' texts make this argument more often than 'Nat' texts.

6.2.3.2 Differentiation According to the Approaches Assessed

The Variable 'Approach' divides the documents of the sample into groups according to the CE approach the document focuses on. The subdivision of values for the various frame elements according to these document groups is difficult, however, as many groups exist, which distributes the number of codings too widely to allow any stable inferences to be drawn. However, some interesting observations shall be imparted here: Firstly, CE is identified as the main problem more often in texts on SRM than in those addressing CDR measures. Secondly, SAI garners a negative evaluation outcome more often than other approaches; DAC and SA garner a positive evaluation outcome more often.

Image 6.13 presents the distribution of different centrally identified problems across document groups. It shows values normalized over the sum of codings for frame element 1 in the specific group. The most commonly mentioned central problem in the sample is climate change, which is true across all document groups. However, segments identifying CE as the central problem are more common, on average, in texts on SRM approaches in general ('SRM—several') and in texts on

[10]Frame element values 'norm – pro' and 'pol – pro' are not represented here as these values appear only in 1 and 2 codings, respectively.

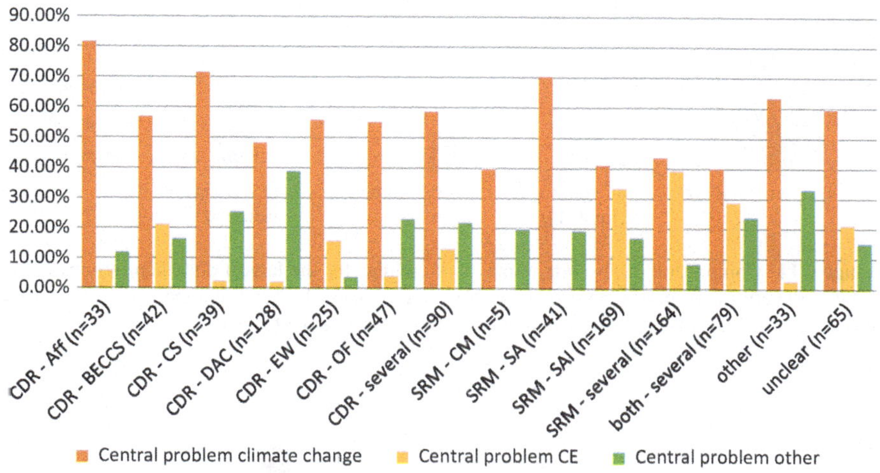

Image 6.13 Comparative distribution of frequencies of various central problem identifications in the document groups relative to number of codings in this group for frame element 1. The numbers below each document group indicate the number of codings in texts of this group for frame element 1. (The numbers do not equal 100% as some segments were coded for several values and value 'none' is left out of the graph.)

the SRM approach SAI than in texts on CDR approaches such as DAC, BECCS or OF. However, SRM approaches SA and CM also do not present segments in which CE is identified as the central problem. The central problem 'other' garnering high coding numbers for various CDR approaches very often refers to research challenges such as unknowns which might prevent DAC or CS from being used on a large-scale.

While definitions of climate change as a security problem and of CE as a complex challenge both have a comparatively high frequency in the sample, they are distributed unevenly: It is useful, at this point, to focus on the three largest document groups 'SRM—SAI' (16.2% of all texts in the sample), 'SRM—several' (13.2%) and 'CDR—DAC' (16.9%). Texts belonging to the group 'CDR—DAC' present no codings of for the definition of CE as a complex problem. This is particularly interesting when assessed in comparison to the other large document group 'SRM—SAI', in which 12 codings represent this value for frame element 1. Furthermore, Image 6.14 shows that none of the texts from group 'SRM—SAI' presents any segments in which the central problem climate change is presented as an economic issue.

Concerning the values for frame element 3 (Evaluation of CE), the document groups present a very diverse picture. As the document groups have largely different sizes, a comparison is difficult. However, Image 6.16 Comparison of share of codings for selected values for frameFrame element 3 and selected document groups (values relative to sum of codings in this document group) 6.15 indicates that the outcomes of evaluation differ quite strongly depending on what the CE approach which is the focus of the document in question is. Documents on the CDR approach OF, the SRM

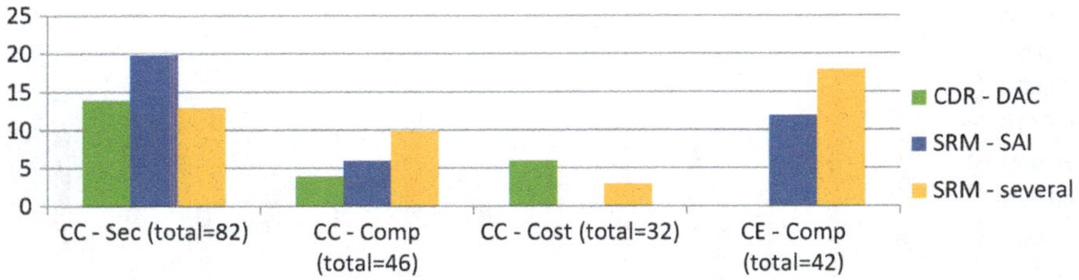

Image 6.14 Comparison of absolute numbers of codings for the largest document groups in the sample concerning the most common values for frame element 1. 'Total' numbers refer to the sum of codings for these values for frame element 1 across all document groups

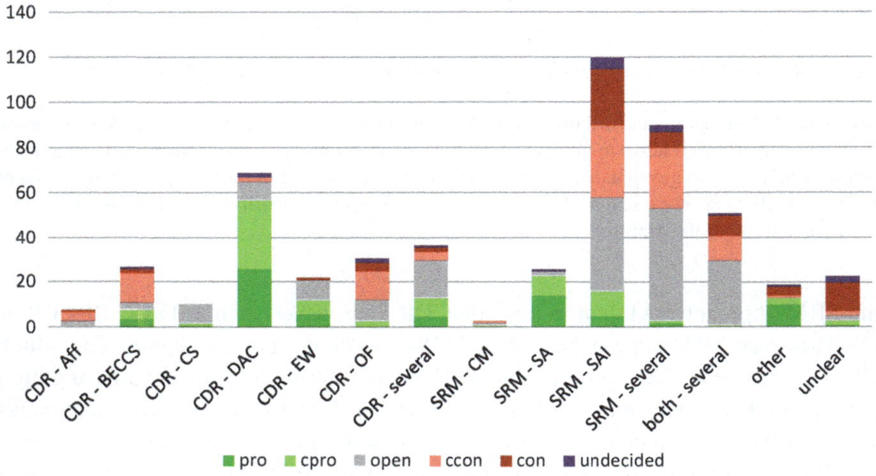

Image 6.15 Comparison of absolute frequencies of codings for various outcomes of codings across the document groups (absolute numbers of codings)

approach SAI or covering several SRM approaches present much more codings for 'counter CE' or 'carefully counter CE' as well as 'open' than for 'pro CE' or 'carefully pro CE'. Documents on the CDR approach DAC and the SRM approach SA (surface albedo modification) present positive evaluation outcomes more frequently (Image 6.17).

Due to the large number of values that frame element 3 (Evaluation of CE) can take on (44), the large number documents groups under variable 'Disciplinary Focus' (14), and the relatively small number of codings for this frame element (536), the decision was made to assess in detail only those values for frame element 3 and only those document groups which, in total, were coded more often than they would have been if all codings were distributed equally across all possible values[11]: More than

[11] For this, values, 'cpro' and 'pro' are aggregated as well as 'ccon' and 'con'; 'open' and 'unclear'. This presents 18 possible values for frame element 3.

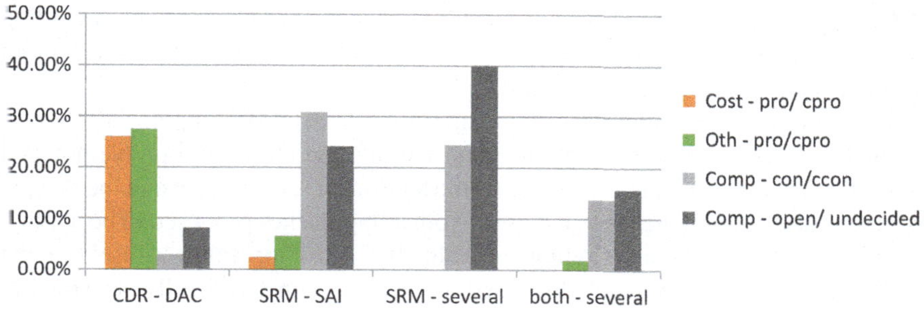

Image 6.16 Comparison of share of codings for selected values for frame element 3 and selected document groups (values relative to sum of codings in this document group)

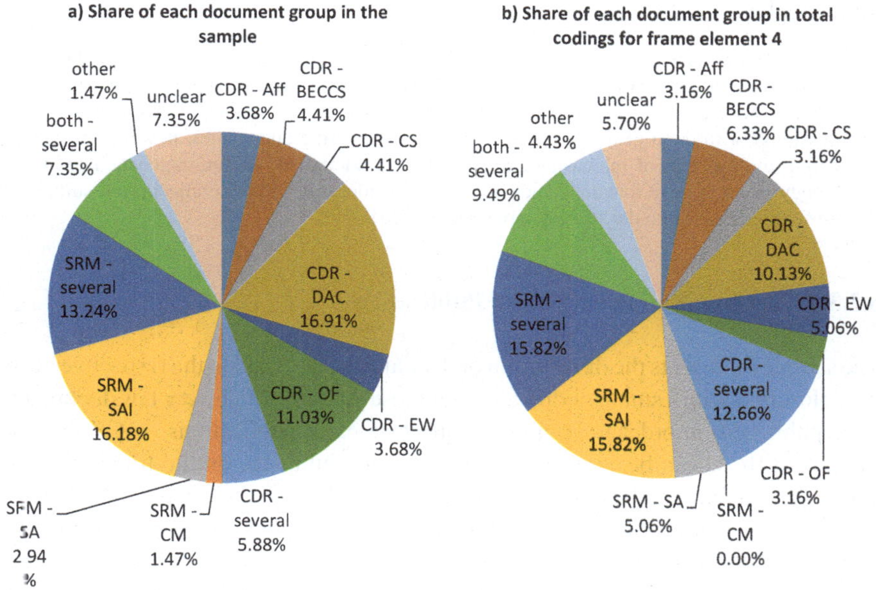

Image 6.17 Comparison of shares of document groups in (**a**) the sample and (**b**) the sum of codings for frame element 4

38 times for document groups and more than 29 times for values of frame element 3. Document groups 'CDR—DAC', 'SRM—SAI', 'SRM—several' and 'both—several' meet this criterion as well as positive evaluations along economic criteria (Cost—pro/cpro), negative, as well as open-ended evaluations using complexity criteria (Comp—con/ccon and Comp——open), and positive evaluations according to other criteria (Oth—pro/cpro) The following graphs show that values for this frame element presenting an evaluation of CE along the lines of complexity are more common in texts on CE in general, on SRM approaches in general or on SAI in particular than, for example, in texts on DAC. Values presenting an evaluation

of CE approaches along the lines of cost and benefit considerations or according to feasibility or effectiveness criteria and coming to a positive outcome are more common, conversely, in texts focusing on DAC.

The remedy coded most often in the sample is research into CE approaches ('CE—research'). This is not surprising when considering that the central problem coded most often is climate change. The second most frequent remedy coded in the sample is 'other—use', which means that there are other approaches proposed to be applied to address central problems. Codings for value 'CE—use' or 'Port——use', i.e. the application of the CE approach discussed in the text—either generally or explicitly as part of a portfolio of measures—is not present at all in the document groups 'CDR-CS', 'CDR-OF' and 'CDR-CM'. In document groups 'CDR—DAC' and 'SRM—SAI' they are present only in the form of statements proposing these approaches explicitly as part of a portfolio of measures to address the central problem in the text, for example:

> Projected anthropogenic warming and increased CO_2 concentration present a twofold threat, both from climate change and from CO_2 directly through increasing the acidity of the oceans. Future climate change may be reduced through mitigation (reductions in greenhouse gas emissions) or through geoengineering. Most geoengineering approaches, however, do not address the problem of increasing ocean acidity. A combined mitigation/geoengineering strategy could remove this deficiency. Here we consider the deliberate injection of sulfate aerosol precursors into the stratosphere (Wigley 2006: 452).

6.2.3.3 Differentiation by Year of Publication

The sample represents the distribution of documents published in the respective years very closely—with a small modification for those years in which very few documents were published, in order to be able to introduce early publications on CE into the sample, as those can be expected to have a strong influence on the framing of the debate.[12] The main findings here are the following: Firstly, the number of segments addressing climate change as the central problem is the highest in comparison to other problems identified across all years of publication. Secondly, while an economic characterization of both the problems of climate change and CE was the first to appear, this characterization theme did not become dominant later on in either case.

Image 6.18 shows numbers of codings for three possible aggregate values for frame element 1 (Problem Definition) over the years. The assessment of numbers of codings by themselves already presented climate change as the most common

[12]This means that, for early years, only one or very few documents are present in the sample, meaning that all codings for these years stem from one and the same document. This needs to be kept in mind when studying the following results. Also, the results can only be indicative, never definite, as a comparatively small number of codings are distributed across a large number of variable values for 'time' and the frame element in question. Thus, only aggregate values for the diverse frame elements are assessed below, not each possible value in itself (e.g., only all values for 'climate change' together, not 'climate change – security' and 'climate change – economy' etc. individually).

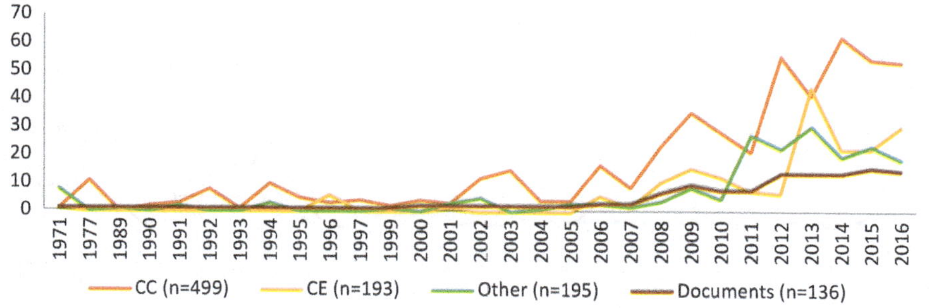

Image 6.18 Absolute numbers of codings for values 'climate change', 'climate engineering' and 'other' for frame element 1 across time; black line indicates absolute numbers of documents in the sample for the respective year; note: no values for years 1972–1976, 1978–1988 and 1998 as no text in sample for these years

problem identified in the sample. The graph indicates, beyond this, how often each of these central problems were presented over the years.

Image 6.19 zooms in on the characterizations of the central problems climate change and CE over time, presenting absolute numbers of codings for each of the values. It shows that in both cases, the characterization of the central problem as an economic one was the first to appear in the discussion and, again in both cases, this characterization did not turn out to be dominant in the later course of the discussion. Where climate change is identified as the central problem, its characterization as a security problem supersedes all other characterizations until very recently, when the

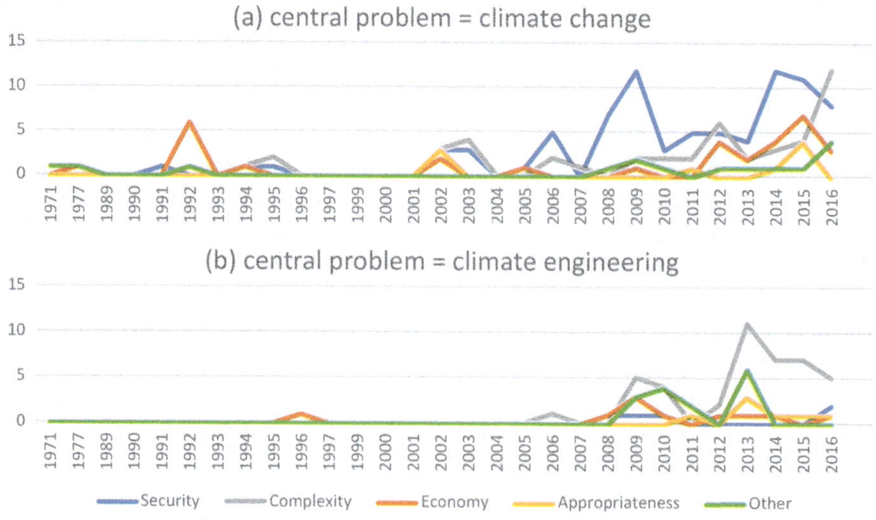

Image 6.19 Absolute numbers of codings for various characterizations of problem climate change (**a**) and climate engineering (**b**) across time; note: no values for years 1972–1977, 1978–1988 and 1998 as no texts in sample for these years

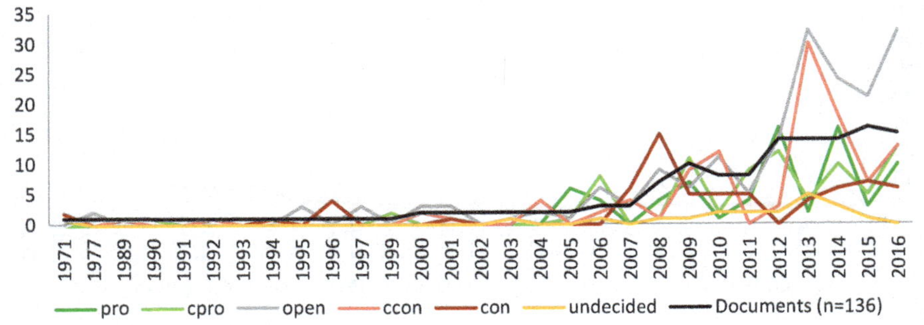

Image 6.20 Absolute numbers of codings for values 'pro', 'cpro', 'open', 'ccon', 'con', 'undecided' for frame element 3 across time; black line indicates absolute numbers of documents in the sample for the respective year; note: no values for years 1972–1976, 1978–1988 and 1999 as no text in sample for these years

characterization as a complex problem becomes more frequent. With regards to CE as the central problem, the characterization as a complex problem is more frequent than any other thematic characterization across the course of the discussion except for 2011, when CE is characterized only in one case, and there as a problem of appropriateness.

Concerning frame element 3 (Evaluation of CE), analysis is rendered very difficult because of a large number of values for this frame element and comparatively small numbers of codings. Values were aggregated to outcomes of evaluation in order to allow for some visualization. Image 6.20 shows the total numbers of codings for various outcomes of evaluation. As a comparison, the number of documents for each year in the sample is again indicated as a black line. All the different outcomes of evaluation show in large fluctuations of the numbers of codings, making the drawing of inferences very difficult. Larger numbers of codings would be needed to allow for assessment concerning the role of the specific approach analyzed in each case. However, Image 6.20 indicates that, while exclusively negative evaluation outcomes peak in 2007, afterwards these outcomes decrease and carefully negative evaluation outcomes—i.e. evaluations in which both pro and counter arguments are considered before the evaluator comes to a negative conclusion—increase. In contrast, positive and carefully positive evaluation outcomes share a similar trend over the years. Text segments presenting arguments for and against CE, but coming to no conclusion as to whether CE should be recommended or not, peak in 2013 as well as 2016, presenting at this point the highest share of codings of all outcomes of evaluation.

Image 6.21 zooms in on the criteria of evaluation applied in those evaluations coming to a, **a** positive or carefully positive, **b** negative or carefully negative, or **c** open or undecided outcome of evaluation.[13]

Concerning frame element 4 (Proposed Remedy), the difficulty of a large number of values and a small number of codings remains. Image 6.22 shows that only very

[13]For these graphs, the numbers of positive and carefully positive evaluation outcomes were aggregated, as well as those for negative and carefully negative evaluation outcomes.

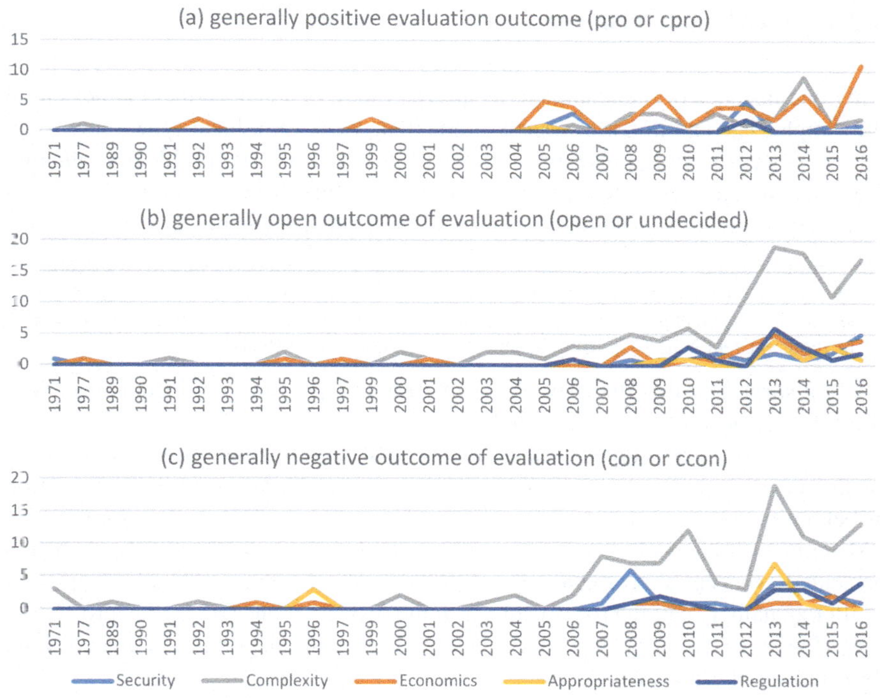

Image 6.21 Absolute numbers of codings for different evaluation criteria across the years presenting, **a** a positive, **b** an open; or **c** a negative evaluation outcome; note: no values for years 1972–1976, 1978–1988 and 1999 as no text in sample for these years

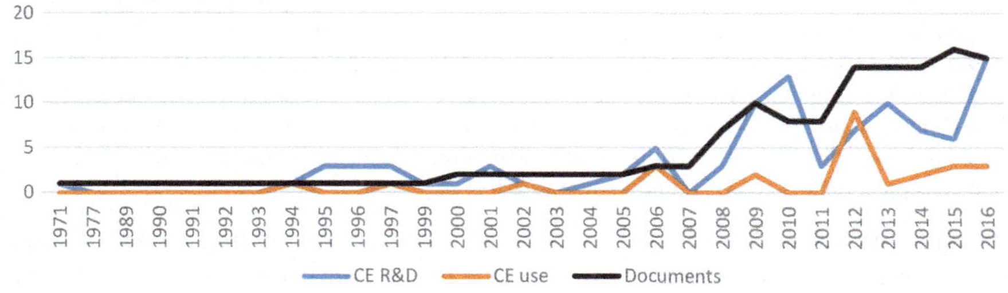

Image 6.22 Absolute numbers of codings for proposals of research and/or development of CE and proposals of use of CE; black line indicates absolute number of documents in the sample for that year; note: no values for years 1972–1976, 1978–1988 and 1999 as no text in sample for these years

few codings for this frame element were made across the sample: The number of documents is higher, for many years, than the number of codings for either the proposal of furthering research and/or development of CE (either individually or as part of a portfolio of climate policies) or the proposal of using CE.

6.2.4 Ideal-Type Frames

Only one instance of one complete ideal-type frame in a text segment is present in the sample: The legitimizing security frame in its weak form Ideal-type frame 'Leg-sec' in its weaker form, presenting climate change as a security problem, evaluating CE as a comparatively safe measure and proposing further research into CE, (CC—sec|safety—pro | Port—dev/res)[14] is present in one text segment:

> Over the long run, the ability to remove CO_2 from the air should be viewed as an essential tool in our kit for managing carbon-climate risks. We therefore need, at the minimum, a serious long-term exploratory research effort to develop air capture along with other direct methods for removing CO_2 from the atmosphere. (Keith 2009: 1654).

It is not surprising that the only ideal-type frame is present in a document of type 'Editorial' instead of a research article.

Of inductively developed frames, the general legitimizing frame 'Leg—none' (without any characterization of climate change as a problem or criteria of evaluation as belonging to one of the ideal-type frames) is present in its complete form on the text segment level, and in only two documents: In one instance climate change is presented as a central problem without further characterization and a CE approach is offered, namely paining roofs white, as remedy (CC—none|Oth—pro/cpro | CE—use). Here, the CE approach at hand, namely surface modification, is evaluated along the criterion of effectiveness:

> With increasing the solar reflectance of urban surfaces, the outflow of short-wave solar radiation increases, less solar heat energy is absorbed leading to lower surface temperatures and reduced out- flow of thermal radiation into the atmosphere. This process of "negative radiative forcing" effectively counters global warming. Cool roofs also reduce cooling-energy use in air conditioned buildings and increase comfort in unconditioned buildings; and cool roofs and cool pavements mitigate summer urban heat islands, improving outdoor air quality and comfort. Installing cool roofs and cool pavements in cities worldwide is a compelling win–win–win activity that can be undertaken immediately, outside of international negotiations to cap CO2 emissions. We review the status of cool roof and cool pavements technologies, policies, and programs in the U.S., Europe, and Asia. We propose an international campaign to use solar reflective materials when roofs and pavements are built or resurfaced in temperate and tropical regions. (Akbari and Matthews 2012: 2).

The same inductively developed frame, in this instance presenting climate change as the central problem and proposing research into soil carbon sequestration and biochar (CC—none | Oth—pro/cpro | CE—res) is present in the following text segment, assessing several approaches, but focusing on soil carbon sequestration in combination with biochar, proposing further research in this area:

> Despite 20 years of effort to curb emissions, greenhouse gas (GHG) emissions grew faster during the 2000s than in the 1990s, which presents a major challenge for meeting the international goal of limiting warming to <2 °C relative to the preindustrial era. Most recent

[14] This signature presents the configuration of codes for the ideal-type frame in the following form: 'Identification of problem – problem characterization (Frame element 1) | Criterion of evaluation – outcome of evaluation (Frame element 3) | Proposed Remedy – Avenue of Activity' (Frame element 4).

> scenarios from integrated assessment models require large-scale deployment of negative emissions technologies (NETs) to reach the 2 °C target. A recent analysis of NETs, including direct air capture, enhanced weathering, bioenergy with carbon capture and storage and afforestation/deforestation, showed that all NETs have significant limits to implementation, including economic cost, energy requirements, land use, and water use. In this paper, I assess the potential for negative emissions from soil carbon sequestration and biochar addition to land, and also the potential global impacts on land use, water, nutrients, albedo, energy and cost. Results indicate that soil carbon sequestration and biochar have useful negative emission potential (each 0.7 GtCeq. yr1) and that they potentially have lower impact on land, water use, nutrients, albedo, energy requirement and cost, so have fewer disadvantages than many NETs. Limitations of soil carbon sequestration as a NET centre around issues of sink saturation and reversibility. Biochar could be implemented in combination with bioenergy with carbon capture and storage. Current integrated assessment models do not represent soil carbon sequestration or biochar. Given the negative emission potential of SCS and biochar and their potential advantages compared to other NETs, efforts should be made to include these options within IAMs, so that their potential can be explored further in comparison with other NETs for climate stabilization. (Smith 2016: 1315).

The section above shows that hardly any complete frames—neither ideal-type nor inductively developed frames—can be found in individual segments in the sample. In order to assess which frames are present in incomplete forms in segments of the sample, a query was run to search for incomplete ideal-type frames—meaning combinations of two frame element values conforming with the configuration required by the ideal-type or inductively developed frames. During this analysis step, I found that legitimizing security, economy and simplicity frames, as well as delegitmizing security, economy, complexity and appropriatentess frames are present in incomplete forms in individual text segments (i.e. at least two frame elements of the respective ideal-type frame co-occur in the relevant text segment). Only the legitiming appropriateness frame is not present in any segment in an incomplete or complete form.

The most common incomplete ideal-type frames are the legitimizing security frame ('Leg–Sec') and the delegitimizing complexity frame ('Deleg—Comp'), while both are still very rare. Aggregated over the entire sample, 51 segments present one of the ideal-type legitimizing frames in incomplete forms and 24 segments present one of the ideal-type delegitimizing frames in incomplete forms.

6.2.4.1 Code Profile of Each Document

While the analysis above is interesting in terms of complete frames on the segment level, it does not unveil frames which are complete, not in one segment, but in one document. This is the case when one frame element is present in one segment (e.g. the introduction) and the other frame elements in another segment (e.g., the conclusion) of the text. Also, as the authors have different styles of writing and the texts are diverse in their length, the amount of each code in the entirety of the sample can be slightly misleading as long texts or texts with very short paragraphs which present code X—or the combination of codes XY—in more than one paragraph might lead to the over-representation of X—or XY—in the sample. In order to account for this problem, aside from quantitative analyses over the entire sample, code signatures for each text were generated on the basis of the codes present in the text: If a code was present in the text, it was added to the code signature, which has the following structure:

> Frame element 1 (Problem Identification) | Frame element 3 (Evaluation of CE) | Frame element 4 (Proposed Remedy).

Nearly all documents in the sample, with very few exceptions, carry segments in which climate change is presented as the central problem. 72 of the texts in the sample (52.94%) do not bear any incomplete or complete ideal-type frames, but only inductively developed frames or individual frame elements from ideal-type frames. 13 documents of the sample (9.55%) present complete frames on the document level. In these, the ideal-type legitimizing appropriateness frame and the delegitimizing economy and security frames are never present in complete forms on the document level in the sample.

Of those documents with complete or incomplete ideal-type frames, most also present frame elements from other ideal-type frames in some location in the text (see table 6.5). Thus, those documents which do carry complete or incomplete frames usually provide diverse framings of CE, not only one specific one. Also, very interestingly, a number of documents bear frame elements or frames, not only from diverse frames, but from both legitimizing and delegitimizing ideal-type frames: 48 of the 64 documents containing a complete or incomplete frame bear also frame elements or incomplete/complete forms of frames with the opposite effect on decision-making—a counter-frame –, thus presenting a contension concerning CE within the same document. One very common combination of diverse frames in one document is the combination of the legitimizing security frame with elements from the delegitimizing complexity frame. Thus, a number of documents present the legitimizing security frame in an incomplete or complete form, but also an evaluation according to criteria of complexity, and coming to a negative evaluation outcome, which is a frame element from the delegitimizing complexity frame (Tjiputra et al. 2016; Tilmes et al. 2013; Steffen et al. 2011a; Vaughan and Gough 2016; Azar et al. 2010; Crutzen 2006; Hidy and Spencer 1994; Keith 2009; Keith and MacMartin 2015; Keith et al. 2010; Robock 2008; Robock et al. 2009; Corner and Pidgeon 2010). Of those 16 documents in which only either legitimizing or delegitimizing frames or frame elements are found—but not both—, 4 present a complete legitimizing frame and 12 an incomplete legitimizing frame. However, for this calculation, frame element 'cc—none' was considered a legitimizing frame element. If this frame element is not considered thus, another 4 documents present only a delegitimizing frame without further legitimizing elements. Those texts with only legitimizing complete or incomplete frames and no frame elements from delegitimizing frames, for the most part, focus on CDR techniques such as Afforestation (Aff), Direct Air Capture (DAC), Carbon Sequestration (CS) or Enhanced Weathering (EW), or on the SRM approach Surface Albedo Modification (SA).

Some documents present legitimizing and delegitimizing elements of ideal-type frames of the same rationale: For example, Corner and Pidgeon (2010); Barrett

Table 6.5 Complete and incomplete ideal-type frames on document level

Ideal-type frame	Leg—sec	Leg—cost	Leg—comp	Leg—norm	Deleg—sec	Deleg—cost	Deleg—comp	Deleg—norm
Documents presenting this frame in complete or incomplete form	24	16	17	2	5	3	23	3
Of these, documents presenting complete frame	5	3	3	0	0	0	2	1
Other complete or incomplete ideal-type frames in the same documents	Leg—cost/comp/norm Deleg—comp/cost	Leg—sec/comp/norm Deleg—comp	Leg—sec/cost/norm Deleg—sec/cost/comp	Leg—sec/cost/comp	Leg—sec Deleg—comp/cost/norm	Deleg—sec/comp	Leg—sec/comp Deleg—sec/cost/norm	Leg—sec Deleg—sec/comp
Occurrences of this frame without other frames or frame elements in the same document	12	4	6	1	1	0	8	1

et al. (2014), Trenberth (2011) npresent climate change as a security problem (legitimizing) and evaluate CE as a security threat (delegitimizing). Only 17 of the texts present legitimizing ideal-type frames (complete or incomplete) in their strong variants (proposing the use of CE). These texts focus on the approaches Surface Albedo Modification, Stratospheric Aerosol Injection, Direct Air Capture, Afforestation, or argue for SRM or CDR in general. This means that no text presents legitimizing ideal-type frames in their strong variants to propose the usage of, for example, Ocean Fertilization. The only text proposing the usage of SAI (Wigley 2006) presents delegitimizing frame elements as well as a complete security frame in favor of a portfolio approach combining mitigation and SAI.

6.3 Summary

The results presented in this chapter indicate that the academic discussion of CE, as represented in peer-reviewed journal publications, presents a very diverse territory to the observer, which, however, is characterized by some very interesting landmarks. Firstly, there is no dominant frame present in the sample. Instead, I find that, on the document level, more than half of the sample does not present any ideal-type frames, not even in incomplete form. The majority of the rest of documents present one or more complete or incomplete frames, but also elements of counter-frames. This means that many documents present either no complete frames—while individual frame elements or frame element fragments are present in every document of the sample—or a combination of counteracting frames which provides for a curious tension inside the individual document. There is only one instance of a complete ideal-type frame on segment level in the sample. All ideal-type frames are present in incomplete form on the document level except for the legitimizing appropriateness frame. At the same time, most text segments in the sample do not carry any coding for a frame. This indicates that there is a high diversity of frames in the discussion, while framing is, in general, very rare.

Secondly, precedence does not ensure dominance when it comes to framing in the academic discussion on CE: While an economic characterization of both problem climate change and problem CE was the first to emerge, climate change was more often, later, characterized as a security problem CE as a complex challenge. Thirdly, CE is identified as the main problem more often in texts on SRM than on CDR measures and SAI garners a negative evaluation outcome more often than other approaches; DAC and SA garner positive evaluation outcome more often. Fourthly, CE is identified as the central problem in social science and humanities texts more often than in natural science texts. Further, climate change is understood as a security problem more often in social science text than in natural science texts. Fifthly, positive evaluations of CE, which are, in general, very rare, take place most commonly according to criteria of cost and benefit of CE. Finally, causal responsibility for any central problem is ascribed only extremely rarely in any of the texts analyze.

References

Akbari H, Matthews HD (2012) Global cooling updates: reflective roofs and pavements. Energy Build 55:2–6

Azar C, Lindgren K, Obersteiner M, Riahi K, Van Vuuren D, Den Elzen KMGJ et al (2010) The feasibility of low CO_2 concentration targets and the role of bio-energy with carbon capture and storage (BECCS). Clim Change 100:195–202

Bais AF, McKenzie RL, Berhard G, Aucamp PJ, Ilyas M, Madronich S et al (2015) Ozone depletion and climate change: impacts on UV radiation. Photochem Photobiol Sci 14:19–52

Barrett S, Lenton TM, Millner A, Tavoni A, Carpenter SR, Anderies JM et al (2014) Climate engineering reconsidered nature. Clim Change 4:527–529. https://doi.org/10.1038/nclimate2278

Corner A, Pidgeon N (2010) Geoengineering the climate: the social and ethical implications. Environ Sci Pol Sustain Dev 52(1):24–37

Crutzen PJ (2006) Albedo enhancement by stratospheric sulphur injections: a contribution to resolve a policy dilemma. Climatic Change 77:211–219

Dumbrell NP, Kragt ME, Gibson FL (2016) What carbon farming activities are farmers likely to adopt? A best-worst scaling survey. Land Use Policy 54:29–37

Goeppert A, Czaun M, Prakash GKS, Olah GA (2012) Air as the renewable carbon source of the future: an overview of CO_2 capture from the atmosphere. Energy Environ Sci 5:7833–7853

Hidy GM, Spencer DF (1994) Climate alteration. A global issue for the electric power industry in the 21st century. Energy Pol 22(12):1005–1027

Keith DW (2009) Why capture CO_2 from the atmosphere? Science 325:1654–1655

Keith DW, MacMartin DG (2015) A temporary, moderate and responsive scenario for solar geoengineering. Nat Clim Change 5:201–206

Keith DW, Parson EA, Morgan MG (2010) Research on global sun block needed now. Nature 463:426–427

Kravitz B, MacMartin DG, Wang H, Rasch PJ (2015) Geoengineering as a design problem. Earth System Dynamics 6:1635–1710

Lenton TM, Vaughan NE (2009) The radiative forcing potential of different climate geoengineering options. Atmos Chem Phys 9:5539–5561

Macnaghten P, Szerszynski B (2013) Living the global social experiment: an analysis of public discourse on solar radiation management and its implications for governance. Global Environ Change 23:465–474. https://doi.org/10.1016/j-gloenvcha.2012.12.008

Marland G, Pielke RA, Apps M, Avissar R, Betts RA, Davis KJ et al (2003) The climatic impacts of land surface change and carbon management, and the implications for climate-change mitigation policy. Clim Policy 3:149–157

NAS (2015a) Climate intervention: carbon dioxide removal and reliable sequestration. The National Academy of Sciences, Washington, D.C.

NAS (2015b) Climate intervention: reflecting sunlight to cool earth. Washington, D.C.: The National Academy of Sciences

Nordhaus WD (1992) An optimal transition path for controlling greenhouse gases. Science 258:1315–1319

Peters GP (2016) The 'best available science' to inform 1.5°C policy choices. Nature Climate Change 6:646–649

Rayner S, Heyward C, Kruger T, Pidgeon N, Redgwell C, Savulescu J (2013) The Oxford principles. Clim Change 121:499–512. https://doi.org/10.1007/s10584-012-0675-2

Rickels W, Klepper G, Dovern J, Betz G, Brachatzek N, Cacean S. et al (2011) Large-scale intentional interventions into the climate system? Assessing the climate engineering debate. Scoping Report conducted on behalf of the German Federal Ministry of Education and Research (BMBF). Kiel, Kiel Earth Institute

Robock A (2008) 20 reasons why geoengineering might be a bad idea. Bull Atomic Scient 64(2):14–18. https://doi.org/10.2968/064002006

Robock A, Marquardt A, Kravitz B, Stenchikov G (2009) Benefits, risks, and costs of stratospheric geoengineering. Geophys Res Lett 36:1–9. https://doi.org/10.1029/2009GL039209

Shepherd J, Cox P, Haigh D, Keith DW, Launder B, Mace G et al (2009) Geoengineering the climate: science, governance and uncertainty. The Royal Society, London

Smith P (2016) Soil carbon sequestration and biochar as negative emission technologies. Glob Change Biol 22:1315–1324

Steffen W, Persson A, Deutsch L, Zalasiewicz J, Williams M, Richardson K et al (2011a) The Anthropocene: From Global Change to Planetary Stewardship. Ambio 40:739–761

Steffen W, Grinevald J, Crutzen P, McNeill J (2011b) The anthropocene: conceptual and historical perspectives. Philos Trans Royal Society A 369:842–867

Tilmes S, Fasullo J, Lamarque J-F, Marsh DR, Mills M, Alterskjaer K et al (2013) The hydrological impact of geoengineering in the geoengineering model intercomparison project (GeoMIP). J Geophys Res 118:11036–11058

Tjiputra JF, Grini A, Lee H (2016) Impact of idealized future stratospheric aerosol injection on the large-scale ocean and land carbon cycles. J Geophys Res: Biosci 121:2–27. https://doi.org/10.1002/2015JG003045

Trenberth KE (2011) Changes in precipitation with climate change. Clim Res 47:123–138

Trenberth KE, Dai A (2007) Effects of Mount Pinatubo volcanic eruption on the hydrological cycle as an analog of geoengineering. Geophys Res Lett 34:1–5. https://doi.org/10.1029/2007GL030524

Vaughan NE, Gough C (2016) Experts assessment concludes negative emissions scenarios may not deliver. Environ Res Lett 11, 10.1088/1748-9326/11/9/095003.

Wigley TML (2006) A combined mitigation/geoengineering approach to climate stabilization. Science 314:452–454

Chapter 7
Climate Engineering as an Instance of Politicization—A Diversity of Frames in the Academic Discussion

Abstract Kreuter offers a much-needed discussion of the role of academic framing of Climate Engineering approaches for the social construction of their meaning and subsequent decision-making in the current situation of little dedicated political discussion of these approaches. As well as discussing the high diversity of framing in the academic discussion of CE, Kreuter explores possibilities of extending the scope of attempts to understand future decision-making concerning CE. Kreuter concludes by considering implications for climate policy, looking at ways in which the diversity of academic framing might be preserved in academic information for political decision-making.

The present thesis provides an account of how, in a situation of limited dedicated high-level political discussion and decision-making, the meaning of CE is constructed in a field in which these approaches are talked about, and which has a specific claim to credibility, namely the academic discussion. Thus, this academic framing can be understood to be relevant to our (socially constructed) reality with regards to CE. Specifically, this study addressed the central research question: *How is the meaning of CE approaches constructed in the academic discussion through frames which legitimize or delegitimize the approaches' use to address anthropogenic climate change?* The following main argument of the thesis summarizes the answer to that question as developed from the findings of the study: *In a situation of little dedicated political deliberation of CE, the intersubjective meaning of these approaches is defined largely in other fora, in particular in the academic discussion on CE. The meaning of CE constructed in the academic discussion through framing is not uniform, but instead composed of a diversity of frames. Two of those frames are particularly frequent in the sample: a security frame legitimizing CE, and a complexity frame delegitimizing these approaches. In this sense, the academic discussion on CE can be understood as an instance of politicization as the diversity of frames may contribute to rendering the issue of CE "the subject of deliberation, decision-making and human agency"* (Hay 2007: 81). *This, I claim, provides a tremendous opportunity for decision-makers: The meaning of CE thus constructed – in particular the diversity of frames of these*

Some ideas expressed in this section have been published in Kreuter (2018).

J. Kreuter, *Climate Engineering as an Instance of Politicization*,
Springer Climate, https://doi.org/10.1007/978-3-030-60340-3_7

approaches – should be preserved if and when a high-level decision-making process is initiated.

In this chapter, I move from the specific to the general and from the known to the potential in drawing inferences from the results: The conclusions are interpretations of the results. The discussion introduces implications of these conclusions and presents a storyline that I chose to put my conclusions into context.

7.1 Conclusions from Theoretical Discussion and Empirical Analysis

The conclusions do not emerge by themselves and automatically from the results. They represent a summary and interpretation of what I found. The following three conclusions are drawn from central points derived from theoretical as well as empirical literature which have gone into the design of the present study[1]:

(1) *Framing of emerging technology can influence decision-making*. According to frame theory, frames can influence decision-making by affecting the legitimacy of political actions such as the introduction or the ban of an emerging technology: Frames "link causal accounts of policy problems to particular proposals for action and facilitate the normative leap from 'is' to 'ought'" (Rein and Schön 1993: 148). Kahneman and Tversky provide early empirical observations of the 'framing effect'—the influence of frames on beliefs and actions—on individual decision-making (Tversky and Kahneman 1981; Kahneman and Tversky 1984). Political scientists have subsequently assessed the framing effect on political decisions (Benford and Snow 2000; Chong and Druckman 2007a, b; Druckman 2011; Jachtenfuchs 1996a; Levy 2003; Nelson et al. 1997; Rein and Schön 1993; Schlipphak 2011; Sniderman and Theriault 2004; Gamson and Modigliani 1989). Frame analyses have been conducted in order to structure media and political debates on technological issues such as nuclear technology (Gamson and Modigliani 1989; Nisbet and Mooney 2007), nanotechnology (Donk et al. 2012) or biotechnology (Nisbet et al. 2003). The framing effect is assumed to be particularly strong in the case of emerging technologies such as CE as.

> interpretative […] flexibility can be assumed to be great in emergent technologies characterized by significant uncertainties, which allows different groups to emphasize different aspects of the technology, like the proverbial blind men examining an elephant. (Luokkanen et al. 2014: 968).

Thus, the early discussion of these technologies can construct the meaning of CE at this stage where these approaches do not (yet) exist as fully-fledged socio-technical systems (Schot and Kanger 2018: 1045), and so pre-define the scope of political decisions concerning these approaches.

[1] These three claims have been found to not be contradicted in the theoretical and empirical literature considered here, but they have not yet been tested empirically.

(2) In situations of little political deliberation, the academic discussion can *influence the space of political decision-making* if and when it occurs. Preceding empirical studies have found that the discussion of CE is, to date, liveliest among academic actors (Matzner and Barben 2018; Buck 2013). These actors do not replace political decision-makers, but rather frame the issue *before* decision-making takes place (Cross 2013: 149; Adler and Haas 1992: 375; Beck 2015: 286), thus defining the conditions of possibility of decisions (Wendt 1998: 106). This influence is particularly strong in situations of high uncertainty on the part of decision-makers, such as environmental phenomena such as climate change (Haas 1992: 13) or issues of emerging technologies (Lee et al. 2005; Donk et al. 2012). The relevance of the influence of academic actors on CE decisions has been noted before:

> Due to the novelty, the uncertainties, and the required scale of any real-world application of geoengineering, researchers play a crucial role in creating the basis for the decisions that determine the future of geoengineering technologies. (Huttunen and Hildén 2014: 5; see also Horton 2015: 2).[2]

In the absence of dedicated political decision-making, the relevance of the academic discussion on CE might, in future, increase further, if informal governance patterns emerge which structure the field of CE (Gupta and Möller 2018; Talberg et al. 2018). Thus, various studies assess inductively how CE is framed in the academic discussion (Bellamy et al. 2012b; Markusson 2013; Bellamy 2014; Huttunen and Hildén 2014; Linnér and Wibeck 2015).

(3) Thematic frames focusing on *security, complexity, economy and appropriateness* can *legitimize or delegitimize* CE as measures to address climate change. The present study builds on the findings of preceding, strongly exploratory studies and gives an account of how the academic discussion of CE is constituted, in particular with regards to whether legitimizing or delegitimizing frames of CE are dominant. IR and STS offer a number of concepts which have been used, for the present thesis, to create a theoretical construct to systematize frames which can either legitimize or delegitimize CE approaches as measures to address climate change. On the basis of Max Weber's differentiation of causes determining social action (Weber 1980 [1922]), securitization theory (Balzacq 2010), economization discourse (Hajer 1997) and the concept of technological fix (Weinberg 1966), the theoretical construct of eight ideal-type frames was developed. The construct comprises frames belonging to four frame themes—security, complexity, economy and appropriateness—of two variants each—one legitimizing, the other delegitimizing CE approaches.

The empirical analysis of this study was conducted on the basis of these theoretical discussions. The current, mainly academic discussion on CE does not provide political decision-makers with a simple answer to the question of whether some or all of these approaches should be considered as measures (among others) to address

[2]As discussed before, I do not assume that the academic discussion is the only factor determining CE decision-making. However, it is a very important one, and thus it can be expected that, if a dominant frame can be found in academic discussion, it will be this very frame which is taken into the governance discussion.

climate change: CE is presented in this discussion in various ways which can both legitimize and delegitimize the consideration of these approaches, both across the discussion and simultaneously by individual authors. The following eight conclusions present the interpretations of the findings from the empirical analysis. (4) *Framing has been found to be rare* across the sample, as more text segments in the sample have not been coded with any frame element code than text segments which have been identified to represent some element of any of the ideal-type frames. Still, the theoretical construct has proven useful to assess the academic framing of CE: Elements of all eight ideal-type frames from the theoretical construct have been identified in the sample. This means that all four frame themes—security, complexity, economy and appropriateness—are present in each of the two variants—legitimizing or delegitimizing CE.

(5) *Framing of CE in the academic discussion* is neither uniform nor clearly polarized, but *diverse*. The framing of CE in the sample has been found to be neither dominated by one specific frame type nor characterized by an antagonism between two clearly differentiated extremes. Elements from each of the ideal-type frames have been identified in the sample, presenting diversity on the level of the sample. Moreover, on the level of the individual document, in many cases elements from both legitimizing and delegitimizing frame variants have been identified in one and the same individual text. Of those documents presenting instances of complete or incomplete ideal-type frames, most also present instances of frame elements from other ideal-type frames in some part of the text. Thus, those documents which carry complete or incomplete frames usually provide diverse framings of CE, not only one specific frame. Also, very interestingly, a number of documents bear not only diverse frame elements or frames, but elements from both legitimizing and delegitimizing variants: 48 of the 64 documents presenting a complete or incomplete frame also bear frame elements or incomplete/complete forms of frames with the opposite effect on decision-making—a counter-frame—, thus presenting a contention concerning CE inside the same document. On the level of the individual text segments, the evaluation of CE approaches conducted in the texts presents positive and negative outcomes to similar degrees. Furthermore, most commonly when an evaluation is conducted in the text, it does not end in a recommendation for or against it, but is left open.

(6) The diversity of framing in the academic discussion on CE can be traced back, partly, to the presence of texts with *various disciplinary foci* in the sample. The results indicate that the diversity of framing present in the sample can be traced back, in part, to the fact that the academic discussion on CE brings together researchers from many different disciplinary backgrounds: Texts assigned to the group 'social sciences and humanities' more commonly present text segments in which CE approaches are discussed as the central problem of the segment than documents from the other groups—instead of, for example, climate change. Further, in documents assigned to the group 'social sciences and humanities', the characterization of climate change as a security problem and the characterization of CE as a complex challenge are more common than in documents assigned to the group 'natural sciences and engineering'. Texts assigned to the document group 'both' more commonly define climate change as a problem of economic cost than texts in the other two groups. Texts in the same

group also more commonly apply economic criteria for evaluation of CE. Many documents in this group jointly address natural science as well as economics questions. These findings indicate that interdisciplinary publications on CE—i.e. journal articles addressing research questions focusing on the subject matter of various academic disciplines—so far might focus on posing natural science together with economics questions. The academic discussion on CE might cover a wider variety of issues if other combinations were to be pursued—for example the combination of natural science and ethics questions.

Further, the analysis shows that the texts assigned to the groups 'natural science and engineering' and 'both' more commonly come to a positive outcome when evaluating CE approaches than texts belonging to the group 'social science and humanities'. One factor that might influence this finding and which could not be controlled in this study due to the limited number of texts analyzed, is the fact that texts belonging to this latter group often do not differentiate between different types of CE approaches, but make general arguments about the whole set. Texts in the 'social sciences and humanities' group more commonly apply security or appropriateness criteria for the evaluation of CE. Concerning proposed remedies in the different text groups, those documents in the groups 'social sciences and humanities' and 'both' more commonly present any remedies to the central problem than the documents in group 'natural science and engineering'. However, the remedies proposed most commonly across all groups are the same: more CE research and approaches other than CE are presented as measures to address the central problem identified. This indicates that there is large agreement between the text groups concerning the best remedy for the problem at hand.

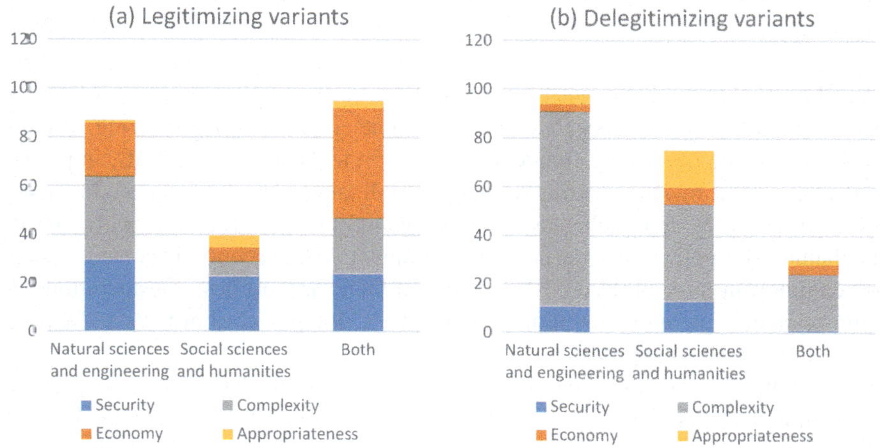

Image 7.1 Absolute numbers of codings for the values of frame elements 1 and 3 belonging to each ideal-type frame of **a** legitimizing variants, and **b** delegitimizing variants in the various document groups (For this graph, all codings for frame elements 1 and 3 belonging to the respective ideal-type frames were aggregated. This means that some coded sections were counted several times—if they carried codings for frame elements of several frame themes or variants.)

Adding all codings for the values of frame elements 1 (problem definition) and 3 (evaluation of CE) belonging to the individual ideal-type frames garners a distribution as presented in image 7.1. The aggregate values indicate that elements of the legitimizing economy frame are more common in texts from the groups ‘natural sciences and engineering’ and ‘both’ than in texts from the group ‘social sciences and humanities’. The group ‘natural sciences and engineering’ further presents the highest number of codings for elements of the delegitimizing complexity frame. Conversely, elements from the legitimizing security frame and the delegitimizing appropriateness frame are more common in the latter group. Accordingly, when the documents from the sample are grouped according to disciplinary focus, patterns emerge in the diversity of framing of CE. This indicates that the presence of a number of different disciplines in the discussion adds to the diversity of framing.[3]

(7) The diversity of framing in the academic discussion on CE can partly be traced back to the presence of texts addressing *various CE approaches* in the sample. The findings show that codings for CE identified as the central problem of the text segment are more commonly found in documents addressing approach SAI or several SRM approaches than in documents focusing on approaches such as Afforestation, CS, DAC, OF or SA.

Further, an evaluation of CE—i.e. a coding for frame element 3—is more common in texts focusing on SRM than texts addressing CDR approaches. When focusing on the outcome of evaluation, I find that texts addressing approaches OF, SAI or several SRM approaches present codings for negative evaluation outcomes more frequently than texts addressing approaches such as DAC or SA. A large amount of positive evaluation outcomes for DAC are based on evaluations according to criteria of cost or of feasibility and effectiveness of the approach. Conversely, negative or open evaluation outcomes of SAI are often based on an evaluation according to criteria of complexity of the approach—e.g., by elaborating on potential detrimental side effects of the approach on physical and/or social systems. Image 7.2 presents aggregated numbers of codings for frame elements 1 (problem definition) and 3 (evaluation of CE) for each of the ideal-type frames. It shows that frame elements of legitimizing variants of the ideal-type frames are frequently applied in texts addressing DAC, in particular with a view towards cost or security considerations. In documents addressing SAI, several SRM approaches or BECCS, delegitimizing variants, in particular of the thematic complexity frame, are highly frequent. While economic considerations are common in texts on DAC, EW or several CDR approaches, considerations of appropriateness are more common in texts on SRM approaches or texts addressing both CDR and SRM approaches.

(8) All in all, elements of legitimizing frame variants are *slightly more frequent* in the sample than elements from delegitimizing frame variants: Firstly, on the level of individual frame elements, climate change is more frequently identified as the central problem of a text segment than CE: 90.44% of all documents in the sample identify

[3]It is important to remember that the image does not represent the relative share of documents belonging to the individual groups in the sample: Texts assigned to the group ‘Natural science and engineering’ make up about two thirds of the sample.

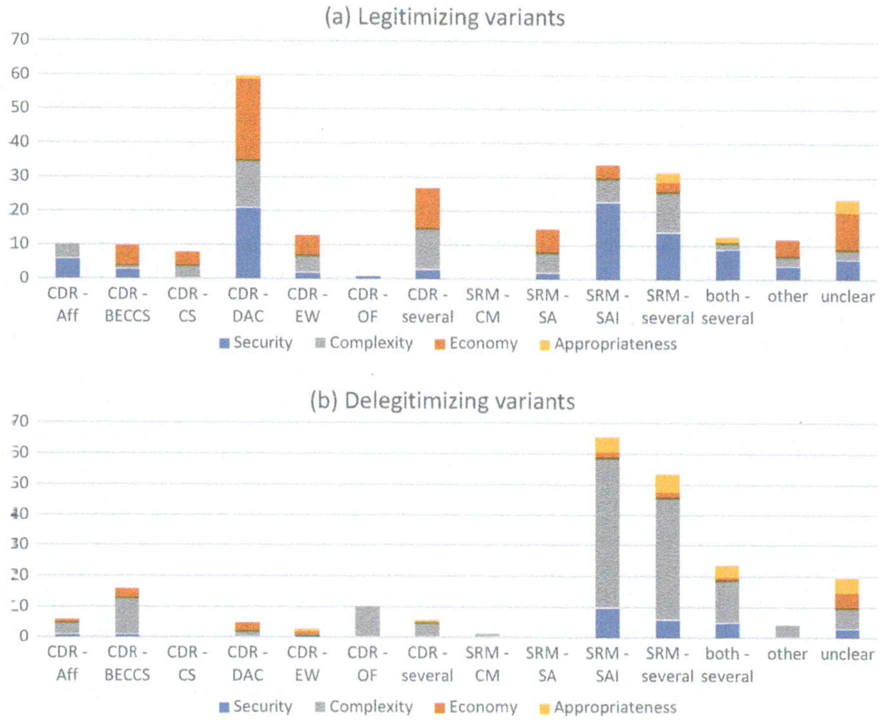

Image 7.2 Absolute numbers of codings for frame elements 1 and 3 of the ideal-type frames, differentiated by approach (For this graph, all codings for frame elements 1 and 3 belonging to the respective ideal-type frames were aggregated. This means that some coded sections were counted several times—if they carried codings for frame elements of several frame themes or variants.)

climate change as the central problem at least once. Secondly, where CE is evaluated, positive outcomes of evaluation are slightly more frequent than negative outcomes of evaluation: 23.53% of all documents present at least one coding for a positive evaluation outcome of CE, 38.24% of texts present at least one coding for a carefully positive outcome—i.e. an evaluation where counter-arguments are considered in the same segment together with pro arguments. In contrast, 35.29% and 22.79% of documents present at least one instance of a carefully negative or negative evaluation outcome, respectively. However, nearly half of all texts present an open evaluation outcome. When the occurrences of legitimizing and delegitimizing values of frame elements 1 and 3 are aggregated, elements of legitimizing frame variants are more common than their delegitimizing counterparts. Similarly, incomplete and complete instances of legitimizing variants of ideal-type frames are more frequently found both inside of individual text segments and across entire documents. It is important to note, however, that legitimizing variants do not present a clear dominance over their delegitimizing counterparts: The differences in frequency are always only slight.

(9) Both the legitimizing and delegitimizing variants of the ideal-type frames develop erratically over time, but *awareness of the duality of high stakes* seems to increase. Global anthropogenic climate change is identified as a central problem early on in the debate, even if the first text present in the sample assessed here—Lamb (1971)—does not discuss GHG emissions, but rather regional water availability issues due to industrial and agricultural overuse of the global water supply. CE approaches are identified as the central problem of a text segment for the first time in 1996, the year of publication of the first special issue in *Climatic Change* dedicated to discussing these approaches. Since then, this central problem has been identified in increasing numbers of text segments. A peak is reached in 2013. This coincides with the foundation of the DFG Priority Program on risks and challenges of CE. Carefully negative outcomes of evaluation of CE also peak in 2013, and positive or carefully positive evaluation outcomes reach a low. All in all, evaluations more and more frequently do not come to any clear conclusion supporting or opposing CE approaches, but instead leave open the outcome of evaluation. Purely negative evaluation outcomes decrease in frequency after 2007. Carefully negative evaluation outcomes increase in frequency after 2012. An assessment of the development of proposed remedies—i.e. the increase and/or decrease of instances of the proposal to research or use CE—is difficult to achieve as remedies are only very rarely proposed in the sample in general. More text data would be required in order to be able to make a statement about these developments.

(10) The two most common types of frames in the sample are the *legitimizing security frame* and the *delegitimizing complexity frame*: On the level of individual frame elements, irrespective of what is identified as the central problem, a characterization of the problem as either a threat (i.e. a security problem) or a complex challenge are most common in the sample. Further, the evaluation of CE is most commonly done according to criteria of complexity or simplicity. 64.71% of all documents present at least one instance of an evaluation of CE approaches according to their complexity, i.e. their unintended side effects and other complex criteria. Instances of incomplete or complete ideal-type frames are generally rare, but, again, the legitimizing security frame and the delegitimizing complexity frame are most frequent, both on the level of individual segments and on the level of the complete document. 24 out of the 136 documents in the sample present at least one complete or incomplete instance of the legitimizing security frame, meaning that all or nearly all elements of this frame are present in some segment of the document; 23 out of 136 documents present at least one instance of a complete or incomplete delegitimizing complexity frame. In 20 text segments in the entire sample, the incomplete instance of the legitimizing security frame is present inside of the segment (i.e., a combination of either frame elements 1 and 3, 3 and 4 or 1 and 4). 16 segments in the entire sample present an incomplete instance of the delegitimizing complexity frame.

(11) Thus, the legitimizing economy frame, while appearing earliest in the sample—in 1977—*does not dominate* over the legitimizing security frame and the delegitimizing complexity frame later on, as the security and complexity frames increase more rapidly in frequency over the following years. Instances of elements of

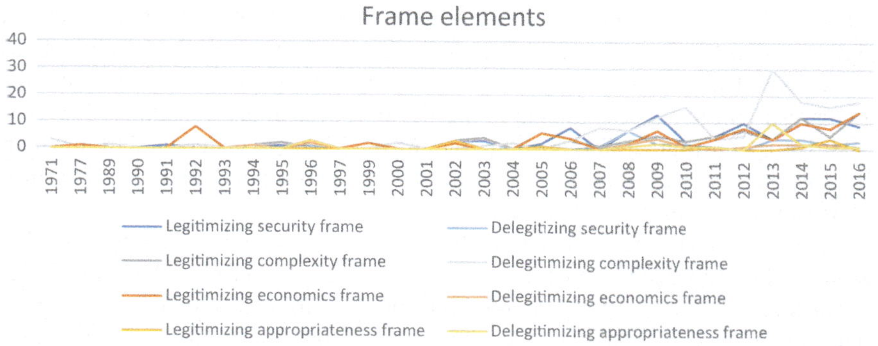

Image 7.3 Absolute numbers of codings of frame elements 1 and 3 belonging to the eight ideal-type frames; note: No values for 1972–1976, 1979–1988, 1998

the legitimizing economy frame surface again and again over the discussion, and positive evaluations in particular are usually conducted on the basis of economic criteria, but this frame does not continue to dominate the discussion. Instances of elements of security frames appear first in 1991. The characterization of climate change as a security problem peaks both in 2009 and in 2014, presenting altogether the most frequent characterization of this problem. Instances of elements of the complexity frame appear first in 1977. While elements of the legitimizing variant of this frame—the notion of CE as a (comparatively) simple fix to the (comparatively) complex problem of climate change—are represented in the sample as slowly increasing numbers of instances, many more sections show instances of the delegitimizing variant. After 2007, open-ended or negative evaluations of CE more commonly apply a complexity criterion than any of the other ideal-type criteria. Both negative and open-ended evaluations according to these criteria peak in 2013, but also keep increasing after 2015. Instances of elements of the appropriateness frames are generally rare.

Image 7.3 shows the aggregated numbers of codings for frame elements 1 (problem definition) and 3 (evaluation of CE) in accordance with each of the ideal-type frames over time. While the legitimizing security frame is present across nearly all years, the delegitimizing complexity frame increases in frequency after 2013.

7.2 Discussion: Context and Possible Implication of Conclusions

In order to be able to put the conclusions into their broader context and to reflect on their implications, the researcher has to take a step back and allow herself, to some degree, to speculate. I will discuss the constitutive and causal relevance (Wendt 1998: 103) of the conclusions of the study, both going backwards—to assess how the conclusions can be explained or understood—and forwards—to speculate upon causal and constitutive implications on other issues. Causal stories aim to "explain *changes* in the state of some variable or system" (Wendt 1998: 105). Causal stories on the framing of CE, then, either aim to explain what has caused the framing of CE to look as it does or to explain which effects this framing has. Constitutive stories, in contrast, aim to "show how the properties of a system are constituted" (Wendt 1998: 106). Constitutive stories of the framing of CE aim to understand from which elements this framing is constructed and which properties it has, and how it, in turn, contributes to constituting other systems.

Accordingly, in this section, the conclusions of this study will be contextualized in two ways: Firstly, possible constitutive understandings and causal explanations of the conclusions will be suggested. Secondly, the implications of these conclusions for the understanding or explanation of other phenomena—such as climate change decision-making, climate policy discourse and securitization or modernization discourses—will be discussed.

7.2.1 Reflecting on Context—Ways of Understanding and Explaining the Findings

Complete frames are rare across the sample, both aggregated on the segment and on the document level. Several reasons can be imagined for this phenomenon: It is conceivable that the generally low degree of framing is due to the specific rules governing the practice of academic publication: In order to serve the ideal of objectivity of information, complex systems of peer-review are in place. In this way, academic journal publication as a practice differs from practices such as political or journalistic publication, which follow different rules. However, it is also conceivable that the low degree of framing is a trait specific to academic publications on CE, and not a feature of academic peer-reviewed publications in general. In other words, it is possible that the academic community discussing CE is particularly careful not to frame the issue in any way supportive or dismissive of these approaches. This might be due to a consciousness that the discussion of CE is characterized by a duality of high stakes—of applying CE and of not applying CE (Linnér and Wibeck 2015). The high frequency of codings for an open outcome of the evaluation of CE across the sample might be considered an indicator that this latter explanation holds true.

The *diversity of framing* of CE in the academic discussion concluded from the empirical analysis can be understood in various ways. The results of the analysis indicate that there are at least two factors which play a role: first, the decision to assess contributions to the discussion from all academic disciplines and second, the definition of the boundaries of the discussion through the umbrella term 'climate engineering'—instead of, for example, assessing only the discussion on SRM or only on CDR or zooming in on one specific approach such as SAI or afforestation. The present study includes documents from all academic disciplines in the analysis as the issue of CE touches upon subject matter of all disciplines (Rickels et al. 2011b: 1). Due, in part, to the large scale of impacts that CE would have—and is intended to have, as otherwise, it would not make a difference in global average temperatures or global precipitation patterns—questions about the physical, chemical, biological, but also social, political, ethical, economic or legal context and implications of CE can be posed. Accordingly, the current, still early discussion about these approaches is conducted in academia between researchers from very diverse disciplinary fields, not rarely in interdisciplinary research projects (Kreuter et al. 2020).[4] Further, the present study addresses documents on all approaches currently included under the umbrella term 'climate engineering' instead of focusing on a specific set or a specific approach. This was done as, despite their numerous differences in scope, point of influence, side effects and other issues, all of these approaches pursue the same goal, namely the large-scale intervention into the climate system, and accordingly, are all highly relevant to climate change decision-making. The results show that the framing of CE found in the academic discussion would have been less diverse if the study had focused on specific approaches or on specific academic disciplines: If, for example, only natural science texts had been studied, the landscape of framing of CE would have been much more uniform. For instance, the share of security frame elements in all legitimizing frame elements would have been much smaller. Similarly, if only texts on DAC had been assessed, the same is true, but with different prevalent values of frame elements: The share of legitimizing frame elements would have been higher.

I suggest that the inclusion of these various approaches and disciplines into what I understand to be the academic CE discussion does not only increase the diversity of frames that I can find in the material because social scientists use other frames than natural scientists, or because researchers discussing DAC use other frames than those discussing SAI. I suppose that the self-conception of the CE research community as a multi- or even interdisciplinary community and the conception of the object of research under the umbrella term can lead to reciprocal inspiration to critically reflect on the frames in communication and frames in thought (Chong and Druckman 2007b: 101) that each researcher uses to make sense of CE approaches and their context and implications. For example, the findings from the present study indicate that the SRM-approach SAI is discussed in a much more critical manner than the CDR-approach DAC: In texts on the former, instances of elements of delegitimizing frames outweigh instances of elements of legitimizing frames. In texts on the latter, the relationship is inverted. This might be due to a much less problematic character and higher potential of DAC—considering DAC as something of a 'silver bullet'—or it might be because the risks and challenges of SAI may have been assessed more intensely than those of DAC in the past—and the meaning of DAC may, accordingly, be constructed as a much less problematic approach. Thus, the communication between researchers assessing SAI and those studying DAC, or between biologists, geographers and sociologists, can allow for the comparison and critical reflection of these frames and inspire the consideration of new or different frames to make sense of these individual

[4]For example projects in the DFG Priority Program Climate Engineering (SPP 1689 2019b) or Harvard's Solar Geoengineering Research Program (2019).

approaches.[5] In this way, frames can be applied intentionally and transparently to define research perspectives and the comparative analysis of frames can help uncover research gaps—e.g. to assess which questions becoming relevant under a security framing have been asked for SAI, but not for DAC or afforestation, and vice versa.

The conclusion of a diversity of framing of CE in the academic discussion might, however, prove a somewhat distorted image of a shared meaning of CE that might make its way into political deliberation: What I did not assess in this study is the role of specific individual actors in contributing to the shared meaning of CE. I assume that each and every instance of communication—each and every segment in a peer-reviewed academic article—has the same weight in contributing to a shared meaning of CE. It is conceivable, however, that some members of the academic CE community are heard more strongly than others, have more clout in the CE discourse than others, in particular with a view to the relevance of academic frames for political decision-making (see, e.g., Buck 2013: 173). In other words, the design of the present study implies that the academic discussion and the political deliberation of CE take place in two entirely dissociate spheres, and that the construction of the shared meaning of CE is concluded in the first before it moves, as a completed and closed entity, to the second.

Furthermore, it is not possible on the basis of this study to ascertain whether the diversity of framing concluded from this project stems from a diversity of framing on the level of the individual actors—meaning that each member of the community applies a variety of frames to describe and discuss CE—or whether it hides a strong polarity of two opposing camps of actors in the discussion, one roughly supporting CE research and potentially even deployment and one opposing these notions. This argument of the existence of two opposing camps or a red team and a blue team in the CE research community has been made before in the literature. Kintisch, for example, argues that.

> [t]he battle lines of geoengineering have begun to take shape. On one side are modern-day romantics, who consider geoengineering an a priori violation of human's role as planetary citizens to let nature be natural and to take a humble place within it. Better to solve the climate problem by reducing our impact on the planet, they say.[...] And then there are the rationalists, who believe that to minimize suffering, it may just be more technological hubris that our species needs. (Kintisch 2010: 233).

Along similar lines, Linnér et al. find that.

> [m]ost peer-reviewed climate engineering literature does not weigh the risks and new, additional, benefits of the various technologies, but emphasizes either the potential dangers of climate engineering or the climate change consequences of refraining from considering the research, development, demonstration and/or deployment of climate engineering technologies. (Linnér and Wibeck 2015: 255).

---

[5]Not only frame analyses offer this potential for critical reflection, of course: Other ways of operationalizing the assessment of discourse are equally useful.

The finding from this study that many documents carrying complete frames—on the document level—also carry elements of frames representing the reverse direction—e.g. an instance of a legitimizing frame and elements from delegitimizing frames—is an indicator against this notion of two clearly delineated camps.

Instances of *legitimizing frame variants* have been concluded to be more frequent, in the sample, than those of delegitimizing frames. It is conceivable that this conclusion can be explained in part by the common practice of defining CE with reference to climate change (Markusson et al. 2016: 1): For example, the Royal Society defines CE as "the *deliberate, large-scale manipulation of the planetary environment to counteract anthropogenic climate change*" (Shepherd et al. 2009: 1). The reference to climate change, in this study, is understood to contribute to an instance of the legitimizing variant of frame element 'problem definition' if climate change is represented as the central problem addressed in a text segment and if this central problem is characterized as a threat, a complex challenge, an economic issue or a normative wrong. As the author of the present thesis, I also decided to use the challenge of climate change as a starting point and provider of relevance to my assessment of the framing of CE, defining CE in reference to this phenomenon.[6] However, it is conceivable to define CE—and to frame CE is academic publications—in a different way, e.g. by identifying it as a set of technologies with potential impacts on social and physical systems. Again another approach is taken by Cairns et al., who argue that CE—or, in their terminology, geoengineering—can be treated as a "discursive phenomenon, the bounds of which are continually being negotiated" (Cairns : 26). This means that they refrain from defining CE in any material way, but explore the "ambiguity and multivalence" of definitions of the term (Cairns 2014: 26). Thus, the reference to climate change is by no means a necessary part of addressing CE, but rather a contingent component of the concept depending on decisions made by the author. Further, the way in which climate change is characterized is up to the decision of the author.

Instances of elements belonging to the *legitimizing security frame and the delegitimizing complexity frame* are slightly higher in frequency in the present study than elements belonging to the other ideal-type frames assessed here. Furthermore, positive evaluations of CE are more frequently conducted with regards to economic criteria than to others. When considered in context with research literature on the discourses of climate change and environmental decision-making and governance, some possibilities for understanding these conclusions emerge—both in the way of understanding the constitution of the academic discussion on CE and understanding its role in constituting other issues. For example, the conclusions from this study can be understood as instances of depoliticizing and politicizing dynamics in the overarching climate discourse of which the CE discussion forms a part. Methmann et al. argue that there were indications for "the contours of an overall pattern" of depoliticization in international climate politics including trends such as the "securitization of climate change conceal[ing] its socio-economic roots", the "reframing of climate change as a matter of economic calculation [which] narrows the scope

[6] See Chap. 1.

of political action" and "the prevailing concentration on technological solutions to climate change" (Methmann et al. 2013a: abstract). The legitimizing security frame, for example, can be understood as an indicator for the depoliticizing trend of securitization mentioned by Methmann et al. Furthermore, the prevalence of economic criteria applied in positive evaluations of CE can be interpreted as instances of an economization trend. Conversely, the delegitimizing complexity frame can be understood as running counter to the technocratization trend identified by Methmann et al. Also, the high frequency of evaluations of CE with an open outcome as well as calls from members of the academic community for political deliberation of these approaches can be interpreted as indicators for a politicizing trend in this discussion. Accordingly, it would be interesting to contextualize the present findings with regards to the larger climate discourse to see whether the inclusion of CE approaches into the discussions strengthens or weakens these depoliticizing trends.

7.2.2 Reflecting on Implications—What Do the Findings Mean for Decision-Making?

The implications of the *slight prevalence of the legitimizing security frame and the delegitimizing complexity frame* for CE decision-making are difficult to ascertain. Firstly, this is the case because the political discourse on which the academic framing is, in this study, assumed to have a causal or constitutive influence, is, to date, very limited. Secondly, as the framing of CE in the academic discussion has been found here to be very diverse and no one single ideal-type frame is dominant, the question remains whether one of these frames will shape political deliberation and decision-making. On the basis of the findings of this study, several tentative hypotheses can be formulated which could provide the starting point for an assessment of the potential causal influence of the academic framing of CE on decision-making:

H1:The more frequently the legitimizing security frame is picked up in political deliberation of CE, the more probable becomes the political decision to include these approaches in the portfolio of measures to address climate change.
H2: The more frequently the delegitimizing complexity frame is picked up in political deliberation of CE, the less probable becomes the political decision to include these approaches in the portfolio of measures to address climate change.

Both hypotheses would allow to assess whether depoliticization processes are taking place with regards to CE. H1 represents the operationalization of technological securitization as outlined in Chap. 3. If CE was found to present a case of technological securitization, climate change would be presented as an existential threat on the basis of which CE, or some specific CE approach, would be introduced as an exceptional measure rendered necessary due to the threatening character of climate change. Whether a case can be made for a securitization of climate change legitimizing CE approaches as extraordinary measures, however, remains to be seen. H2 represents the operationalization of a process running counter to that of technological solutionism as outlined in Chap. 3: In a process of technological solutionism, it would be assumed that 'messy' social problems without clear solutions could be reduced to 'crisp' technological problems to which simple technological fixes could be identified (Weinberg 1966). The delegitimizing complexity frame, conversely, presents CE approaches as messy issues introducing many challenges and side-effects—both on physical and on social systems. Accordingly, CE is, according to the findings of this study largely *not* presented as a technological fix in the academic discussion, but rather as a complex challenge in its own right. Whether this understanding will prevail in political deliberation and influence decision-making, however, remains to be seen. While economy or appropriateness frames have been found, in this study, to be less prevalent in the academic discussion, it is conceivable that they play a role in shaping political deliberation and decision-making. The appropriateness frames in particular might increase in relevance if the central framing of the IPCC 1.5 Report can be understood as an indicator for a future discursive trend, as mentioned above.

Further, the findings of the present study in concert with the illustration through the case of the early discussion of nuclear weapons technology indicate that it might make a difference who decision-makers choose to listen to when they start considering CE approaches: If a small number of individual academic actors are chosen to contribute their views to a political decision-making process of CE, the meaning of CE that they convey might be highly different, for instance, than if the entire academic community of CE is given a voice. In the latter case, the diversity of framing of CE could be preserved (see Kreuter 2018).

7.3 Standing on the Shoulders of Giants—Added Value and Limitations of the Present Study[7]

This study contributes to various literatures and improves understanding with regards to IR theory, political science research method, empirical information about the case of CE, and practical politics. At the same time, the present study is, like any research endeavor, subject to various limitations: There are a number of questions to which the study cannot give an answer, and a number of conditions for the relevance and functioning of the approach espoused here.

Concerning *theoretical considerations*, this study offers insights for the *analysis of emerging as well as existing technology in international relations*. A number of technologies play a significant role for international relations and global governance such as nuclear power technology (see Kaplan 1983; R. L. Brown and Kaplow 2014), genetic engineering (Lander et al. 2019), and networking technologies underlying systems such as the Internet (M. Moore 2018; Hansen and Lim 2018; Ulbricht et al. 2018). Technology scholar Wiebe E. Bijker argues: "Technology matters: it matters to people, planet, and profit; it also matters to policy-making and politics; and should

[7] Some ideas expressed in this section have been published in Kreuter (2018).

thus matter to political studies" (Bijker 2006: 682). The present thesis contributes to social constructivist IR theory by making use of insights from STS beyond the material turn based on ANT. The notion of co-production posits that technology and society are intertwined and technology is not a necessary effect of physical laws, allowing for the discussion of emerging technology as an issue of deliberation and decision-making. The Multi-Level Perspective differentiates several layers of interaction of technology and society and provides a framework to understand how technologies can emerge (Schot and Kanger 2018). The Epistemic Communities approach (Cross 2013: 149; Haas 1992) can be informed by STS to take into account the internal dynamics of academic groups without relativizing academic knowledge on climate change, and to be able to consider also academic groups in political processes which do no intend to actively influence political decision.

Furthermore, the present study contributes to a growing social constructivist IR literature by highlighting the use of frame theory to be able to discuss not only constitution, but also *causal power of meaning, in a social constructivist perspective* (see also Oels and Carvalho 2012; Grauvogel and Diez 2014; Balzacq 2010; Jachtenfuchs 1996b). Frame theory offers a specific way to conceptualize how meaning can become causally relevant. The operationalization of discursive moves such as securitization as a type of frame allows for a structured and systematical assessment of empirical information because frame theory offers both an account of the effect of frames on decision-making and action (Druckman 2011; Chong and Druckman 2007a), and a clear and easy-to-operationalize tool (Entman 1993: 52). The present study offers an instance of how insights from frame theory can be used in order to study meaning-making under the assumption of their causal power as well as highlighting further assumptions that need to be made in order to allow for this operationalization.

Moreover, and most importantly, the present thesis offers insights into understanding how an issue, and specifically *an emerging technology issue, can become political*: The academic discussion is not dominated by one frame, but diverse with regards to framing. It does not present one clear meaning construction of CE, but various, partly opposing constructions, thus opening up the issue for deliberation, and "making previously unpolitical matters political" (Zürn 2019: 977). This understanding of the academic CE discussion as an instance of politicization is interesting also in the light of the claim that "the rule of technological circumstances in the modern era [would] in fact supplant other ways of building, maintaining, choosing, acting and enforcing which are more commonly considered political" (Winner 1977: 237). Quite in contrast to this claim, the present thesis indicates that the construction of meaning of emerging technologies can also increase the relevance of and strengthen the role of political decision-making and action. The contribution of this thesis to politicization literature is threefold: Firstly, I offer a middle ground reconciling foundational and post-foundational outlooks on climate change (Pepermans and Maeseele 2016), by taking into account *both* the contributions of academic research to our knowledge of phenomena in the world, and its contributions to the construction of social meaning of phenomena. This might provide a way to transcend the opposition between two normative perspectives; one regarding the politicization of climate change as a chance—for instance as a process to "revive democratic debate and citizenship" (Pepermans and Maeseele 2016: 478), and the other understanding it as a problem—for example when "the inherent uncertainty of science is emphasized to cast doubt on scientific consensus" (Bolsen et al. 2019). Secondly, I contribute to diversifying the scope of actors that are studied as relevant in processes of politicization by assessing the *academic* discussion on CE, thus moving beyond actor groups such as "individual citizens, non-governmental organizations (NGOs), parties, lobby groups, and governmental bodies" (Zürn 2019: 985). Thirdly, the present study introduces a fourth type of depoliticization processes in the climate debate, assessing not only security, economy and complexity frames, but also appropriateness frames (Methmann et al. 2013c). This last type of frames might become more relevant in the discourse in future. For instance, the authors of the IPCC 1.5 °C report stress that "[e]thical considerations, and the principle of equity in particular, are central to this report" (Allen et al. 2018: 51).

With regards to limitations, it is important to note that the approach espoused here is contingent on the acceptance of a number of theoretical assumptions. Chapter 3 states the hierarchy of assumptions informing the present research approach, and an individual explication of each of them in order to render them plausible to the reader. First among these is the assumption that the academic discussion on CE is in some way relevant to political decision-making concerning these approaches. If this assumption is denied, the relevance of the entire research endeavor dwindles as a result. This assumption is not empirically tested for the present study, but assumed to apply on the basis of theoretical considerations. This is due to the observation that the empirical case of CE decision-making has, to date, not progressed far enough to assess empirically whether the assumption does apply. If the next few years shows an increase in political deliberation and decision-making on CE approaches, an empirical test of the assumption can be conducted in future research projects.

Methodically, the study contributes to further developing the integration of inductive and deductive approaches as well as qualitative and quantitative methods in political research: *Inductive and deductive approaches*, in everyday research work, always permeate each other. This renders a systematic and transparent model for the integration of the two approaches highly relevant. Such concrete illustrations are useful supplements to abstract methodological literature as they provide models which can be referred to for future research endeavors (Egner 2019). In the present study, the selection and definition of the empirical case to be studied—i.e., framing in the academic discussion of CE in peer-reviewed journal articles between 1971 and 2016—was conducted on the basis of a combination of inductive and deductive approaches: Both empirical information about the discussion on CE—derived from an unsystematic exploration of the literature—and theoretical concepts—in particular the EC concept—informed this selection and definition. Further, the selection of salient theoretical concepts to apply to the case was based both on the findings of earlier exploratory studies on representations of CE and on theoretical concepts derived from IR and STS theory. The analysis of empirical data in this thesis was conducted following a largely deductive approach by applying a pre-defined theoretical construct developed from IR and STS theory. However, inductive impacts have been incorporated: The construct was kept open to allow data to 'speak for itself' in cases where the categories of the construct did not capture the data. Also, the data was not analyzed in order to test general statements derived from theory as would have been done in a clearly deductive approach (Behnke et al. 2012: 31). Instead, theory was applied to systematize empirical data (Spindler and Schieder 2010: 23).

Qualitative and quantitative approaches differ in a number of aspects, such as the number of cases considered, the degree of openness in data generation, the treatment of data prior to analysis, and the methods of data analysis. The present thesis contributes to the *integration of qualitative and quantitative research methods* by offering a structured example of a 'complementarity model' ("*Komplementaritätsmodell*", (Behnke et al. 2012: 44, translated by JK) of the mixed methods approach. In this model, qualitative and quantitative elements complement each other in the pursuit to achieve a common research objective.[8] In the present thesis, both of these types of steps are required in order to answer the research question by combining qualitative elements of data generation (close reading) with both qualitative (assembly of frame elements and frames) and quantitative (revision of data) elements in data treatment. Further, qualitative (circularity) and quantitative elements (number of cases, analysis of frequencies) were incorporated in data analysis.

By applying this model in a study using the well-established method of manual empirical content analysis, the present thesis provides a *starting point for automated or semi-automated text mining approaches* to analyze CE in text content (Grimmer and Stewart 2013). For example, the findings of the present study could be translated into a dictionary—i.e., a list of words—dedicated to frames of CE technology in order to "classify documents into categories or to measure the extent to which documents belong to particular categories" (Grimmer and Stewart 2013: 274). Alternatively, the coding applied in this study can be used to inform supervised learning methods which would allow the coding of much larger amounts of text automatically (Waldherr et al. 2016). Such a dictionary or a 'training set' does not exist to date for the theoretical construct applied here, as the construct has been created specifically for the use of this study. As well-defined dictionaries or manually coded portions of the corpus are the foundation of many automated text mining approaches, the work conducted in the present thesis can contribute to the prerequisites of such studies to be conducted in the future. Furthermore, the present thesis contributes to specifying content analysis methods for the political science analysis of discursive representations by *operationalizing the concept of 'frame'* as comprising several categories—which correspond to the four elements of a frame—for a codebook for content analysis. In this, I make useful an approach from communication and media studies (Matthes and Kohring 2008: 264; Donk et al. 2012: 12; Scholte et al. 2013) for moderate social constructivist IR research.

---

[8]In contrast, in the ‚phase model ' („Phasenmodell ", (Behnke et al. 2012: 45, translated by JK), qualitative and quantitative elements are applied in separation in different phases of research procedure.

Concerning limitations, in attempting to combine insights from qualitative and quantitative content analysis, this thesis might be criticized as committing too little to acquiring either adequate qualitative depth or sufficient quantitative scope. The approach pursued here combines qualitative close reading of each individual text with aggregation of these results quantitatively in order to uncover possible trends across document groups. The codebook guiding close reading has been purposefully designed to remain open to new categories and has been tested through inter-coder and intra-coder reliability testing in order to prevent the generation of artifacts. The possibility of mis-codings and misinterpretation of quantitative results is, through these steps, not entirely precluded, however.

Empirically, the present thesis connects descriptive findings from preceding exploratory studies on the CE discourse with theoretical insights on *potential political implications* of the constitution of this discourse: A number of exploratory studies have found various frames, metaphors, imaginaries or other discursive elements to be prevalent in the discussion on CE. The present thesis connects these findings, through the theoretical framework, with potential effects on political decision-making. Thus, we gain insight not only into what is out there, but also a first glimpse of what this might imply for future climate policy decisions. Doing so, I build upon endeavors by, for example, Asayama (2014); Huttunen and Hildén (2014); Luokkanen et al. (2014); Corner and Pidgeon (2015); Bellamy et al. (2012b) which connect specific discursive representations with impacts on policy- or decision-making. I take these approaches at face value and contextualize them in a comprehensive social constructivist account of different frame types with different effects on decision-making and their presence or absence in the academic discussion.

In this, the present thesis contributes to *extending the scope of attempts at understanding future decision-making concerning CE*. Thus, it supplements the findings from two sets of literature: On the one hand, deductive studies aim to model future sociopolitical behavior based on assumptions derived from rational choice theory. For example, game theoretical approaches are applied to assess how future international CE decision-making might look (A. Harding and Moreno-Cruz 2016: 573; see, e.g., Goeschl et al. 2013; Ricke et al. 2013; Quaas 2017; Singer and Matchett 2015; Manoussi and Xepapapdeas 2014) and arguments about the future impacts of CE are made based on the assumption of rational decision-makers (Keith and Irvine 2016). On the other hand, exploratory literature (see, e.g., Corner et al. 2011; Bellamy and Lezaun 2015; Bellamy et al. 2016, 2017) aims to understand "concerns and values regarding future risks and challenges" of CE which might go into decisions on the basis of empirical engagement and deliberation work with "scientific, policy, and civic stakeholders" (Low and Schäfer 2019: 5). The present thesis contributes to extending the "bounds of the debate" (Low and Schäfer 2019: 8) of future CE decision-making by bringing to bear the modernist social constructivist perspective. This means that I avoid assumptions of (bounded) rationality of decision-makers and "choice as a maximization process" (Tversky and Kahneman 1987: 67) or the existence of a given set of preferences, "motivations and constraints facing international actors" (Low and Schäfer 2019: 6). Instead, I accept the assumptions and concurrent pitfalls of modernist social constructivism and STS—such as the role of discursive representations in shaping preferences and the co-production of technology and society—[9] in order to be able to ascertain "how [influential] ideas shape state interests and behavior" (Horton and Reynolds 2016: 455) in CE decisions. Thus, it offers one way—among many—to achieve "a description of the behavior of real people" instead of "a normative model of an idealized decision maker" (Tversky and Kahneman 1987: 67), thus taking seriously the criticism voiced against assumptions of rational choice even in the discipline of economics. Thus, I attempt to make a first step towards answering Horton and Reynold's call to IR scholars to "offer insights into [...] the emerging international politics of climate engineering" (Horton and Reynolds 2016: 438). The two authors make clear that "Constructivism has much to contribute" (Horton and Reynolds 2016: 455) to the clarification of the preconditions and implications of CE for international politics, for example by assessing the influence of "expert networks and expertise" (Horton and Reynolds 2016: 455), as I do, on the constitution of the meaning of CE.

Furthermore, I offer an empirical basis to ascertain to what extent two opposed concerns voiced in academic as well as other fora resonate with how the representation of CE in the academic discussion is constituted empirically: Both concerns apply to *how current CE research should be evaluated, in particular with regards to its role in determining a specific path for the future of CE*. On the one hand, the concern is voiced that CE might constitute a case of a 'slippery slope' which inadvertently leads from research to deployment (Zürn and Schäfer 2013: 272; Hulme 2014: 69). If this hypothesis was true, any consideration of CE approaches—even decidedly restricted research endeavors such as SPP 1689—might prove too much consideration (Rayner 2014). On the other hand, the concern has been voiced that CE constitutes an "uphill struggle" which might require governance to "responsibly incentivize[...]" research (Bellamy and Healey 2018: 1) as otherwise, research might not be conducted at all. The present thesis offers one piece to this puzzle by assessing whether, discursively in the academic discussion, the foundations for a slippery slope process are laid through dominant legitimizing framing of CE or, conversely, the foundations of a hill to be struggled up through dominant delegitimizing frames. The sample studied here does not present clear dominance of frames either discursively legitimizing or delegitimizing CE. Instead, the diversity of frames and the increasing relevance of the complexity frame can be read as an indication that the academic community is engaged in building a complex landscape with slopes and hills and a number of other features. In order to further understanding concerning the role of current early CE research, i.e. to understand in what ways the concerns of slippery slope and uphill struggle are justified, future studies are required such as analyses of other discursive spheres, as well as assessments of material decisions on CE.

It is important to note that the decision to conduct a study of the constitution—instead of the causal mechanisms—of CE framing and the chosen focus on framing in the academic discussion operationalized through peer-reviewed journal articles sets clear limits to what can be learned about the case in question: Firstly, and related to what has been outlined above, the present thesis does not assess the comparative

[9] See Chap. 3.

causal influence of the academic framing on decision-making compared to other factors, but instead assumes that there is some influence and moves further from there. Secondly, the present thesis does not provide a constitutive account of the entire discourse on CE—across discursive arenas such as media, academia, politics, the public etc., across geographical locations, across groups of speakers etc.—but instead focuses on a specific arena—academic publication—and a specific set of actors—researchers from diverse academic disciplines. Further, the study does not take into account diverse elements of discourse—such as imaginaries, metaphors, storylines, arguments etc.—but focuses on frames in order to render discursive moves visible. Thus, the findings of the present thesis need to be ascertained in conjunction with those of other studies assessing other elements or the entirety of the discourse.[10]

7.4 Avenues for Future Research

Further research could, based on this study, follow at least three general directions: First, the theoretical conclusions developed above could be problematized further as well as empirically tested. Second, conclusions drawn from empirical analysis about patterns in the discussion could be contextualized through studies on different cases in order to assess whether they present general patterns. Third, explanations and interpretations of the conclusions drawn from the empirical analysis conducted here could be theoretically more deeply rooted and empirically tested in order to assess why the patterns are there.

The first direction concerns the critical reflection and empirical assessment of the theoretical conclusions drawn in this thesis. For instance, concerning the influence of academic framing on political decision-making, one might ask: Can an academic and a political 'sphere' really be differentiated this clearly, or does the co-constitution of the two need to be taken into account in order to represent the empirical information more precisely? Following Max Weber (1980 [1922]), for the present study I differentiated an ideal type of academic actor—who offers answers to research questions—from an ideal type of political actor——who decides on how to address wicked problems—in order to be able to identify potential deviations from these types in empirical reality. Starting from such a differentiation, one can assess whether and, if so, how academic actors are involved in the decision-making concerning wicked problems. It is important to differentiate between these types in order to be able to assess how processes of influence take place. To follow up on this, future research might apply a contrasting theoretical conceptualization drawing, for example, on the notion of boundary work (Gieryn 1983; Hoppe 2005) in order to unearth how various actors define and negotiate the boundary between the academic and the political, as has been done for other issue areas (see, e.g., Schot and Schipper 2011). The assumption of relevance of academic discussion for decision-making could be tested empirically. One might, for instance, conduct interviews with the representatives

[10]See Chap. 2.

involved in supporting or rejecting the Swiss draft resolution on CE assessment at the 2019 UNEA meeting (IISD Reporting Services 2019: 13) in order to assess which role the academic discussion of CE played for their decisions. Further, the researcher might turn to other cases such as Artificial Intelligence (Kiggins 2018), CRISPR/Cas (Lander et al. 2019) or nuclear technology (Lederer and Kreuter 2018) in order to make a general argument about the role of academic framing for decision-making on emerging technology or to identify differences between the cases.

Closely related to this issue, future research might consider the relevance of relative power of various academic actors to construct the meaning of CE: If we assume that academic framing can influence decision-making, it is conceivable that each academic actor involved in the discussion does not have exactly the same weight in constituting the meaning of CE, and accordingly influence decision-making. This means that it is possible that a specific frame is more relevant for the constitution of the meaning of CE when it is applied by actor A than when applied by actor B. In an approach focusing more strongly on actors, future research might ask: Which academic actors or institutions are particularly influential in the constitution of the meaning of CE and how do they frame CE? A research endeavor proceeding in this direction might map out the exchange of discursive elements such as frames, but also the exchange of material resources such as research funds in order to sketch a network of power relationships in the field of CE, between actors from the academic, the political and other spheres. From this, channels of influence could be ascertained in order to assess from where decision-makers draw their knowledge and ideas. Such an approach would probably require extensive data gathering, e.g. through interviews.

There are indications that some members of the CE research community are heard more than others. For example, in an assessment of a sample of articles on CE in global English-language newspapers, Buck finds that communications from researchers such as David Keith, Ken Caldeira, John Shepherd and Paul Crutzen top the list of assertions on CE cited in those articles (Buck 2013: 173). This means that for the readers of these papers, those researchers shape the meaning of CE more profoundly than other actors. Furthermore, projects such as the C2G2/C2G decidedly aim to "expand the conversation from the scientific and research community to the global policy-making arena" (C2G2 2019).[11] Arguably, the frames that members of projects like this choose in order to make this connection will find their way more easily into the political debate than others.

Another venue for future research based on critically reflecting on the theoretical conclusions drawn in this thesis is the empirical assessment of the idea that the ideal-type frames identified here really do have their assumed impact on decision-making. While, on the basis of earlier empirical studies and theoretical literature, these impacts seem probable, this thesis did not empirically test that assumption. To do so, future research might either experimentally test the frames or assess their actual impact on decisions, e.g. by operationalizing the two hypotheses developed above.

[11] Interestingly, C2G2 recently changed their name to C2G, at least in part in order to be able to achieve this aim, because the term 'geoengineering' "hampered our ability to engage with society, and to catalyse the learning processes necessary to take informed decisions." (Pasztor 2019).

In the former case, the experiment would need to be set up in a way as to indicate whether respondents are more likely to support CE approaches after being acquainted with a legitimizing frame or more likely to oppose it after being acquainted with a delegitimizing frame than a control group. While such experiments have been made for various discursive representations of CE (see, for example, Corner and Pidgeon 2014, 2015; Asayama et al. 2017), the specific frames defined here have not, yet, been applied in experimental analysis. In the latter case, future research would aim to draw a causal connection between the frames applied in the CE discourse and the decisions made with regard to these approaches. The causal influence of framing implied in these considerations would need to be carefully tested against a number of control variables covering the potential role of, for example, political entrepreneurs or possible windows of opportunity for specific policy contents (Kingdon 2011).

The second direction of future research concerns the contextualization of empirical conclusions drawn here to find out whether they capture patterns generalizable beyond the case of CE. This might be done empirically, by extending the basis of material, e.g. on other technologies or from other discursive realms. Further, theoretically, the present case could be embedded in the theoretical discussions around the larger climate or environmental discourses. Methodically, patterns in framing could be compared to patterns in other discursive elements such as metaphors or imaginaries (Low 2017) or patterns in practices adding to the constitution of meaning of CE (Low and Schäfer 2019). Empirically, future research might assess whether the academic discussion on CE is characterized by an exceptionally low degree of legitimizing or delegitimizing framing, and if so, why; or whether the patterns of prevalent frames can be replicated with regards to other emerging technologies or other discursive arenas. In order to do so, the academic discussion on other emerging technologies, such as CRISPR/Cas or AI could be studied applying the same theoretical construct as in the present case, in order to be able to compare these discussions. Alternatively, the media or political discussion on CE could be assessed through this construct, in order to assess whether framing is significantly more common in those arenas. Also, through such a comparative study, it might be ascertained whether the two frames found to be slightly more common in the academic discussion than the rest of the ideal-type frames, namely the legitimizing security frame and the delegitimizing complexity frame, are also prevalent in other discursive arenas.

To connect the results of the present study to theoretical debates and assess whether they represent general patterns, insights from literatures on the securitization (see, e.g., Grauvogel and Diez 2014; Corry 2012; Lucke et al. 2014, 2016; Oels and von Lucke 2015; Trombetta 2008), the economization (see, e.g., Hajer 1997; Levy and Spicer 2013; Anshelm and Hultman 2015) and the technocratization of climate change or the environment (see, e.g., Corner and Pidgeon 2010; Borgmann 2012; D. Scott 2012; Nerlich and Jaspal 2012; Morozov 2014) could be brought together and systematized to form the basis for an analysis of the comparative relevance of these individual trends and their influence on the construction of the meaning of CE and other climate policy measures. The present study has contributed a first attempt at systematizing the assessment of these different trends by operationalizing them through frame theory. This can serve as a starting point. Future research could aim to provide either a broader understanding of this discourse—by including larger numbers of texts from a variety of sources—or a more detailed understanding of this discourse—by including close reading methods or differentiating sub-types of the frame types assessed here, e.g. on the basis of the differentiation of three security frames conceptualized by Grauvogel and Diez (2014), the differentiation of processes of riskification and securitization as presented by Corry (2012) or on the basis of the differentiation of different economic 'discourses' presented by Anshelm and Hultman (2015). The former approach could address questions such as whether a security framing is more common in the CE discussion than in the general climate change discourse and what this means for climate policy, and whether an appropriateness framing might increase in relevance after the publication of the IPCC 1.5 Report which arguably applies an appropriateness frame by stressing in its executive summary that "[e]thical considerations, and the principle of equity in particular, are central to this report" (Allen et al. 2018: 51).

Through such a study embedding CE as a policy measure in the broader discourse on climate policy, furthermore, we could draw nearer an understanding of the role of real world events such as UNFCCC COP meetings and the publication of IPCC Assessment Reports for fluctuations in the academic framing of CE. For example, in the years 2009 and 2014, peaks in the number of text segments identifying climate change as the central problem coincide with the striking of the Copenhagen Accord at COP15 and the preparation of the fifth Assessment Report of the IPCC, respectively. Similarly, the peak in instances of carefully negative evaluations of CE coincides, in 2013, with the foundation of the DFG Priority Program on CE. It remains to be seen whether these coincidences can be drawn back to causal or constitutive relationships between the academic discussion on CE and those real-world political events.

In the third direction for future research starts from the conclusions, theoretical and empirical work could be done in order to explain and understand the patterns identified here. For instance, in the preceding section, I suggest that interaction between researchers, both of those working on different approaches of CE and those from different disciplinary backgrounds, might diversify framing and the constitution of meaning of CE. For future research, analyses of a larger sample or of communications in different media—e.g. through newspaper interviews or on social media—could broaden the empirical data base. A theoretically more profound concept of 'discipline' could build a theoretical foundation for the testing of such a supposition. In order to empirically test this suggestion, future research projects might, for example, assess the framing applied in oral or written discussions (e.g. at conferences or on social media) on CE and compare cases of mono- with cases of multidisciplinary discussion contributions or, similarly, cases focusing on one specific approach with cases addressing various CE approaches. Alternatively, an experimental setting could provide insight into whether the use of frames becomes more moderate after the researcher has had contact with researchers from other disciplines or other CE approach research contexts.

Future research endeavors could, further, shed light onto why the academic discussion is constituted in the way found here, and why there is currently only limited high-level political decision-making on CE. Why is the academic discussion on CE so diverse, why is there no dominant frame? On the basis of this study, the role of individual academic actors could be assessed more closely in order to identify directions of influence of researchers on each other, for instance through network analysis. It would be very interesting to see whether the network is structured around a clear focus, or whether there are several important clusters, and which role large research projects play in this structure. A larger-scale network analysis could move beyond academic actors alone and assess the interactions between them and political decision-makers in order to assess whether and, if so, where there are power struggles. Process tracing might be applied to assess whether and, if so, how specific frames move around between individuals, e.g. through academic reports.

7.5 Insights for Practical Policy-Making

The present study, I claim, offers some valuable insights useful for practical politics. Firstly, I offer decision-makers a—surely not uncontroversial—account of *potential implications of a lack of dedicated political decision-making on CE.* This account might contribute to a discussion of the situation characterizing current climate decision-making, in which CE approaches are implicitly assumed to be available. Specifically, I claim that not making explicit political decisions with regards to CE approaches is a decision in its own right which implies accepting that these approaches will develop according to their inherent dynamics or according to decisions made by other actors. Gupta and Möller (2018) and Talberg et al. (2018) provide two arguments of how these dynamics currently play out: Talberg et al. argue that the '[g]overnance-by-default' to which CE is subject.

> is characterized by uneven regulation from existing multilateral agreements established for other purposes, an absence of regulation specifically focused on geoengineering, guidance from an international ambition to hold global average warming below 2 °C and to achieve net-zero emissions in the second half of the century, and strong normative engagement by the research community (Talberg et al. 2018: 229).

Similarly, Gupta and Möller claim that CE is a case of 'de facto governance' "discernible in an ordering of this nascent field of inquiry by unacknowledged sources of steering", such as "high-level 'authoritative assessments' of CE" (Gupta and Möller 2018: 480). Furthermore, Chap. 4 of this thesis offers an account of how specific framing can play a role in legitimizing an emerging technology with daunting consequences. The present thesis does not aim to predict the future of CE, but it can give first indications that should be considered when deciding whether or not to make political decisions on CE approaches.

Secondly, the present thesis highlights that there is, at this moment, *a unique opportunity to actively and critically shape the way in which CE approaches are considered*—if at all. This is the case because these technologies have not yet entered our reality on a large scale, and further, because the meaning of these approaches in the academic discussion is, at this point, not yet settled or sedimented. Instead, the academic discussion on CE is characterized by a high degree of frame contestation and frame diversity (see also Linnér and Wibeck 2015), which led me to argue, earlier, that it presents an instance of politicization of the issue of CE: It opens up the issue for debate and dispute. Thus, the scope of political decision-making is not pre-defined by a stable intersubjective academic understanding of CE, but instead is highly malleable, offering an arena for the inclusion of a number of positions and interests. It is important to remember that there are a large number of decisions to be made, and that, in making these decisions, the global community can shape its own future. This thesis highlights the political character of these decisions as well as the opportunity provided by the highly reflexive academic discussion currently taking place. The present thesis provides an access point which systematizes this multifaceted discussion without narrowing it down to one specific way of framing CE.

Thirdly, the findings from this study indicate that a political decision-making process for CE should draw on *a large and diverse range of knowledge production endeavors* to inform decisions. In the 2015 Paris Agreement, the signatories decided to take mitigation, adaptation and monitoring actions in accordance with "the best available science" (UNFCCC 2015: Art. 4, 7, 14). As an example of an expert organization which includes not only 'science' in the narrow sense of the term—commonly denoting natural sciences—but also "being inclusive of different disciplines and knowledge-systems" (Borie and Hulme 2015: 487), the Intergovernmental Platform on Biodiversity and Ecosystem Services (IPBES) is discussed. The inclusion of social sciences and humanities into policy-relevant knowledge bodies to inform CE decision-making provides insights into the issue from different angles as well as answers to different questions. For instance, with regards to uncertainties of CE, Matzner and Barben (2018: 160) find that social uncertainties such as potential future political conflict are assessed in more depth, differentiation and detail by social sciences, which is hardly surprising. Low and Schäfer (2019) argue that different "epistemologies, objectives, and user communities surround [different] research practices, highlighting that different modes of constructing and interpreting evidence about an unformed future yields different kinds of results and signals for actions to be taken." Similarly, the findings of the present thesis indicate that including knowledge from social sciences and humanities as well as natural and engineering sciences draws a completely different picture of CE than if only some of these disciplines are considered because members of various disciplines frame CE in different ways. Different types of knowledge should, I argue, be complemented when generating policy-relevant expertise.

In order to prevent the threat of knowledge relativism, open discussions are required about truth claims and evaluation mechanisms. One example to attempt to bring together knowledge from a diverse set of disciplines, based on a wide range of theoretical assumptions and methodological approaches, attempting to address a wide array of questions concerning CE, is the SPP-CE (SPP 1689 2019a; Kreuter et al. 2020). In the course of the six year-program, the researchers involved not only produced knowledge about specific questions concerning CE from their own disciplinary perspective, but also interacted at common meetings and in publications, aiming to make their findings and assumptions speak to each other. This is a highly strenuous endeavor in its own right, and one that has to take place on top of disciplinary research work. In spite of this, decision-makers should make use of the entirety of knowledge at their disposal when turning to deliberating the future of CE.

On the basis of the above reflections, I suggest that, if the diversity of the make-up of the academic community discussing CE can be preserved in academic information for political decision-making—e.g. by ensuring that decision-makers draw upon expertise from diverse academic disciplines and addressing diverse approaches subsumed under the umbrella term 'climate engineering'—then the dominance of one frame type and, thus, the inadvertent narrowing of political deliberation can be prevented. Similar recommendations have been made before, for example by Caldeira and Keith, who suggest a red team/blue team approach to SRM research:

> [...] research efforts should take a red team/blue team approach, wherein one team is tasked with showing how an approach can be made to work, and another team is tasked with showing why the approach cannot produce a system that can actually diminish environmental risk at an acceptable cost. (Caldeira and Keith 2010: 61).

I go further than Caldeira and Keith, however: Not only researchers from two opposing camps—one supportive of CE, one opposing these approaches—should be included in the group of academics informing decision-makers about CE, but a diversity of researchers with regards to disciplinary expertise and the approach they address would be required. It is conceivable that such an inclusive approach could help address two opposing concerns connected with CE decision-making: The first concern is designated by the term 'slippery slope' and points out the "possibility that permitting research on CE could itself be a step onto a slippery slope, making development and eventual deployment of a technology more likely." (Oldham et al. 2014: 2) The opposing concern is that the ability to conduct some type of CE is implicitly relied on in global climate decision-making without an explicit decision to provide this ability through research and development. Arguments relating to this concern are exemplified by Rayner's "magical thinking" (Rayner 2016) or Rockström et al.'s "the world's biggest gamble" (Rockström et al. 2016). Both concerns address the implications of omitting dedicated political deliberation on CE: The first identifies potential 'runaway' research on CE as the problematic implication, the second the failure to research CE enough in order to be able to use it to the degree implied in climate decisions. The inclusions of researchers from diverse backgrounds into policy advice might prevent situations as occurred with regards to the question of using nuclear technology in the Interim Committee in the USA in 1945.[12]

A political decision-making process on CE drawing on academic expertise in a way as outlined above could, then, help to realize two aims in the relation between research and political decision-making: It could help ensure that the 'wickedness' of CE as a problem is kept in mind while, at the same time, provide the basis for a system in which political decisions are based on the findings of academic research.

---

[12]See Chap. 4.

Academic research—be it in the social sciences, humanities, natural or engineering sciences—can answer a lot of very important questions surrounding CE in a very precise way. For example, it can provide information on what the probable impact of large-scale application of specific stratospheric aerosols is on the photosynthesis rate of plants or whether humans can be expected to uphold mitigation endeavors when they have learned about the possibilities and limitations of SAI. However, problems such as whether CE approaches should be considered as parts of a portfolio of climate measures are wicked problems that do not have a clear right or wrong solution (Hulme 2014: 114; Zhang and Posch 2014: 392). Instead, different approaches to address the problem produce diverse implications, advantageous to some, disadvantageous to others. For example, while academic research can provide, with a specific probability and a specific degree of uncertainty, information on which effect a given degree of albedo modification will have on the average precipitation in China or France, or average temperatures in India or South Africa, it cannot define which level—if any—of albedo modification should be chosen. Thus, it can inform a decision, but this does not, usually, preclude a decision. In this, the issue of CE is not different from other decisions concerning the global climate or environment. While, for example, the IPCC can and does inform decision-makers concerning the implications of various levels of climate stabilization, this body explicitly recognizes "the importance of governance, ethical dimensions, equity, value judgements" for "[e]ffective decision-making to limit climate change and its effects" (IPCC 2013b: 17). Its results are meant to be policy-relevant, but not policy-prescriptive (IPCC 2018b).[13] Thus, lessons learned from knowledge-producing bodies such as the IPCC or the Intergovernmental Platform on Biodiversity and Ecosystem Services (IPBES) might be used in order to establish institutional arrangements which ensure the inclusion of a diversity of types of academic expertise into informing decision-making. The IPBES in particular applies a framework to ensure the representation of different kinds of knowledge in assessments (Turnhout et al. 2015: 69; Borie and Hulme 2015). The findings of the present thesis support the claim that such an approach is relevant and useful also in the case of CE.

Similar to climate and environmental governance in general, not making a decision concerning CE is, I claim, a decision in its own right: If GHG emissions are allowed to continue unabated, this has consequences for various groups in society, just as much as deciding on rigorous mitigation measures does. Not making dedicated political decisions on CE implies accepting that these approaches will develop according to their own dynamics. This means that they might disappear due to lack of research funding, or that they might be developed in a way that does not consider the interests and needs of all those affected, but instead serves certain particular interests and goals. While dedicated political deliberation does not guarantee that the best decision for all those concerned will be made, it can provide a tool to draw near this ideal (SPP 1689 2019a: 63). As pointed out above, CE approaches exist, so far, mainly in the world of ideas, but have not been implemented on a large scale in the material

[13]For a critical discussion of whether policy-relevant knowledge can ever achieve this aim of not being policy-prescriptive, see Turnhout et al. (2015).

world. These approaches, then, are not (yet) embedded in a highly complex and far-reaching socio-technical system (Schot and Kanger 2018) entangling extensive infrastructures, artifacts, investments and large amounts of livelihoods depending upon them. If they were, this would make decision-making even more complex than it already is, as well as costly in terms of the need to re-allocate values in society. The current situation offers the opportunity to actively and critically shape the way in which CE develops in the future, and to put into place a process as hinted at above, in which decisions are made on the basis of a diverse academic discussion while keeping in mind the wicked problem character of CE.

7.6 Concluding Remarks

At this moment in time, the discussion of CE turns it into a political issue, which means that it opens up for deliberation and debate. This means that, right now, we are at a crossroads: Decisions will be made, and soon, concerning which way we want to proceed. Not making a decision is a decision in its own right. This moment is one of great opportunity, but also of great uncertainty. Researchers have the chance, at this moment, to assess this issue not in disjunction from each other, but in cooperation in order to collect and combine as many perspectives as possible on the context and implications of CE before one individual perspective starts to dominate all others, shaping a specific future of CE and sidelining alternatives (Low 2017). Decision-makers, in turn, have the opportunity to include a diversity of frames, and more broadly speaking, a diversity of types of policy-relevant knowledge into decision-making processes. The constitution of the academic discussion itself offers the opportunity to critically reflect on the constructedness of meanings concerning climate change and CE approaches, while at the same time upholding the relevance of knowledge derived through academic research. Through this instance of politicization, it might be possible to learn how to reconcile insights about the role of power in the realm of academic research with taking seriously the findings from academic research about climate change and CE and their implications for our future. Thus, as UN Secretary-General António Guterres announced in September 2018: "Climate change is the defining issue of our time—and we are at a defining moment." (UN Secretary-General 2018). How and through which processes we choose to define the role of CE approaches at this moment in time will be decisive for our future (Tables 7.1 and 7.2).

Table 7.1 The meaning of CE according to the eight ideal-type frames making up the theoretical construct applied in this study

Variant/ Theme	Legitimizing CE	Delegitimizing CE
Security	CE as a measure to address the existential threat of anthropogenic climate change	CE as an existential threat
Complexity	CE as a simple fix to the complex challenge of anthropogenic climate change	CE as a complex problem
Economy	CE as a low-cost solution to the expensive implications of anthropogenic climate change	CE as an expensive measure
Appropriateness	CE as a way to right the wrongs incurred through anthropogenic climate change	CE as normatively wrong

Table 7.2 Contributions and limitations of the present study

	Contributions to the respective literature/knowledge base	Limitations/challenges
Theory	• Contributing to social constructivist IR theory by applying concepts from STS and frame theory • Offering a way to make use of STS concepts in IR study in order to be able to assess political role of technology • Further developing application of concept of framing for political analysis • Contributing to securitization theory by offering an approach to simultaneously assess various types of legitimizing moves in discourse • Contributing to the de-/politicization debate by offering a middle ground between conventional and critical studies of climate change politicization	• Relevance and functionality of approach highly dependent on acceptance of theoretical assumptions • Does not empirically test the theoretical assumption that framing of CE can influence political decision-making
Method	• Further developing integration of deductive and inductive research approaches • Further developing the integration of qualitative and quantitative methods • Providing a starting point for automated corpus analyses of the discussion on CE • Operationalization of concept 'frame' for qualitative and quantitative content analysis	• Neither deeply qualitative nor largely quantitative study: danger of misinterpreting numbers generated through qualitative coding

(continued)

Table 7.2 (continued)

	Contributions to the respective literature/knowledge base	Limitations/challenges
Empirical knowledge about the case	• Connecting findings from preceding exploratory studies with potential political implications • Extending the scope of attempts at understanding future decision-making concerning CE • Offering empirical basis to ascertain concerns of 'slippery slope' and 'uphill struggle' of CE	• Does not assess relevance of academic framing in influencing political decision-making in comparison to other influential factors • Does not provide a constitutive account of the entire discourse on CE

References

Adler E, Haas PM (1992) Conclusion: epistemic communities, world order, and the creation of a reflective research program. Int Organ 46(1):367–390

Allen MR, Dube OP, Solecki W, Aragón-Durand F, Cramer W, Humphreys S et al (2018) Framing and context. In: Masson-Delmotte V, Zhai P, Pörtner HO, Roberts D, Skea J, Shukla PR et al (eds) Global warming of 1.5 °C. An IPCC Special Report on the impacts of global warming of 1.5°C above pre-industrial levels and related global greenhouse gas emission pathways, in the context of strengthening the global response to the threat of climate change, sustainable development, and efforts to eradicate poverty. Geneva, IPCC, pp 49–92

Anshelm J, Hultman M (2015) Discourses of global climate change. Apocalyptic framing and political antagonisms. Routledge, London

Asayama S (2014) Catastrophism toward 'opening up' or 'closing down'? Going beyond the apocalyptic future and geoengineering. Curr Sociol 63(1):89–93. https://doi.org/10.1177/0011392114559849

Asayama S, Sugiyama M, Ishii A (2017) Ambivalent climate of opinions: tensions and dilemmas in understanding geoengineering experimentation. Geoforum 80:82–92

Balzacq T (2010) Constructivism and securitization studies. In: Cavelty MD, Mauer V (eds) The Routledge Handbook of Security Studies. Routledge, London, pp 56–72

Beck S (2015) Science. In: Bäckstrand K, Lövbrand E (eds) Research handbook on climate governance. Edward Elgar, Cheltenham et al., pp 286–296

Behnke J, Baur N, Behnke N (2012) *Empirische Methoden der Politikwissenschaft* (2. aktualisierte, Auflage. Ferninand Schöningh, Paderborn

Bellamy R (2014) Safety first! Framing and governing climate geoengineering experimentation. Climate Geoengineering Governance Working Paper Series, 014

Bellamy R, Healey P (2018) Slippery slope' or 'uphill struggle'? Broadening out expert scenarios of climate engineering research and development. Environ Sci Policy 83(May):1–10. https://doi.org/10.1016/j.envsci.2018.01.021

Bellamy R, Lezaun J (2015) A review of deliberative public engagements with climate engineering. Climate Geoengineering Governance Working Paper Series, 019

Bellamy R, Chilvers J, Vaughan NE, Lenton TM (2012) A review of climate geoengineering appraisals. WIREs Clim Change 3(6):597–615

Bellamy R, Chilvers J, Vaughan NE (2016) Deliberative mapping of options for tackling climate change: citizens and specialists 'open up' appraisal of geoengineering. Publ Understand Sci 25(3):269–286. https://doi.org/10.1177/0963662514548628

Bellamy R, Lezaun J, Palmer J (2017) Public perceptions of geoengineering research governance: an experimental deliberative approach. Global Environ Change 45:194–202

Benford RD, Snow AD (2000) Framing processes and social movements: an overview and assessment. Ann Rev Sociol 26:11–39
Bijker WE (2006) Why and how technology matters. In: Goodin RE, Tilly C (eds) The Oxford handbook of contextual political analysis. Oxford University Press, New York, pp 681–706
Bolsen T, Palm R, Kingsland JT (2019) Counteracting climate science politicization with effective frames and imagery. Sci Commun 41(2):147–171. https://doi.org/10.1177/1075547019834565
Borgmann A (2012) The setting of the scene: technological fixes and the design of the good life. In: Preston CJ (ed) Engineering the climate: the ethics of solar radiation management. Lexington, Plymouth, UK, pp 189–200
Borie M, Hulme M (2015) Framing global biodiversity: IPBES between mother earth and ecosystem services. Environ Sci Policy 54:487–496
Brown RL, Kaplow JM (2014) Talking peace, making weapons: the iaea technical cooperation and nuclear proliferation. J Confl Res 58(3):402–428
Buck HJ (2013) Climate engineering: spectacle, tragedy or solution? A content analysis of news media framing. In Methmann C, Rothe D, Stephan B (eds) Interpretive approaches to global climate governance. (De)constructing the greenhouse. London, Routledge, pp 166–180
C2G2 (2019). C2G2 mission. https://www.c2g2.net/c2g2-mission/. Accessed 22 Mar 2019
Cairns R (2014) Climate geoengineering: issues of path-dependence and socio-technical lock-in. WIREs Clim Change 5:649–661. https://doi.org/10.1002/wcc.296
Caldeira K, Keith DW (2010) The need for climate engineering research. Issues Sci Tech, Fall 2010:57–62
Chong D, Druckman JN (2007a) A theory of framing and opinion formation in competitive elite environments. J Commun 57(1):99–118
Chong D, Druckman JN (2007b) Framing theory. Ann Rev Polit Sci 10:103–126. https://doi.org/10.1146/annurev.polisci.10.072805.103054
Corner A, Pidgeon N (2015) Like artificial trees? The effect of framing by natural analogy on public perceptions of geoengineering. Clim Change 130:425–438. https://doi.org/10.1007/s10584-014-1148-6
Corner A, Pidgeon N (2010) Geoengineering the climate: the social and ethical implications. Environ Sci Pol Sustain Dev 52(1):24–37
Corner A, Parkhill K, Pidgeon N (2011) Experiment earth: reflections on a public dialogue on geoengineering. Working Paper
Corner A, Pidgeon N (2014) Geoengineering, climate change scepticism and the 'moral hazard' argument: an experimental study of UK public perceptions. Philos Trans Royal Soc A 372(2031):1–14. https://doi.org/10.1098/rsta.2014.0063
Corry O (2012) Securitization and 'riskification': second-order security and the politics of climate change. Millennium: J Int Stud 40(2):235–258
Cross M, A. D. (2013) Rethinking epistemic communities twenty years later. Rev Int Stud 39(1):137–160. https://doi.org/10.1017/S0260210512000034
Donk A, Metag J, Kohring M, Marcinkowski F (2012) framing emerging technologies: risk perceptions of nanotechnology in the German press. Sci Commun 34(1):5–29. https://doi.org/10.1177/1075547011417892
Druckman JN (2011) What's it all about? Framing in political science. In: Keren G (ed) Perspectives on framing. Psychology Press, New York et al., pp 279–301
Egner B (2019) Methoden der Politikwissenschaft. UVK, Eine anwendungsbezogene Einführung (UTB Basics)
Entman RM (1993) Framing: toward clarification of a fractured paradigm. J Commun 43(4):51–58
Gamson WA, Modigliani A (1989) Media discourse and public opinion on nuclear power: a constructionist approach. Am J Sociol 95(1):1–37
Gieryn TF (1983) Boundary-work and the demarcation of science from non-science: strains and interests in professional ideologies of scientists. Am Sociol Rev 48(6):781–795
Goeschl T, Heyen D, Moreno-Cruz J (2013) The intergenerational transfer of solar radiation management capabilities and atmospheric carbon stocks. Environ Resour Econ 56(1):85–104

Grauvogel J, Diez T (2014) Framing und Versicherheitlichung: Die diskursive Konstruktion des Klimawandels. Zeitschrift Für Friedens- Und Konfliktforschung 3(2):203–232

Grimmer J, Stewart BM (2013) Text as data: the promise and pitfalls of automatic content analysis methods for political texts. Polit Anal 21:267–297

Gupta A, Möller I (2018) De facto governance: how authoritative assessments construct climate engineering as an object of governance. Environ Polit. https://doi.org/10.1080/09644016.2018.1452373

Haas PM (1992) Introduction: epistemic communities and international policy coordination. Int Organ 46(1):1–35

Hajer M (1997) The politics of environmental discourse: Ecological modernization and the policy process. Clarendon Press, Oxford

Hansen I, Lim DJ (2018) Doxing democracy: influencing elections via cyber voter interference. Contemp Polit 25(2):150–171. https://doi.org/10.1080/13569775.2018.1493629

Harding A, Moreno-Cruz J (2016) Solar geoengineering economics: from incredible to inevitable and half-way back. Earth's Future 4(12):569–577. https://doi.org/10.1002/2016EF000462

Harvard's solar geoengineering research program (2019) Projects. https://geoengineering.environment.harvard.edu/projects#. Accessed 2 May 2019

Hay C (2007) Why we hate politics. Polity, Cambridge, MA

Hoppe R (2005) Rethinking the science-policy nexus: from knowledge utilization and science technology studies to types of boundary arrangements. Poeisis Prax 3(3):199–215

Horton JB (2015) The emergency framing of solar geoenengineering: Time for a different approach. Anthropocene Rev Prepublication. https://doi.org/10.1177/2053019615579922

Horton JB, Reynolds J (2016) The international politics of climate engineering: a review and prospectus for international relations. Int Stud Rev 18:438–461

Hulme M (2014) Can science fix climate change? A case against climate engineering. Polity Press, Cambridge, UK

Huttunen S, Hildén M (2014) Framing the controversial: geoengineering in academic literature. Sci Commun 36(3):3–29

ISD Reporting Services (2019) Summary of the fourth session of the United Nations environment assembly. Earth Negot Bull 16(153)

IPCC (2013b) Summary for policymakers. In Stocker TF, Qin D, Plattner GK, Tignor M, Allen SK, Boschung J, Nauels A, Xia Y, Bex V, Midgley PM (eds) Climate Change 2013: The Physical Science Basis. Contribution of Working Group I to the Fifth Assessment Report of the Intergovernmental Panel on Climate Change. Cambridge, UK, and New York, USA, IPCC

Jachtenfuchs M (1996a) Frames and learning. In: Jachtenfuchs M (ed) International policy-making as a learning process? The European Union and the greenhouse effect. Ashgate, Aldershot et al., pp 23–43

Jachtenfuchs M (1996b) A methodology for frame analysis. In: Jachtenfuchs M (ed) International policy-making as a learning process? The European Union and the greenhouse effect. Ashgate, Aldershot et al., pp 43–60

Kahneman D, Tversky A (1984) Choices, values, and frames. Am Psychol 39(4):341–350

Kaplan FM (1983) The wizards of armageddon. Stanford University Press, Stanford, CA

Keith DW, Irvine PJ (2016) Solar geoengineering could substantially reduce climate risks—a research hypothesis for the next decade. Earth's Future 4:549–559. https://doi.org/10.1002/2016EF000465

Kiggins RD (2018) Big data, artificial intelligence, and autonomous policy decision-making: a crisis in international relations theory? In: Kiggins RD (ed) The political economy of robots. Prospects for prosperity and peace in the automated 21st century. London, Palgrave Macmillan, pp 211–234

Kingdon JW (2011) Agendas, alternatives, and public policies (Revised second, edition. Longman, Boston

Kintisch E (2010) Hack the planet. Science's best hope—or worst nightmare—for averting climate catastrophe. Wiley, New York, NY

Kreuter J (2018) Climate engineering and political decision-making: the importance of polarity in the academic debate. INoGOV Policy Brief

Kreuter J, Matzner N, Baatz C, Keller DP, Markus T, Wittstock F et al (2020) Unveiling assumptions through interdisciplinary scrutiny: observations from the German priority program on climate engineering (SPP 1689). Clim Change. https://doi.org/10.1007/s10584-020-02777-4

Lamb HH (1971) Climate-engineering schemes to meet a climatic energy. Earth Sci Rev 7:87–95

Lander ES, Baylis F, Zhang F, Charpentier E, Berg P, Bourgain C et al (2019) Adopt a moratorium on heritable genome editing. Nature 567:165–168

Lederer M, Kreuter J (2018) Organising the unthinkable in times of crises: will climate engineering become the weapon of last resort in the Anthropocene? Organization 25(4). https://doi.org/10.1177/1350508418759186.

Lee C-J, Scheufele DA, Lewenstein BV (2005) Public attitudes toward emerging technologies: examining the interactive effects of cognitions and affect on public attitudes toward nanotechnology. Sci Commun 27(2):240–267. https://doi.org/10.1177/1075547005281474

Levy JS (2003) Applications of prospect theory to political science. Syntheses 135:215–241

Levy DL, Spicer A (2013) Contested imaginaries and the cultural political economy of climate change. Organization 20(5):659–678

Linnér B-O, Wibeck V (2015) Dual high-stake emerging technologies: a review of the climate engineering research literature. WIREs Clim Change 6:255–268

Low S (2017) The futures of climate engineering. Earth's Future 5:67–71. https://doi.org/10.1002/2016EF000442

Low S, Schäfer S (2019) Tools of the trade: practices and politics of researching the future in climate engineering. Sustain Sci, Online First. https://doi.org/10.1007/s11625-019-00692-x

Lucke F, v., Wellmann, Z., & Diez, T. (2014) What's at stake in securitizing climate change? Towards a differentiated approach . Geopolitics 19(4):857–884

Lucke F, v., Wellmann, Z., & Diez, T. (2016) Klimakämpfe: Eine komparative Studie der Versicherheitlichung von Klimawandel. Zeitschrift Für Internationale Beziehungen 23(2):112–143. https://doi.org/10.5771/0946-7165-2016-2-112

Luokkanen M, Huttunen S, Hildén M (2014) Geoengineering, news media and metaphors: framing the controversial. Public Understand Sci 23(8):966–981. https://doi.org/10.1177/0963662513475966

Manoussi V, Xepapapdeas A (2014) Cooperation and competition in climate change policies: mitigation and climate engineering when countries are asymmetric. Clim Change Sustain Dev

Markusson N (2013) Tensions in framings of geoengineering: constitutive diversity and ambivalence. Climate Geoengineering Governance Working Paper Series, 003

Markusson N, Venturini T, Laniado D, Kaltenbrunner A (2016) Contrasting medium and genre von Wikipedia to open up the dominating definition and classification of geoengineering. Big Data & Society, July—December 2016, 1–17

Matthes J, Kohring M (2008) The content analysis of media frames: toward improving reliability and validity. J Commun 58:258–279

Matzner N, Barben D (2018) Verantwortungsvoll das Klima manipulieren? Unsicherheit und Verantwortung im Diskurs um climate engineering. In: Janich N, Rhein L (eds) Unsicherheit als Herauforderung für die Wissenschaft. Peter Lang, Berlin et al., pp 143–179

Methmann C, Rothe D, Stephan B (2013a) Interpretive approaches to global climate governance. (De-)constructing the greenhouse. London, Routledge

Methmann C, Rothe D, Stephan B (2013c) Reflections. In: Methmann C, Rothe D, Stephan B (eds) Interpretive approaches to global climate governance. (De)constructing the greenhouse. London, Routledge, pp 248–256

Morozov E (2014) To Save Everything, Click Here. Penguin, London

Nelson TE, Oxley ZM, Clawson RA (1997) Towards a psychology of framing effects. Polit Behav 19(3):221–246

Nerlich B, Jaspal R (2012) Metaphors we die by? Geoengineering, metaphors, and the argument from catastrophe. Metaphor Symbol 27(2):131–147

Nisbet MC, Mooney C (2007) Framing science. Science 316(5821):56

Oels A, Carvalho A (2012) Wer hat Angst vor 'Klimaflüchtlingen'? Wie die mediale und politische Konstruction des Klimawandels den politischen Handlungsspielraum strukturiert. In: Neverla I, Schäfer MS (eds) Das Medien-Klima. Fragen und Befunde der kommunikationswissenschaftlichen Klimaforschung. Wiesbaden, Springer VS, pp 253–276

Oels A, von Lucke F (2015) Gescheiterte Versicherheitlichung oder Sicherheit im Wandel: Hilft uns die Kopenhagener Schule beim Thema Klimawandel? Zeitschrift Für Internationale Beziehungen 22(1):43–70

Oldham P, Szerszynski B, Stilgoe J, Brown C, Eacott B, Yuille A (2014) Mapping the landscape of climate engineering. Philos Trans Royal Soc A, 372(2031). https://doi.org/10.1098/rsta.2014.0065

UNFCCC (2015) Adoption of the Paris agreement. In U. N. F. C. o. C. Change (ed), (vol. FCCC/CP/2015/L.9/Rev.1). Paris, United Nations

Pasztor J (2019) What's in a name? Why we became C2G. C2G Blog

Pepermans Y, Maeseele P (2016) The politicization of climate change: problem or solution? WIREs Clim Change 7:478–485. https://doi.org/10.1002/wcc.405

Quaas MF (2017) Are there reasons against open-ended research into solar radiation management? A model of intergenerational decision-making under uncertainty. J Environ Econ Manag 84:1–17. https://doi.org/10.1016/j.jeem.2017.02.002

Rayner S (2014) To know or not to know? A note on ignorance as a rhetorical resource in geoengineering debates. Climate Geoengineering Governance Working Paper Series, 010s

Rein M, Schön D (1993) Reframing Policy Discourse. In: Fischer F, Forester J (eds) The argumentative turn in policy analysis and planning. Duke University Press, Durham et al., pp 145–166

Rickels W, Klepper G, Dovern J, Betz G, Brachatzek N, Cacean S et al (2011b) Large-scale intentional interventions into the climate system? Assessing the climate engineering debate. Scoping Report conducted on behalf of the German Federal Ministry of Education and Research (BMBF). Kiel, Kiel Earth Institute

Ricke K, Moreno-Cruz J, Caldeira K (2013) Strategic incentives for climate geoengineering coalitions to exclude broad participation. Environ Res Lett 8:1–8

Rockström J, Schellnhuber HJ, Hoskins B, Ramanathan V, Schlosser P, Brasseur GP et al (2016) The world's biggest gamble. Earth's Future 4(10):465–470. https://doi.org/2016EF000392

Schlipphak B (2011) Was ist ein Frame?—Zu den Grundlagen des Frame-Ansaztes in den Sozialwissenschaften. In B. Schlipphak (ed) Framing ideology. Die Kommunikation ideologischer Positionierungen zwischen Parteien, Wählern und Medien. Baden-Baden, Nomos, pp 21–51

Scholte S, Vasileiadou E, Petersen A (2013) Opening up the societal debate on climate engineering: how newspaper frames are changing. J Integr Environ Sci 10(1):1–16

Schot J, Kanger L (2018) Deep transitions: emergence, acceleration, stabilization and directionality. Res Policy 47:1045–1059

Schot J, Schipper F (2011) Experts and the European transport integration, 1945–1958. J Eur Public Policy 18(2):274–293. https://doi.org/10.1080/13501763.2011.544509

Scott D (2012) Insurance policy or technological fix? The ethical implications of framing solar radiation management. In: Preston CJ (ed) Engineering the climate: the ethics of solar radiation management. Lexington, Plymouth, UK, pp 151–168

Shepherd J, Cox P, Haigh D, Keith DW, Launder B, Mace G et al (2009) Geoengineering the climate: science, governance and uncertainty. The Royal Society, London

Singer C, Matchett L (2015) Climate action gaming experiment: methods and example results. Challenges 6:202–228

Sniderman PM, Theriault SM (2004) The structure of political argument and the logic of issue framing. In Saris WE, Sniderman PM (eds) Studies in public opinion. Attidues, nonattitudes, measurement error, and change. Princeton and Oxford, Princeton University Press, pp 133–165

Spindler M, Schieder S (2010) Theorien in der Lehre von den internationalen Beziehungen. In: Schieder S, Spindler M (eds) *Theorien der Internationalen Beziehungen* (3, Auflage. Barbara Budrich, Opladen, pp 9–38

SPP 1689 (2019a) Climate engineering und unsere Klimaziele—eine überfällige Debatte. In: Schwerpunktprogramm C 1689 der Deutschen Forschungsgemeinschaft "Climate engineering: risks, opportunities?" (SPP 1689) (ed) Kiel, Kiel Earth Institute

SPP 1689 (2019b) Projekte. https://www.spp-climate-engineering.de/index.php/die-projekte.html. Accessed 02 May 2019

Talberg A, Christoff P, Thomas S, Karoly D (2018) Geoengineering governance-by-default: an earth system governance perspective. Int Environ Agreements 18:229–253. https://doi.org/10.1007/s10784-017-9374-9

Trombetta MJ (2008) Environmental security and climate change: analysing the discourse. Cambridge Rev Int Aff 21(4):585–602

Turnhout E, Dewulf A, Hulme M (2015) What does policy-relevant global environmental knowledge do? The cases of climate and biodiversity. Curr Opin Environ Sustain 18:65–72

Tversky A, Kahneman D (1987) Rational choice and the framing of decisions. In: Hogarth RM, Reder MW (eds) Rational choice: the contrast between economics and psychology. University of Chicago Press, Chicago, pp 67–94

Tversky A, Kahneman D (1981) The framing of decisions and the psychology of choice. Sci New Ser 211(4481):453–458

Ulbricht L, Haunss S, Hofmann J, Klinger U, Passoth J-H, Pentzold C et al. (2018) Dimensionen von big data: eine politikwissenschaftliche systematisierung. In: Kolany-Raiser B, Heil R, Orwat C, Hoeren T (eds) Big data und gesellschaft. Eine multidisziplinäre Annäherung, Technikzukünfte, Wissenschaft und Gesellschaft). Wiesbaden, Springer, pp 151–232

UN Secretary-General (2018) Secretary-general's remarks on climate change [as delivered]

Waldherr A, Heyer G, Jähnichen P, Niekler A, Wiedemann G (2016) Mining big data with computational methods. In: Vowe G, Henn P (eds) Political communication in the online world. Theoretical approaches and research designs. New York, Routledge, pp 201–217

Weber M (1980) [1922]). Wirtschaft und Gesellschaft. Grundriss der verstehenden Soziologie (5. Revidierte, Studienausgabe. Mohr, Tübingen

Weinberg AM (1966) Can technology replace social engineering? Bull Atomic Scient 22(10):4–8

Wendt, A. (1998). On constitution and causation in international relations. Rev Int Stud 24(Special Issue):101–117

Winner L (1977) Autonomous technology. Technics-out-of-control as a theme in political thought. MIT Press, Cambridge, MA

Zhang Y, Posch A (2014) The wickedness and complexity of decision making in geoengineering. Challenges 5:390–408. https://doi.org/10.3390/challe5020390

Zürn M (2019) Policization compared: at national, European, and global levels. J Eur Publ Policy 26(7):977–995. https://doi.org/10.1080/13501763.2019.1619188

Zürn M, Schäfer S (2013) The paradox of climate engineering. Global Policy 4(3):266–277

Index

J. Kreuter, *Climate Engineering as an Instance of Politicization*, Springer Climate, https://doi.org/10.1007/978-3-030-60340-3

The manufacturer's authorised representative in the EU is Springer Nature Customer Service Centre GmbH, Europaplatz 3, 69115 Heidelberg, Germany. If you have any concerns regarding our products, please contact ProductSafety@springernature.com

Printed and bound by CPI Group (UK) Ltd, Croydon, CR0 4YY
30/09/2026
02225287-0008